EINSTEIN IN SPAIN

Einstein in Spain

Relativity and the Recovery of Science

THOMAS F. GLICK

PRINCETON UNIVERSITY PRESS

PRINCETON, NEW JERSEY

Published by Princeton University Press, 41 William Street,
Princeton, New Jersey 08540
In the United Kingdom: Princeton University Press,
Guildford, Surrey

Library of Congress Cataloging in Publication Data will be
found on the last printed page of this book

ISBN 0-691-05507-6

Publication of this book has been aided by a subsidy from the
Program for Cultural Cooperation Between Spain's
Ministry of Culture and United States' Universities

This book has been composed in Linotron Bembo

Clothbound editions of Princeton University Press books
are printed on acid-free paper, and binding materials are
chosen for strength and durability. Paperbacks, although
satisfactory for personal collections, are not usually
suitable for library rebinding

Printed in the United States of America by
Princeton University Press,
Princeton, New Jersey

To my parents,
Lester and Ruth Glick,
in recollection of the
landscape of Castile,
1954

To my parents,
Lester and Ruth Glick,
in recollection of the
landscape of Castile,
1954

Contents

List of Illustrations and Tables   viii

Preface   xi

Introduction   3

CHAPTER 1. Spanish Science and the Reception of
Relativity   17

CHAPTER 2. The Einstein Phenomenon   74

CHAPTER 3. Barcelona: Einstein and Catalan
Nationalism   100

CHAPTER 4. Madrid: The Two Aristocracies   123

CHAPTER 5. The Debate over Relativity in the 1920s   150

CHAPTER 6. Relativity and Spanish Engineers: The
Scientific Middle Class   188

CHAPTER 7. The Slave at the Sermon: Einstein and the
Spanish Intelligentsia   238

CHAPTER 8. Flow and Transformation of Ideas   276

CHAPTER 9. After Einstein's Visit   302

Appendixes
1. Einstein's Travel Diary for Spain, 1923   325
2. German Consular Reports of Einstein's Trip to Spain   327
3. Einstein's Madrid Lectures on Relativity   332

Bibliography   357

Index   377

List of Illustrations and Tables

ILLUSTRATIONS

1. Receptions of Darwinism, Relativity, and Psychoanalysis in Spain.   84
2. Reception for the Einsteins at the City Hall of Barcelona.   114
3. Einstein at the Industrial School of Barcelona.   119
4. Latin menu presented to guests at Rafael Campalans's banquet for the Einsteins.   120
5. Einstein departing Barcelona.   121
6. Einstein arriving in Madrid.   123
7. Einstein at Cabrera's laboratory.   125
8. King Alfonso XIII and Einstein at the Academy of
   Sciences, March 4, 1923.   129
9. Einstein and Ortega in Toledo, March 6, 1923.   136
10. Rector José R. Carracido and Einstein (in his aca-
    demic hood) at the University of Madrid, March 8,
    1923.   140
11. Einstein and members of the Faculty of Sciences of
    the University of Madrid, March 8, 1923.   141
12. Einstein at the Madrid Athenaeum, March 8, 1923.   142
13. "A Relativity Paradox."   207
14, 15. Newspaper cartoons on Einstein's Madrid lec-
    tures.   295
16. Ramón Pérez de Ayala, Einstein, and A. S. Yahuda
    in Belgium, April 1933.   304

Photo Sources: 1-3, Instituto Histórico Municipal, Barcelona; 4, Terradas
Archives, Institut d'Estudis Catalans, Barcelona; 7, Faculty of Sciences,
University of Madrid; 9, Fundación José Ortega y Gasset, Madrid; 10,

LIST OF ILLUSTRATIONS AND TABLES

16, *ABC*, Madrid; 13, *Nature*, 110 (1922), 844. Reproduced with permission. All other photos were taken from defunct newspapers and are in the public domain.

TABLES

1. Guests at a Dinner for Levi-Civita, Madrid, 1921   66
2. Spanish Society of Physics and Chemistry (Membership in 1920, by Category)   190
3. Scientific Middle Class (Selected Spanish Scientific Societies)   191
4. Popular Responses to Einstein   240

Preface

THIS book is a contribution to the history of civil discourse in matters of science in an ideologically polarized society, Spain during the first thirty-five years of this century. By civil discourse I mean the process wherein a divided elite agrees to place in abeyance in mutually agreed upon areas the habit of making all ideas serve the ends of ideology. In Spain around the turn of the century such conditions came to prevail in the areas of science and technology for the purpose of modernizing the country, since its scientific backwardness was identified by all political sectors as one of the primary reasons for Spain's defeat in 1898. In this context, civil discourse is viewed as the central mechanism whereby a broad base of support for pure science was built and a climate of opinion that viewed science positively was fashioned. The specific institutional context out of which civil discourse in science emerged around 1900 is the subject of the Introduction.

Einstein's trip to Spain in 1923 served to sharpen the image of pure science in Spain, and an examination of the repercussions of his visit tells much about the nature of the scientific enterprise in Spanish society. A glance at the index of this volume, moreover, brings immediate attention to the fact that Einstein touched a significant portion of the Spanish intelligentsia of the 1920s: Araquistáin, Baroja, Fernández Flores, Gaziel, Gómez de la Serna, Machado, Maeztu, Ors, Ortega y Gasset, Pérez de Ayala, Sagarra, Soldevila, Unamuno—the vital core around whom, even more than scientists themselves, civil discourse flowed. Such persons had analogues within the scientific community, the protagonists of the present volume, with whom they interacted on a daily basis: Cabrera, Carracido, De Buen, Marañón, Novoa Santos, Plans, Terradas. The interpenetration of scientific and lay intelligentsias accounts for the broad scope of the present investigation, which may be characterized as a study of scientific popularization, writ large. That is, I intend to show the multilevel discussion and impact of a specific set of scientific ideas and to explore relationships among various levels or domains of discourse, in order to illuminate the process of civil discourse in Spain and in Spanish science. I am more concerned here with the social appropriation of scientific ideas than with the ideas themselves. I am a social historian, not a physicist, and this study complements my parallel studies of the receptions of Darwinism and Freudian psychology in Spain.

MANY institutions and individuals have aided me in this project since its inception. The Fundación Juan March, the Aula de Cultura of the Caja de Ahorros of Alicante and Murcia, the University, Autonomous University, and Polytechnic University of Barcelona and the Departament d'Economia of the Generalitat de Catalunya all sponsored lectures based on this material. Joan Badal, of the last-named entity, was unstinting in his encouragement. Many persons provided me with specific information: Michel Biezunski, Antonio de Castro, José A. García Diego, Judith Goodstein, and Dirk Struik. Family tradition accounts for my interest in Einstein's 1921 trip to Cleveland where he was feted in the home of my grandparents, Anna and Isadore Rothstein. My father, Lester Glick, researched this episode for me. The encouragement and help of my colleagues at the University of Valencia, J. M. López Piñero, Víctor Navarro, and Francesc Bujosa, were basic to the enterprise. The entire manuscript was read by Stanley Goldberg, John Stachel, Antoni Roca, and J. M. Sánchez Ron, whose help in evaluating and explaining numerous points of physics that were beyond my understanding ultimately made this book possible. Joan C. Ullman's valued comments are embodied in the Introduction. Permission to consult and reproduce materials contained in the Einstein Archive, then in Princeton, New Jersey, and now at the Hebrew University in Jerusalem, was kindly granted by the late Dr. Otto Nathan.

PREFACE

In the text of the book, names of Catalan personages are represented in that language; their Castilian publishing names can be found in the index or bibliography.

All translations from Spanish and Catalan into English are mine.

EINSTEIN IN SPAIN

Introduction

SCIENCE IN MODERN SPAIN

Spain had been infertile soil for the physical sciences since the late eighteenth century, when the Bourbon monarchs and their enlightened ministers made a valiant and partially successful attempt to import modern physics and chemistry from France and England. In particular, the banner of scientific modernity was borne by an extraordinary group of navigators, under the aegis of Admiral Vicente Tofiño, two of whom were elected to membership in the English Royal Society (José Mendoza y Ríos in 1793, and Felipe Bauzá in 1819). At the same time, modern chemistry was introduced by the French chemists Louis Joseph Proust and François Chavaneau. Symbolic of chemistry's progress in Enlightenment Spain were the identification of two new elements by Spaniards: tungsten by Juan José Elhuyar and vanadium by Andrés Manuel del Río. The Bourbon effort was reflected in scientific institutions of high quality: the three botanical expeditions to Spanish America, the astronomical observatories of Montevideo and Bogotá, and, in Madrid, the Museum of Natural Sciences, the Royal Museum of Machines, the Royal Botanical Garden, and the Chemical Laboratory of the Ministry of State, among others.

The Napoleonic invasions caused the nearly complete ruin of the scientific gains of the previous period. Convenient symbols are the lack of success that all three botanical expeditions had in publishing their findings or the ruin of the Highway School and machine collection and the flight of their founder Agustín de Betancourt to Imperial Russia. Nor was there any palpable recovery over the next half century, when university science instruction (except in medicine) was virtually dormant and new scientific ideas were met with formidable prejudice.

EDUCATIONAL POLITICS IN THE
NINETEENTH CENTURY

In order to understand how civil discourse could emerge out of a political community sharply polarized over the goals (and content) of state-directed education, we must first review the nature of higher education in nineteenth-century Spain and the positions of major political groups. In 1857 the Minister of Development (*fomento*, the ministry in which control over education was vested until the establishment of a separate Ministry of Public Instruction in 1901) instituted an educational reform whose main lines have subsisted until very recently. Following a centralist model of French inspiration, the Moyano Plan established a system of university education that was, in the words of the educational reformer José de Castillejo, "doctrinaire, secular and regalist, uniform and hierarchical," and, in the characterization of two recent historians, "an iron mold which would oppress public instruction for more than a century."[1] The law put the university system under the direct control of the ministry, which had dictatorial powers over the curriculum and textbooks and could remove professors at will and gave diocesan officials a virtual right of censorship over textbooks and course content. The minister was advised by an appointed council in whose deliberations university professors had only minimal participation.

In January 1867, Manuel de Orovio, development minister in the reactionary government of General Narváez, issued a decree setting forth criteria for the destitution from their chairs of professors who espoused "erroneous doctrines" of a moral, religious, or political nature. The government then moved against the reformist academic leadership of liberals and "Krausists" (followers of the German philosopher Karl Krause), specifically Julián Sanz del Río, Fernando de Castro, and Nicolás Salmerón, all professors at the Central University

---

[1] José de Castillejo, *War of Ideas in Spain* (London: John Murray, 1937), p. 86; Mariano and José Luis Peset, *La universidad española (siglos XVIII y XIX)* (Madrid: Taurus, 1974), p. 465.

of Madrid who soon became the intellectual heroes of the Revolution of 1868. This revolution, which led to the short-lived First Republic of 1873, provided an opening for new ideas, such as Darwinism, and for new scientific and educational institutions that functioned outside of the rigid university system.

With the Bourbon restoration of 1874, Orovio returned to the cabinet of the conservative prime minister Antonio Cánovas de Castillo and again promulgated a decree forbidding the teaching of anti-Catholic doctrines. Although the targets were ostensibly the same Krausists as before, Julio Caro Baroja has argued that now the real target was Darwinism, a more potent threat to Catholic dogma than Krausism ever was. Indeed, the first two university professors to resign rather than comply with the order were Augusto González de Linares (1845–1904), a zoologist whose open espousal of Darwinism at the University of Santiago de Compostela had attracted threats of violence by neo-Catholic students who accused him of heresy and of contradicting Linnean and Cuverian doctrine on the fixity of species, and the mineralogist Laureano Calderón (1847–1894), also a Darwinian.[2] Another "separated" professor, the Krausist philosopher and pedagogue Francisco Giner de los Ríos (1839–1899) responded to Catholic pressure by founding (in 1876) the Institución Libre de Enseñanza (Free Institute of Education) in order to inculcate secular, religiously neutral primary and secondary education on the model of the English "public" schools. Education was progressive—there were no examinations and student notebooks generally replaced the use of texts. The Institución taught the values of liberal democracy, social equality, and modern science and inspired an

---

[2] Julio Caro Baroja, "El miedo al mono o la causa directa de la Cuestión Universitaria, en 1875," in the collective volume, *En el centenario de la Institución Libre de Enseñanza* (Madrid: Tecnos, 1977), pp. 23–41. On the University crisis generally, see Pablo de Azcárate, ed., *La cuestión universitaria, 1875* (Madrid: Tecnos, 1967), and Alberto Jiménez [Fraud], *Historia de la universidad española* (Madrid: Alianza, 1971), pp. 354–374.

entire generation of Liberal and Republican politicians and public figures who had studied there.

The core of the Institute's staff were the professors who had resigned during the university crisis the previous year, including the natural scientists González de Linares, Laureano Calderón, and his brother Salvador (1853–1911), also a mineralogist. Science instruction followed John Tyndall's concepts, using simple experiments, with more emphasis on scientific method and less on memorization.[3]

The Count of Toreno, Orovio's successor, argued in the Cortes that academic freedom (*libertad de la ciencia*) could never put the professoriate above religion. Academic freedom was a central issue whose nature and limits were debated by parliamentary political groups over the next twenty-five years.[4] The real issue in these debates was the amount of clerical control over education that civil society would tolerate. Although Spain was officially a Catholic country, the Church had been progressively losing its grip on higher education as successive educational reforms expanded the prerogatives of the state and diminished those of the Church. As in France, higher education was viewed officially as being wholly the state's province, a position that the Church's defenders frequently attacked as "Jacobin." The Church's position on academic freedom was, simply, that error (as the Church defined it) should not have the same privileges as truth. Its spokesman favored a return to the educational ideals of the "Catholic unity" of knowledge that informed Spanish education prior to the Enlightenment. The Church's supporters in the Cortes, the "neo-Catholic"

[3] On the Institución Libre de Enseñanza, see Yvonne Turín, *La educación y la escuela en España de 1879 a 1902* (Madrid: Aguilar, 1967), pp. 175–225; María Dolores Gómez Molleda, *Los reformadores de la España contemporánea* (Madrid: C.S.I.C., 1966); and Vicente Cacho Viu, *La Institución Libre de Enseñanza, I. Orígenes y etapa universitaria (1860–1881)* (Madrid: Rialp, 1962). On the Institución's scientists, see Antonio Jiménez-Landi, "Científicos de la Institución Libre de Enseñanza," in *En el centenario de la Institución Libre de Enseñanza*, pp. 89–101.

[4] The following section is based on Turín, *Educación en España*, pp 103–174.

6

deputies of the far Right led by the Marqués de Pidal and Joaquín Sánchez de Toca, developed a political strategy that was at odds with its conviction that the only truth was that defined by the Church. Since the prevailing opinion in both Conservative and Liberal parties was that the right of members of religious orders to teach should be sharply restricted, the neo-Catholics sided with the Republican deputies of the left—their ideological arch enemies—in defending academic freedom as a means of preserving a role for churchmen in higher education. Academic freedom diminished in importance as an educational issue of the moment because the rest of the political spectrum was more inclined to accept this right as one already acquired and no longer in doubt. The official spokesmen of the Conservatives and Liberals, however, applied this ideal differently. Conservatives, of course, wanted to preserve the greatest possible influence of the Church, which they regarded as an ally in braking social progress. Liberals, on the other hand, tended to view the Church as an impediment to the "Jacobin" program of the complete laicization of education. As the Liberal party developed its educational policies, it eventually backed away from a wholly interventionist policy and came to favor the right of individual professors to set their own curricula without state interference and to oppose compulsory religious instruction. The Republicans had the same objectives but sided with neo-Catholics in protecting the religious orders' right to teach, because they did not like to see the precedent of particular ideological groups being saddled with some civil disability, having already learned from experience that a willful minister could exclude them. Both neo-Catholics and Republicans favored the autonomy of the university, believing that their ideas would fare better with decentralization than under the tutelage of the state, as mainline Conservatives and Liberals held.

Government ministers of the 1870s and 1880s, whether Conservative or Liberal, upheld the state's right to determine university policy in consonance with their political views. Such a situation both stimulated political polarization over ideas, including scientific ones, and created a climate in which the espousal of controversial ideas, such as Darwinism, was imprudent, given the fact that any university professor could lose his chair if his ideas were found not of sufficient orthodoxy to please the minister.

The Emergence of Civil Discourse

By the end of the century, nevertheless, a new consensus had emerged concerning education among Conservative and Liberal politicians. The two first ministers of public instruction, the conservative Antonio García Alix (1900–1901) and the liberal Count of Romanones (1901–1902), both agreed that the state's educational policy had to be politically neutral (which was almost certainly not the view of Cánovas and his associates). Romanones had long been feared by the far Right as the leading educational "Jacobin," and García Alix was uncharacteristically open-minded and impartial for a conservative education minister, consulting with specialists drawn from a broad political spectrum, such as the Institucionista zoologist, Ignacio de Bolívar. In concert, they introduced modest reforms that somewhat mitigated the strangulating effect of the Moyano Plan, broadening the membership of the ministry's advisory council to include more academics and representatives of all political tendencies. This political depolarization of education at the official level merged with other currents newly present in the broader culture—such as the widespread feeling that Spain's defeat in the war of 1898 was due to its scientific and technological backwardness—to create a climate that was propitious for the development of science and favorable to an open discussion of scientific ideas, without their automatically being appropriated as weapons in the ideological struggle between Left and Right. From this time, until the outbreak of the Spanish Civil War in 1936, civil discourse prevailed in the scientific arena, creating an ambience in which scientific modernization became a general goal of the society.

The participation of ideological enemies in the common

task of modernization in a climate of civil discourse can be seen in the functioning of the Junta para Ampliación de Estudios (Board for the Extension of Study), drafted by Amalio Gimeno as Minister of Public Instruction and established by Royal Decree in June 1907 as an autonomous organ of the Ministry of Public Instruction, administered by a board of twenty trustees under the presidency of the Nobel laureate Santiago Ramón y Cajal, with the energetic Castillejo as permanent secretary. The Junta's most noteworthy function was the administration of postdoctoral and other grants for scholars to broaden and perfect their education in foreign centers in the hopes that they would invigorate their disciplines when they returned. (Grants in the physical sciences are discussed in chapter 1.) The Junta also administered, among other research and educational programs, the National Institute of Physical Sciences, the Association of Laboratories, and the Residence for Students (Residencia de Estudiantes), all established in 1910, and an experimental secondary school, the Institute-School, founded in 1918. The Residence was an influential residential college that, among its other noteworthy attributes, had seven laboratories in various disciplines.[5]

Trustees of all political tendencies were represented on the Junta's governing board, aspiring, in the words of a Junta brochure of 1911, "to make it a technical organism, with neutrality as a permanent element, beyond political oscillations, that it can promote the social function of education above all differences of opinion, belief or party."[6] Board members took the

[5] On the Junta para Ampliación de Estudios, see Castillejo, *War of Ideas*, pp. 118–133; Jiménez, *Universidad española*, pp 396–402; José Subirá, "La Junta para Ampliación de Estudios," *Nuestro Tiempo*, 24¹ (1924), 23–55; 24² (1924), 69–85, 166–184; and Pierre París, "Junta para Ampliación de Estudios e Investigaciones Científicas," *Bulletin Hispanique*, 18 (1916), 114–131. The Residencia's laboratories, all established between 1912 and 1919, were of microscopic anatomy, general chemistry, physiological chemistry, general physiology, neurology, histology, and serology (Jiménez, *Universidad española*, p. 492).

[6] Cited by París, "Junta para Ampliación de Estudios," p. 118, n. 1. Subirá ("Junta," p. 29) says "with absolute independence from the interests of the political parties."

charge of neutrality seriously and, in spite of the disparity of political opinions of its members, reached all decisions by unanimity.

The composition of the Junta's first board is indicative of the spirit of civil discourse informing it. There were Liberal (Eduardo Vincenti) or Republican (Gumersindo de Azcárate) politicians associated with the Institución Libre and neo-Catholic members such as the ideologue Marcelino Menéndez y Pelayo and the politician Joaquín Sánchez de Toca. Scientist members included the mathematician José de Echegaray, the neurologist Luis Simarro, the physiologist (and Liberal politician) Amalio Gimeno, the entomologist Bolívar, chemists like José Casares Gil and José Rodríguez Carracido, the astronomer Victoriano Fernández Ascarza, and the inventor Leonardo Torres Quevedo. I might add that Cajal, Carracido, Gimeno, and Simarro were articulate evolutionists; Carracido and Simarro had engaged in polemics on the pro-Darwin, anti-Catholic side, while Gimeno, no less a Darwinian, argued evolution's compatibility with Catholicism at a time (1875) when it was daring to do so. Sánchez de Toca, for his part, had defended the Catholic position against the social views of Darwin and his followers in 1879, when there was no possibility of composition between Right and Left on that key issue.[7] We may furthermore

---

[7] Simarro had been expelled from school for reading Darwin and later adopted an evolutionary approach both to morphology and to social theory. See Temma Kaplan, "Luis Simarro, Spanish Histologist," in *Actas, II Congreso Nacional de Historia de la Medicina*, 3 vols. (Valencia, 1971), II, 527–528. Carracido criticized the Catholic approach to evolution in a polemic with Ceferino Cardinal González in 1889; see "Las ciencias físico-naturales en el Congreso Católico," in his *Estudios histórico-críticos de la ciencia española*, 2nd ed. (Madrid: Imprenta Alrededor del Mundo, 1917), pp. 313–349. On Gimeno, see my discussion in "Spain," in T. F. Glick, ed., *The Comparative Reception of Darwinism* (Austin: University of Texas Press, 1974), pp. 325–326. On Sánchez de Toca's anti-Darwinism, see his article, "La doctrina de la evolución de las modernas escuelas científicas," *Revista Contemporánea*, 21 (1879), 55–91, 273–303, where he argues against the applicability of biological models to the study of society and asserts that the histories of human language and civilization both refute the notion of progress as a mechanism of human history. In later years the geolo-

note that Sánchez de Toca, who even in 1879 had admitted that scientists would have to decide the merits of Darwin's biological arguments, apparently acquiesced in the teaching of the evolution of species in the science curriculum of the Institute-School.[8] The Junta was attacked by the clerical Right, but was, in turn, defended by the numerous priests who had been the beneficiaries of its grants. José Subirá, a supporter of the Junta, confirmed that its role was "to carry out purely scientific and cultural functions, in which neither interests of party or the sectarian intransigence of one or the other side ever played a role, but rather there prevailed exclusively a feeling of solidarity in favor of anything that might benefit universal culture or national science or art."[9]

FACULTIES OF SCIENCE AND TECHNICAL SCHOOLS

At mid-century there were no freestanding faculties of sciences in Spanish universities. What few chairs existed (including only two of mathematics among all Spanish universities in 1850) were scattered through faculties of philosophy. Other science chairs were based outside of the university system, in such institutions as the Royal Botanical Garden (botany), the Museum of Natural Sciences (zoology and mineralogy), and

gist Domingo Orueta, whose Darwinian and Spencerian library was famous, was elected to the board; see Jiménez, *Universidad española*, p. 429.

[8] See Subirá, "Junta," p. 50.

[9] Ramón Ruiz Amado, "La Junta para Ampliación de Estudios y sus instituciones," *Razón y Fe*, 68 (1924), 80–94; and Subirá, "Junta para Ampliación de Estudios," pp. 172–178, 184. Ruiz Amado alleged that the Junta was a nest of *institucionistas* and that elements hostile to the Institución were excluded from the board. The newspaper *El Debate*, which reflected neo-Catholic views in the twentieth century and which directed much of its editorial invective against the Institución Libre, admitted in 1923 that the Junta had been respected by Liberal and Conservative politicians alike (May 22, 1926). In 1930 it praised the reelection of board members J. M. Torroja Miret, Inocencio Jiménez Vicente, Juan La Cierva, and José Ortega y Gasset, "without inquiring as to whether they were of the left or right; they were professors, men of value, and this was enough reason for us to applaud their reelection" (March 8, 1930).

the Royal Observatory (astronomy), all in Madrid, together with chairs of exact and physical sciences in the engineering schools.

The Moyano Plan of 1857 provided for the separation of philosophy and letters on the one hand, and the sciences on the other, divided into three sections—physical-mathematical sciences, chemistry, and natural sciences (although only the Central University offered the doctorate in all three). In some universities, such as Zaragoza, the faculty of science served merely as a preparatory school for the faculty of medicine. It did not gain independent existence there until 1893, after an unsuccessful attempt by the medical faculty to suppress it completely. The science faculties began to function effectively when they were restructured in 1900 into four sections—exact, physical, chemical, and natural sciences.

Modernization of the medical schools came earlier, in 1843, when the old colleges of medicine, surgery, and pharmacy were abolished. In their place, full medical faculties were established in Madrid, Barcelona, Santiago, Valencia, and Cádiz, and independent pharmacy faculties were introduced. As was the rule in all disciplines, textbooks were fixed by the government, in consultation with the Public Instruction Council, and all curricula were carefully controlled from Madrid. As a general rule, science professors did not do research until around 1875, with medicine constituting somewhat of an exception.[10]

"Special schools" are the schools of engineering, where scientific subjects were taught, although the term also refers to other institutions of higher learning such as music conservatories and fine arts academies. The oldest of the engineering

[10] Peset and Peset, *La universidad española*, pp. 513, 625, 628, 659, 665; Antonio Alvarez de Morales, *Génesis de la universidad española contemporánea* (Madrid: Instituto de Estudios Administrativos, 1972); José María Albareda, *Creación de facultades universitarias y producción científica en el pasado siglo* (Granada: Universidad de Granada, 1950), pp. 46–47; and Antonio Rumeu de Armas, *Ciencia y tecnología en la España ilustrada: La Escuela de Caminos y Canales* (Madrid: Turner, 1980), p. 503.

schools was the Escuela de Caminos (School of Highways and Canals) in Madrid, founded in 1802 and reestablished after a period of decadence in 1834, which during the 1840s struggled with physics and chemistry laboratories that were so rudimentary that the students had to perform experiments elsewhere.[11] Only in the 1870s did the level of instruction improve substantially.

Three other important government engineering schools were founded in 1835, on the model of the Highway School: the Schools of Mines, of Geographical (topographical) Engineers, and of Forestry (the Escuela de Ingenieros de Montes, which did not actually function until 1843). Although all these schools flourished in the last quarter of the century, the Forestry School was probably the most productive in terms of original research in a wide range of disciplines including descriptive and microscopic botany, geology, and hydraulics. The schools of industrial engineering were established by an 1850 decree in Madrid, Barcelona, Seville, and Vergara. Those in Seville (and another of slightly later foundation in Valencia) were abolished in 1865, and the Royal Institute of Madrid followed them into oblivion two years later for want of students. For practical purposes the Escola d'Enginyers Industrials (School of Industrial Engineers)[12] of Barcelona had a monopoly on higher education in industrial engineering until 1899, when a viable school was established in Bilbao (and another in Madrid soon after). The roster of important state engineering schools is completed with the Academy of Military Engineers, established in 1803 and installed at Guadalajara in 1833. Military engineers were conspicuous in Spanish mathematics and physics circles in the last quarter of the nineteenth century, and are so in the present. (Don Carlos, a character in Clarín's novel, *La Regenta* [1884], represents a type of military engineer

[11] Rumeu de Armas, *Escuela de Caminos*, p. 494

[12] Not to be confused with the Escola Industrial [Industrial School] of Barcelona, founded in 1904. This was a multifaceted technical school that provided instruction on many levels, from basic skills for textile workers to applied chemistry and mechanical applications.

who was a devotee of physics and mathematics and, through them, philosophy. Such persons were avid members of scientific societies. Artillery officers were also interested in higher mathematics.)[13]

Among private institutions, the most important was the Instituto Católico de Artes e Industrias (Catholic Institute of Arts and Industries), founded in Madrid in 1908. It specialized in electrical engineering, a newer field not adequately covered in the older schools, and it linked up with a number of other influential Jesuit scientific institutions, such as the Ebro Observatory. To the extent that the special schools served the interests of the political elite directly, its faculty members were well-placed to promote a positive image of science and technology.

ENGINEERING AND IDEOLOGY

Of particular relevance to the argument sustained in this book concerning the central role played by engineers in the support of pure science in general and relativity in particular are a number of facets of the professional ideology of this group as it developed in the last quarter of the nineteenth century. According to Ramon Garrabou, the "sacralization" of science and attachment of a scientific ethos to a program for social and technological change were hallmarks of the industrial engineering community of Catalonia. These engineers had a common historical perspective according to which the decadence of Hapsburg Spain was the result of the abandonment of science and technology and only by exalting science could Spain enter the stream of modern civilization. As early as 1865 the engineer Pau Sans i Guitart (d. 1900) indicated that it was only

[13] On the special schools, see Peset and Peset, *La universidad española*, pp. 451–460 (quotation from *La Regenta*, chap. 6, pp. 458–459); *Enciclopedia Universal Ilustrada*, 28: 1473–1489, s.v. "Ingeniero"; Rumeu de Armas, *Escuela de Caminos*; Alvarez de Morales, *Génesis de la universidad española*, pp. 458–472; and Ramon Garrabou, *Enginyers industrials, modernització económica i burgesia a Catalunya* (Barcelona: L'Avenç, 1982).

the practical spirit of engineering that had freed Spanish physics, chemistry, and metallurgy from the constraints of academic routinization. The engineer, said another, was a missionary, "the priest of modern science." Such views—broadly positivistic—were constantly reiterated in popular engineering and business journals and from the podia of engineering associations; they helped to create a public mentality more favorable to science and technology, and engineers lobbied openly for a national research capability. In 1879, observing the lack of laboratories and research centers, Josep Vallhonesta Vendrell noted the need to mould opinion and convince the public of the need to create "these sanctuaries for science."[14]

A second aspect of engineers' ideology was their apoliticism, which was so strongly felt that it became a dogma. In order to enhance their status as professionals, engineers cultivated an aversion to ideology and were openly critical of politicians, whose ethos was considered antithetical to their own. This was a common posture of European engineers in the late nineteenth century, but the aversion to politicians was heightened in Spain because of the political power of agrarian oligarchs indifferent or hostile to industrial research and development.[15] I have noted the emergence of a consensus favoring civil discourse in science as a means toward facilitating the modernization of the country. This consensus emerged around 1900, after a full quarter-century of agitation on behalf of the scientific ethos by an engineering community making steady gains in status and prestige. The constant stream of public utterances by engineers could not have failed to contribute positively to a climate of opinion favorable to modernization. During Einstein's visit, their voice was heard again on the social value of scientific research.

THE FOREGOING discussion provides the minimal background for understanding the broader social, cultural, and political

[14] Garrabou, *Enginyers industrials*, pp. 226–236.
[15] Ibid., pp. 280–284.

15

contexts in which the debate over relativity took place. As with Darwinism previously and Freudian psychology contemporaneously, the scientific and public debates were framed by certain ground rules—tight ideological control in the Darwinian polemics of the 1870s and 1880s, civil discourse in the Freudian and Einsteinian debates of the 1920s—which guided public opinion, informed the course of reception, and suggested a mode for contemporary evaluation of the events that form the subject matter of this book.[16]

[16] Cf. Glick, "Spain," pp. 307–345, and "The Naked Science: Psychoanalysis in Spain, 1914–1948," *Comparative Studies in Society and History*, 24 (1982), 533–571.

ONE

Spanish Science and the Reception of Relativity

AROUND 1900 a political consensus favoring vigorous academic freedom emerged at the same time when Spaniards were reassessing national goals in the wake of the Spanish-American War. Modernization was one of the answers to the nation's most pressing problems, and young academics of the "Generation of 1914," such as José Ortega y Gasset, took up the cry eagerly. For Ortega, science was the way to regeneration, and, since where there was shouting there would be no science, civil discourse was a necessary precondition for modernization.[1] In the first decade of this century, when the young Ortega was formulating his program for national regeneration, Spanish mathematics, physics, and astronomy—the three scientific disciplines on which the theory of relativity was to have an impact toward the end of the decade—were all undergoing a process of modernization and renewal. These disciplinary developments form the immediate context of the reception of relativity in Spain.

THE REVITALIZATION OF MATHEMATICS

According to Julio Rey Pastor, the mathematics of nineteenth-century Europe was not introduced in Spain until 1865 when José de Echegaray (1832–1916), who taught mathematics at the Escuela de Caminos (Civil Engineering School) in Madrid, published two volumes on plane and analytical geometry. Be-

---

[1] On Ortega's early political program, see Robert Wohl, *The Generation of 1914* (Cambridge, Mass.: Harvard University Press, 1979), pp. 122–159, esp. p. 134.

fore then, mathematics instruction was based entirely on compendia of eighteenth-century mathematics, texts made obsolete after analysis had been completely remade by Gauss, Abel, and Cauchy and after Riemann had created the modern theory of functions.

Spanish mathematicians before and contemporaneous with Echegaray believed that the field had been fully developed and that there was nothing left to solve except a few intractable problems, such as the quadrature of the circle and the trisection of the angle. Echegaray put an end to the circle-squarers in 1886, when he divulged Carl Lindemann's 1882 research on the transcendence of *pi* which demonstrated the impossibility of quadrature. Later he finished off the trisectors as well. At the Escuela de Caminos, he introduced the higher geometry of Michel Chasles and initiated a profound renovation of Spanish mathematics, centered in the engineering schools, which resulted in the assimilation of European mathematics (for example, that of Cauchy and von Staudt) of the first half of the century by around 1890. In this task he was aided by a few other figures, such as Eduard Torroja (1847–1918), Echegaray's successor at the Escuela de Caminos after the former accepted the chair of mathematical physics at the University of Madrid in 1878, who introduced Christian von Staudt's projective geometry; Ventura Reyes Prosper (1863–1922) and Josep M. Bartrina i Capella, two of the few Spanish mathematicians who were interested in non-Euclidean geometry; and Zoel García de Galdeano (1846–1924). Torroja's signal contribution was the initiation of original research in mathematics; his best work was the synthetic study of the curvature of surfaces.[2] Reyes Prosper, significantly, was unable to gain a university chair and always taught in secondary schools. His interests extended to mathematical logic, and he introduced the work of such logicians as Charles Sanders Pierce and Christine Ladd-Franklin in Spain. Of all Spanish mathematicians of his

[2] See discussion by José M. Plans, "Las matemáticas en España en los últimos cincuenta años," *Ibérica*, 25 (1926), 172.

generation, Galdeano was most in contact with European, particularly German, colleagues, and he was a regular participant in European mathematical congresses. Primarily a pedagogue, he introduced Cauchy's work on the functions of a complex variable.[3]

This is where matters stood—squarely in the early nineteenth century—when Rey Pastor sketched the recent history of Spanish mathematics in 1915. "Mathematical ideas," he wrote in summary, "arrive [in our country] when they have given of themselves all they have to give, when it is virtually impossible to continue the exploitation of the lode, that is, when they have crystallized into a book. The history of our mathematical culture is not the history of ideas, nor even the history of mathematicians; it is the history of text-books."[4] Both the textbooks and curricula of Spanish mathematics around the turn of the century were based on antiquated French models, the texts following pre-1850 French manuals both in format and contents.

The next phase in the history of Spanish mathematics, directed by Rey Pastor (1888–1962) himself, witnessed the assimilation of contemporary research. In 1915, at Rey Pastor's behest, the Junta para Ampliación de Estudios created the Mathematical Seminar and Laboratory (possibly modeled after Vito Volterra's seminar in Rome) where Rey Pastor trained the first generation of modern Spanish mathematicians whose work was fully contemporaneous with that of research-front mathematics. Rey Pastor's own research covered much of the ground of contemporary mathematics: functions of real and complex variables, theory of groups, and conformal map-

[3] See ibid., p. 172, and Gino Loria, "Le matematiche in Spagna ieri ed oggi: I matematici moderni," *Scientia*, 25 (1919), 443–444.

[4] Julio Rey Pastor, "El progreso de España en las Ciencias y el progreso de las Ciencias en España," in Ernesto and Enrique García Camarero, eds., *La polémica de la ciencia española* (Madrid: Alianza, 1970), pp. 458–478; Santiago Garma Pons, "José Echegaray y Eizaguirre," "Ventura Reyes Prosper," and "Eduardo Torroja Caballé," in J. M. López Piñero, et al., *Diccionario de la Ciencia Moderna en España* (Barcelona: Ediciones Península, 1983), s.v.

19

ping, as well as a whole series of problems in projective geometry. Research topics for his disciples were carefully chosen by Rey Pastor and were designed to overcome the derivative nature of previous Spanish work in the field: "The choice of research topics appropriate to the capacity of each, of not excessive difficulty so as to avoid discouragement, of previously established novelty so that the research might be fruitful and useful, represents, without doubt, the professor's most important task."[5] Research had to be novel, to ensure that the students' efforts were not wasted and to contribute to progress in the discipline as a whole.

By the end of the war decade Rey Pastor had created a "Valuable nucleus of mathematicians. . . . Pedro Pineda, Olegario Fernández Baños, Pere Puig Adam, José M. Lorente Pérez—he, above all, the best—Roberto Araujo, José Maria Orts, etc., and, among the youngest, Tomás Rodríguez Bachiller, Fernando Lorente de Nó, Teófilo Martín Escobar . . . all of them my students."[6] These young men worked on problems of current interest: Lorente de Nó and Puig Adam on relativity, Fernández Baños on complex spaces of n dimensions, Pineda on conformal mapping, Orts on Dirichlet's problem, and so forth.

Virtually all of these men had studied abroad, most of them in Italy, an all but compulsory requisite to their assignment to original and current problems. In 1917 Pineda received a Junta grant to study the theories of functions and groups with Hermann Weyl in Zurich. The following year Araujo also went to Switzerland to study higher geometry, and Fernández Baños, with another grant from the Junta, traveled to Switzerland and Italy. Lorente de Nó began to study harmonic curves (those with equations satisfying Laplace's equation) at the Mathematical Seminar in 1918, but the next year he was in Italy studying relativity with Tullio Levi-Civita. At a meeting of

[5] Quoted in Sixto Ríos et al., *Julio Rey Pastor, matemático* (Madrid: Instituto de España, 1979), p. 31.

[6] Ramiro Ledesma Ramos, "El matemático Rey Pastor," *La Gaceta Literaria*, II, 30 (March 15, 1928), 1. I have provided given names.

the Mathematical Society in Madrid the same year, Josep Maria Plans read an extract sent by Lorente on the movement of a point in an Einsteinian field, "which is a preview of research carried out in Rome under the direction of Prof. Levi-Civita and which, in its final form, will be published in the *Rendiconti* of the Reale Accademia dei Lincei."[7] Plans's letters to Levi-Civita over the following decade mention Lorente frequently. "Together," Plans remarked in 1922, "we speak frequently—and with all due praise—of the Italian mathematicians." In another letter, Plans notes that Vito Volterra had been "very friendly" to Lorente when the latter was studying in Rome with Levi-Civita. Indeed, when Volterra visited Madrid in 1932, it was noted in the press that Lorente had attended his classes in Rome in 1919 and that "to him is owing the first accounts of the original theories of Volterra, which he divulged after his return from Rome; thereafter he continued to correspond with his two great professors."[8] More of Rey Pastor's disciples followed, as Plans mentioned in a letter to Levi-Civita on March 9, 1926: "A number of times those young men who have had the good fortune to study [in Italy] (T. Martín Escobar, [Angel] Saldaña, etc.) speak of you in their letters to me." Others of Rey Pastor's group were indirectly in touch with the Italians: through letters, Puig Adam consulted Levi-Civita about his dissertation on the theory of relativity, and, following Plans's advice, Fernando Peña sent copies of his articles.[9] To publish in Italy was a mark of distinction; for example, Cabrera, Comas Solà, Plans, Rey Pas-

[7] Junta para Ampliación de Estudios, *Memorias, 1916–1917* (Madrid: 1918), pp. 42, 53, 187; *Revista Matemática Hispano-Americana*, 1 (1919), 224. I have been unable to locate Lorente's paper.

[8] Plans to Levi-Civita, October 23, 1921, April 20, 1922, and May 2, 1925. These and the following letters are cited from copies deposited in the California Institute of Technology Archives, Pasadena, by the Accademia dei Lincei, Rome.

[9] Plans to Levi-Civita, October 23, 1921, January 3 and September 9, 1924. Plans also sent José María Orts to Levi-Civita in 1924, but only for a recommendation to an Italian colleague with whom the Spaniard could study probability theory.

tor, and Terradas all published in the influential journal *Scientia* between 1917 and 1936.

In welcoming Federigo Enriques to Buenos Aires in 1927, Rey Pastor gave, in passing, a clue to his understanding of the value of the Italian connection for Spanish mathematicians. Enriques, he noted,

> also knew how to cast the net of inspiration, capturing the imaginations of those who could aspire, later on, to a deeper pedagogical experience than was contained in the more ephemeral one of his lectures. It is much to be desired that Argentinian students, for whom Europe is Paris, would share their preferences with the universities of Italy, where more intimate contact with professors makes study there more fruitful. With reference to the exact sciences, it is difficult to find at any one European university figures of the stature of Levi-Civita, Volterra, Enriques, Severi, Bompiani, Fermi, Amoroso, Gini, Amaldi, Vacca . . . under whose tutelage one can not only learn the present state of any theory and contribute to its advance, but also acquire that higher synthetic vision of science which only philosophical minds can achieve and without which scientific learning is mere erudition, falling short of cohesive understanding.[10]

This revealing passage suggests a number of observations. There is, first, the derogatory reference to Paris, symbol of French culture and science, against whose former dominance over Spanish science Rey Pastor was reacting. Second, he emphasizes the highly personalized pedagogical style of Italian professors and their ability to create affective bonds with their students. I believe there is an implicit comparison here with

[10] Julio Rey Pastor, "Federico Enriques," clipping from *La Nación*, 1927. I discuss the connection of Spanish mathematics with Italy more fully in "Einstein, Rey Pastor y la promoción de la ciencia en España," *Actas, I Simposio sobre Julio Rey Pastor* (Logroño, 1985), pp. 79–90. In the same volume, Giorgio Israel notes that there were also significant cognitive affinities between the two schools.

German pedagogy, characterized by the distance between student and teacher. Given the lack of a strong scientific tradition in Spain, Rey Pastor wanted his students to avail themselves of a more personal and nurturing relationship with the professoriate.

Rey Pastor's group was the most cohesive nucleus of scientists to receive the theory of relativity in Spain. Those physicists and astronomers who participated in the process did so more as individuals than as members of an articulated disciplinary network of discussion. The primary institutional focus of reception, beyond the Mathematical Seminar, was the Mathematical Society of Madrid (founded in 1911) whose discussions can be followed in the *Revista Matemática Hispano-Americana*, founded by Rey Pastor in 1919. García de Galdeano's pioneering journal, *El Progreso Matemático*, had failed in the 1890s because of the lack of quality material available for submission. The Mathematical Society's *Revista*, the predecessor of Rey Pastor's journal, suffered the same fate. Rey Pastor's 1915 attack on senior colleagues and his call for the Europeanization of Spanish mathematics had the effect of making many Spanish mathematicians, ostensibly those of the old school, unwilling to publish at all. The *Revista*'s editor, Cecilio Jiménez Rueda, complained in an editorial the same year: "When all Spanish mathematical papers are accused of discovering things already discovered, one hears talk of semi-obscurity in which, like bats, a group of Spanish mathematicians enjoys a comfortable position, when it is alleged that the Spanish Mathematical Society has lived long enough already and we must incorporate ourselves into the European scientific movement, we can understand the fear that many worthy persons display when we ask them to contribute to our journal."[11] The *Revista*'s successor—still extant today—was wholly dominated by the "European" group of the Mathematical Seminar.

This group, in spite of its dynamism in the 1920s, did not produce sustained research of a highly original nature. Many

[11] *Revista de la Sociedad Matemática Española*, 5 (1915), 4.

of these mathematicians, after completing competent theses, ended up as secondary-school teachers and authors of textbooks (Puig Adam, who wrote his dissertation on relativity, is a good example). But the role that those textbooks played in the revitalization of Spanish mathematics (their excellence as didactic tools is emphasized by present-day Spanish mathematicians and physicists who learned from them) should not be underestimated.

The protagonist of Francisco Vera's novel *El hombre bicuadrado*, a mathematician who worked as a bureaucrat in an office in Madrid, was accustomed, when suffering insomnia, to calm his nerves by reading Rey Pastor's *Introducción a la matemática superior*. His choice of reading material was not adventitious. Rey Pastor's book, which explicated contemporary mathematical theory by breaking it down into three principle families of ideas—sets, functions, and groups—was recognized at the time as having contributed notably to the inculcation of interest in modern mathematics among Spanish students. Vera, who had himself been a student of Rey Pastor, explained through his fictional alter ego that mathematics, "all order and method, is the best remedy for overstimulated nerves, the best antidote for *obscure* ideas." *Oscuridad* (darkness) was a code word that Rey Pastor invoked to characterize the routine textbook mathematics of the preceding generation, as well as its retrograde appeal to "traditional values" to oppose modern science—an excuse, in Rey Pastor's view, for its inability to come to terms either with modern science or with modern society.[12]

How successful, then, was the movement of institutionalization centered around the Mathematical Seminar and the Mathematical Society? In retrospect, and in the context of the present volume, the high quality of the members' education, their international connections, and their strong internal cohe-

[12] Plans, "Las matemáticas en España," p 174; Francisco Vera, *El hombre bicuadrado* (Madrid: Páez, 1926), pp. 88–89; Rey Pastor, "El progreso de España en las Ciencias," p. 459.

sion make intelligible the swift and nearly unanimous acceptance of relativity by the Spanish scientific community. But the leadership was uncertain of its own success. For Esteve Terradas, there was insufficient interest in higher mathematics among prospective students to permit the launching of the discipline, in spite of his best efforts. As he explained to Levi-Civita in 1922: "What a pity it is that interest in mathematics is so little developed among us! I do all I can to awaken it, although I recognize the insufficience of my own mind. But just like one who makes it a duty to instruct persons with little aptitude for music in the marvelous refinement of a Brahms, I strive to sow seeds which perhaps might flower in a not too distant future."[13] Scientists, of course, fully participated in the myth that Spain provided infertile soil for the cultivation of science. Plans, more optimistic, reached somewhat similar conclusions, although his analysis was sociological, not cultural. In summing up Spanish progress in the field since the 1880s, he wrote:

> Mathematics has made a great leap in Spain; the lack of phase with respect to other countries has almost been overcome. What is lacking is to increase research, which is still in diapers and restricted to a few, very few, privileged minds. Foreign journals are now read here, but one still notes in them the absence of Spanish names. How to remedy this? It is mainly a question of numbers. There can be no abundance of research papers so long as there is no regular number of persons who devote themselves to it.[14]

Einstein's visit, along with others by Levi-Civita, Jacques Hadamard, and European figures of similar stature, contributed to the crystallization of the discipline, at least as far as its self-image was concerned. But it is noteworthy that these visits took place not even a full decade after Rey Pastor's 1915

---

[13] Terradas to Levi-Civita, March 22, 1922.
[14] Plans, "Las matemáticas en España," p. 174.

speech, at a time when the vital core of the discipline was composed in large part of men in their twenties or early thirties,
few of whom held—or would ever hold—chairs in major universities. It was this discrepancy between promise and pros-
pect that made it appear that the undeniable achievements of
the Mathematical Seminar still rested on shifting sand.

The Renewal of Physics

In physics (and astronomy) the pattern of historical development just observed was repeated: a rudimentary level of instruction in the late nineteenth century, with very poor material facilities, followed by a leap forward during the first decade
of the twentieth century. The lowly state of Spanish laboratories was a leitmotif of nineteenth-century scientific literature.
The chair of biochemistry in Madrid, which served the faculties of medicine, pharmacy, and sciences, had no laboratory
budget at all since its establishment in 1887 until 1901. As a result, according to José R. Carracido, the subject was taught "as
if it were metaphysics." Finally, the sum of 6,000 pesetas, approximately the cost of a Huefner spectrophotometer, was al-
located for a laboratory. The total laboratory budget for all five
university faculties was also 6,000 pesetas annually, or 38.25
pesetas per chair per trimester, according to Carracido's calculations![15]
  The result of such impoverishment, in physics as well as in
chemistry, was not only a practical lag but a theoretical one as
well. Without laboratories it was impossible to demonstrate
recent theoretical advances. At the close of the nineteenth century the University of Madrid had but one laboratory of physics. As José M. Madariaga complained in 1902, "In what physics laboratory can the Zeeman effect be observed? In how
many is Verdet's constant in rotary magnetic polarization familiar? Where has it been possible to secure an accurate value

[15] José R. Carracido, *Estudios histórico-críticos de la ciencia española*, 2nd ed. (Madrid: Imp. Alrededor del Mundo, 1917), pp. 388–390.

for the length of Hertzian waves?"[16] Manuel Moreno-Caracciolo recalled "the picturesque chair of higher physics in the Central University [in the 1890s], from which its incumbent advised us not to believe what was said of X-rays, explicable, so he said, as a photographic trick." The early appreciation of the nature of X-rays, discovered by Roentgen in 1896, was dependent upon having access to the equipment. In contrast to the situation in Madrid, Eduard Fontserè and Eduardo Lozano gave a public demonstration of the obtainment of X-rays from a cathode tube at the Academy of Sciences of Barcelona in 1896, and the following year Bernabé Dorronsoro, professor of pharmacy in Granada, demonstrated in his own course photographs taken in May 1897 with X-ray apparatus he had acquired in Paris with his own funds.[17]

Renewal began after the turn of the century, and new laboratories sprang up in the years just preceding World War I—new chemistry laboratories in particular, because these were directly linked to the interests of major industries. Thus the assay laboratory (Laboratori General d'Assaigs) of the Mancomunitat (regional authority) of Catalonia was established in 1908, and the Laboratory of General Chemistry of the Residencia de Estudiantes in Madrid in 1912. By the early 1920s there was a well-established network of industrially oriented laboratories, housed in private industrial concerns, in engineering schools, or under the auspices of various public entities.[18]

[16] José María Albareda, *Creación de facultades universitarias y producción científica en el pasado siglo* (Granada: Universidad de Granada, 1950), p. 45, citing Blas Cabrera; José M. Madariaga, "Exposición de algunas consideraciones sobre la explicación de ciertos fenómenos y de sus relaciones con los de luz (1902), cited by Antonio Lafuente, "La relatividad y Einstein en España," *Mundo Científico*, no. 15 (June 1982), 585.

[17] M[anuel] Moreno Caracciolo, "El Laboratorio de Investigaciones Físicas," *El Sol*, September 10, 1920; Albareda, *Creación de Facultades Universitarias*, pp. 47–48

[18] See, for example, *Guia de les institucions científiques i d'ensenyança*, pp. 58 (Laboratori General d'Investigacions i Assaigs), pp. 156–157 (chemistry laboratory of the Escola de Teneria), pp. 159–160 (Laboratori d'Estudis Superiors

Exhibits of laboratory hardware were an important feature of the meetings of the Spanish Association for the Progress of Science (founded in 1908), and the War provided a direct stimulus for locally produced laboratory equipment. With the products of Zeiss and other German glassware suppliers cut off, the Spanish Army, as well as private and public chemistry and medical laboratories, had to replace equipment locally. By 1919 both optical equipment and laboratory glassware were being produced by Spanish companies.[19]

The most striking advances in instrumentation, however, were centered in a unique institution, the Laboratory of Automation, founded in 1906 in the Palace of Industry of the old Hippodrome of Madrid, and enlarged in 1914. There the director, Leonardo Torres Quevedo (1852–1936), best known for his spectacular inventions such as the automatic chess player and the aerial tramway at Niagara Falls, built a whole series of scientific instruments requested (and frequently designed) by various Spanish scientists to fit their own research requirements. Thus he built microtomes for Ramón y Cajal's neurohistology group; cardiographs for the physiologist José Gómez Ocaña; a device for determining the constricting or dilating action of different substances passing through blood vessels for Gómez Ocaña's successor Juan Negrín; a seismograph for Eduardo Mier y Mirva; a spectrograph and many other devices for magnetochemical measurements for Blas Cabrera; a Michelson-style interferometer for Manuel Martínez Risco, and so forth.[20]

---

de Química); Anon., "Los laboratorios de la Residencia [de Estudiantes]," *Residencia*, 5, no. 1 (1934), 26–30; *Ibérica*, 11 (1919), 101 (Laboratorio Químico Industrial of the Escuela de Minas); ibid., 18 (1922), 146–147 (Laboratorio Químico de Análisis Industriales y Agrícolas, Zaragoza), etc. *Ibérica* featured frequent articles on laboratories, particularly industrial ones.

[19] Diego de Imaz, "Los vidrios científicos. Nueva industria española," *Ibérica*, 11 (1919), 221–223.

[20] On the Laboratorio de Automática, see José García Santesmases, *Obra e inventos de Torres Quevedo* (Madrid: Instituto de España, 1980), pp 301–306. A list of apparatus built for specific scientists is in the exhibition catalog, *Leo-*

In the 1920s, Spanish laboratories were at least able to support experimentation in a limited range of subdisciplines, relying on the resourcefulness of individual researchers and a mechanical genius like Torres Quevedo. But outside of a few well-funded organizations, such as Cabrera's physics Institute and the six medical laboratories of the Residencia de Estudiantes, the infrastructure, particularly in the universities, was weak. Study abroad continued to be a basic requirement for advanced research that, in the experimental sciences, could not be undertaken on the basis of the limited facilities available in Spain. The problem of university instruction was critical. As late as 1927 the chemist Enric Moles (1883–1953) was threatening to resign his chair of chemistry at the University of Madrid because of the laboratory problem:

> . . . the installation of laboratories is indispensable—to the point where I told the Director General [of higher education] that if I don't secure their installation during the next academic year I will consider that I have failed and will resign my chair. Believe me, there is no arrogance in this. It is just a matter of my conviction that every effort is sterile without them. The teaching of chemistry in Spain has been a bad joke. It has been explained on blackboards, more or less as grammar is. All this is fruitless and would be considered not only laughable in any European country, but inconceivable.[21]

Perhaps the most important change, then, was one of values. In the nineteenth century, according to Carracido, there had been no "public conception" regarding the laboratory needs of chemists or other scientists. By the 1920s, the image

---

nardo Torres Quevedo (Madrid: Colegio de Ingenieros de Caminos, Canales y Puertos, 1978), pp. 121–122. On instrumentation for Negrín, see Gonzalo R. Lafora, "El Congreso Internacional de Fisiología en Paris," *El Sol*, August 10, 1920. On Cabrera, see Moreno Carraciolo, "El Laboratorio de Investigaciones Físicas."

[21] "Una conversación con el químico Sr. Moles," *El Sol*, June 20, 1927.

had clearly changed. The protagonist of Ramón Gómez de la Serna's science-fiction novel, *El dueño del átomo* (1926), is a physicist who works in a Madrid laboratory seemingly inspired by that of Blas Cabrera, "a recondite institute hidden in the outskirts of the city." He is, moreover, an experimentalist "who works in an important laboratory with new apparatus, extraordinary clocks, machines to weigh the imponderable and a lot of glass in spirals and retorts, and all this teratology of small transparent monsters that complexify laboratories."[22] There, after patient experimentation, he splits an atom.

Scientists themselves, and those with scientific education who popularized their vocation, were responsible for changing the public image of the scientist. To Luis Urbano, a Dominican philosopher who had studied physics in Madrid with Plans and spent long hours of research in Cabrera's Institute, laboratory researchers were to be admired: "They are solitary workers in the shop of science, who labor through vocation and without philosophical bias."[23]

Practically speaking, modern physics in Spain dates to 1910, when the Junta para Ampliación de Estudios created the Institute of Physical Research in the Hippodrome of Madrid, with Blas Cabrera as its director. Cabrera (1878–1945) received his doctorate at Madrid in 1901 and had pursued research mainly on the properties of electrolytes. He had already established himself as the country's leading experimental physicist by 1910. The subsequent development of Spanish physics was inextricably linked to Cabrera's Institute.

In an evocative article written ten years after the founding of the Institute, Moreno-Caracciolo noted that those who, like himself, had passed through the University of Madrid's science faculty in the last years of the nineteenth century, with its "oral classes of physics and chemistry, with their undigestible lessons, repeated from memory," could scarcely believe that

[22] Ramón Gómez de la Serna, *El dueño del átomo* (Buenos Aires: Losada, 1945), p 11.

[23] Luis Urbano, *Einstein y Santo Tomás: Estudio crítico de las teorías relativistas* (Madrid-Valencia: La Ciencia Tomista, [1926]), pp. xxv–xxvi.

the institute was an official center of instruction.[24] Of course, in reality it was not an official center because it was not dependent on a university but rather upon the Junta, which supplied its budget (40,000 pesetas in 1919). Cabrera had total financial control, and Caracciolo stressed the connection between budgetary independence, the independence of work carried out at the Institute, and, in a broader perspective, the consequent reinvigoration of Spanish physics and the overthrow of rote instruction. The independence of research was crucial. As an example, Santiago Piña de Rubíes, who had studied in Geneva and held an assistantship in the physics department there, received 300 pesetas monthly to support his spectrographic research. But Piña had no Spanish academic title and thus, Caracciolo notes ironically, "he could serve only as a beadle in an official educational institution." The Institute's independence attracted the very best researchers, as they were able to work there without having to waste time attempting to ascend the tortuous Spanish academic ladder, a fact that caused considerable resentment among the established professoriate.

Assaying the practical accomplishments of the Institute's first decade, Moreno-Caracciolo noted that it had educated a whole generation, first, of university professors (Jerónimo Vecino at Zaragoza was the leading figure) and, second, of secondary school teachers (Vicente García Rodeja in Oviedo and José de la Puente in Barcelona, among others).[25] Beyond that considerable accomplishment, its investigations "have corrected scientific theories universally held to be certain and have exposed errors committed by foreign scholars which have given rise to the deduction of false consequences." Below I will comment on the contents of this research, but here I stress the

[24] The following section is based on Moreno-Caracciolo, "El Laboratorio de Investigaciones Físicas."

[25] Both Vecino and de la Puente participated in the reception and diffusion of relativity. On Vecino's role, see chapter 4. De la Puente was the translator of P. Kirchenberger, *¿Qué puede comprenderse sin matemáticas de la Teoria de la Relatividad?* (Barcelona: Juan Ruiz Romero, 1923).

seemingly ingenuous, or rather modest, nature of Caracciolo's criterion for scientific excellence. For a Spanish scientist—he implies—the ability to criticize mainstream science intelligently is goal enough. In fact, the ability to criticize is intimately linked to the effectiveness of scientific communication and is a necessary complement of the capability for independent research.

Around 1920, nearly all the significant research in Spanish experimental physics was concentrated in the Institute. There Cabrera, the chemist Moles, and their disciples worked on magnetism—Moles on atomic weights, Miguel Catalán and Piña on spectrography, and so forth. In 1926 the Institute received a Rockefeller (International Education Board) grant to upgrade its physical plant. This concentration of energy produced astounding results: of all articles on physics published in the *Anales de la Sociedad Española de Fisica y Quimica*, from the founding of the Institute in 1910 through 1937, this single center (and the successor National Institute of Physics and Chemistry, established in 1931) accounted for 40.9 percent of all authors and no less than 72.3 percent of all articles.[26]

THE RECEPTION OF RELATIVITY BY SPANISH SCIENTISTS

Esteve Terradas

The first Spaniards to grasp the significance of relativity and its place in modern physics were Esteve Terradas and Blas Cabrera. Terradas (1883–1950) had attended elementary school in Berlin-Charlottenburg, as a result of which he became the best Germanist among Spanish physical scientists. As a university student, according to his Madrid roommate Enric de Rafael, he never read the assigned texts but would prepare himself with some better volume in German. For the second-level course in analysis, accordingly, Terradas prepared for the ex-

---

[26] Manuel Valera Candel, "La producción española en física a través de los Anales de la Sociedad Española de Física y Química, 1903–1937," doctoral dissertation, University of Murcia, 1981.

amination by reading Eugen Netto's *Vorlesungen über Algebra*. Terradas spoke German to a fault and his mastery of the language was legendary.[27]

Because of his easy access to German culture, especially scientific culture, Terradas was probably the first Spaniard to learn of special relativity. Most likely, he had read Einstein's article soon after publication because he subscribed to (or owned many numbers of) the *Annalen der Physik* from 1903 on, and in 1905 he had corresponded with the editor Paul Drude on a problem related to his doctoral dissertation on the absorption of light by pleochroitic crystals.[28] Terradas received his doctorate in physics from the University of Madrid in 1905 and was auxiliary professor there in 1906 while obtaining another doctorate in mathematics in order to compete for a chair of Rational Mechanics at the University of Zaragoza, which he held in 1906–1907. In April 1907 he won the chair of Acoustics and Optics at the University of Barcelona, where he was also interim professor of Electricity and Magnetism.

Special relativity made its first appearance in Spain in papers by Terradas and Cabrera at the first congress of the Spanish Association for the Progress of Science, held in Zaragoza in 1908. Terradas, in presentations on black-box radiation and theories of the emission of light (and again the following year in his discourse of reception in the Barcelona Academy of Sciences) alluded to the Special Theory only as a new deduction of the "principle discoverd by Lorentz," adding that Einstein and Laub "have recently applied it to establish more general laws of Electrodynamics."[29]

---

[27] Enrique de Rafael, "Juventud y formación científica de Terradas," in Real Academia de Ciencias Exactas, *Discursos pronunciados en la sesión necrológica en honor de . . . Esteban Terradas e Illa* (Madrid, 1951), p. 6; on Netto (1848–1919), see *Dictionary of Scientific Biography*, 16 vols. (New York: Charles Scribner's Sons, 1970–1980), X, 24.

[28] De Rafael, "Juventud de Terradas," p. 8; on Drude, see *Dictionary of Scientific Biography*, IV, 189–193.

[29] Terradas, "Teorías modernas acerca de la emisión de la luz," in Asociación Española para el Progreso de la Ciencia, Zaragoza Congress, Sección segunda, *Ciencias Físico-Químicas* (Zaragoza, 1908), pp. 1–21; "Sobre la emisión de ra-

At the University of Barcelona, the young Terradas impressed his students with his innovative teaching methods as well as his grasp of the new physics. One of his early students, Julio Palacios, remarked that the usual approach to physics encountered by Spanish university students in the first decade of the century was that "science was already something finished and closed, to which nothing remained to be added."[30] By contrast, Terradas made clear to his students his total receptivity to new ideas. Palacios recalled:

> Around the middle of my [undergraduate] career, I had the good luck to have Terradas as a professor, and the contrast [with the former view] could not have been more brusque. He had no method, no text, not even a syllabus, and the first day he left us speechless when he asked if we wanted to learn classical wave optics or if we preferred (in 1910!) Planck's quantum theory. If that wasn't enough, every day he assigned books and journals in English and German, assuring us that, with but a little effort, we would be able to understand them.[31]

By 1910, therefore, Terradas had incorporated quantum physics into his university teaching, and in 1915 he gave a course on quantum theory entitled "Discrete Elements in the Matter of Radiation" at the Institut d'Estudis Catalans. In 1912 he published a detailed, discursive review of Max von Laue's book on special relativity, in which he noted that "The principle of relativity is accepted today by almost everyone. In chairs of phys-

diaciones por cuerpos fijos o en movimiento" (Barcelona, 1909), and discussion by Antoni Roca, "La incidència del pensament d'Einstein a Catalunya," in *Centenari de la naixença d'Albert Einstein* (Barcelona: Institut d'Estudis Catalans, 1981), pp. 170–172.

[30] Julio Palacios, "Terradas físico," in Real Academia de Ciencias Exactas, *Discursos*, p. 16. This was a stereotypical view expressed by leading physical scientists prior to 1905—for example, Albert Michelson's assertion (in 1903) that "the most important fundamental laws and facts of physics have all been discovered"; cited by Lewis Feuer, *Einstein and the Generations of Science* (New York: Basic Books, 1974), p. 253.

[31] Palacios, "Terradas físico," p. 16.

ics its language has been generally adopted."[32] We can assume that Terradas had in mind not only his own pedagogy but also direct information from European physics departments—German, most likely—which testified to the reception of the new physics. At this juncture, according to Antoni Roca, relativity was for Terradas still wholly consistent with the theory of electromagnetism, a new language for thinking about electricity. He did not yet see it as the basis for a new mechanics.[33]

By the middle of the decade, when he was following developments in general relativity, Terradas had abandoned the Lorentzian framework and he now presented relativity in completely Einsteinian terms. Then, in the winter of 1920–21, he gave a thirty-session course, "Relativity and the New Theories of Knowledge," given under the auspices of the Mancomunitat of Catalonia. The text of the course is not preserved, but de Rafael published his notes from a number of Terradas's lectures, broken down into six themes: Galilean relativity, the aether, the Michelson-Morley experiment, some generalizations regarding aether-drift experiments, the Lorentz contraction, and local time.[34] In contrast to many Spanish commentators, Terradas was unequivocal in his acceptance of special relativity, which had disproved the existence of aether, an imponderable substance that supposedly filled space and served as a medium for the propagation of light waves: *Aether does*

[32] Victor Navarro Brotons, "Esteve Terradas e Illa," in *Diccionario Histórico de la Ciencia Moderna en España*, s.v. Terradas, "Sobre'l principi de relativitat," *Arxius de l'Institut de Ciències*, 1 (1912), 94.

[33] Roca, "Incidència del pensament d'Einstein," pp. 172–173. As proof of this contention, Roca notes that the first mention of Einstein in the *Enciclopedia Universal Ilustrada* is in the article "Electricidad," which Terradas wrote around 1914. There is no article on Einstein. Terradas's 1912 review of von Laue is "Sobre'l principi de la relativitat," *Arxius de l'Institut de Ciències*, 1 (no 2), 84–94.

[34] Enrique de Rafael, "De relatividad (Apuntes con ocasión de las conferencias de E. Terradas en el Institut)," *Ibérica*, 15 (1921), 89–91, 218–221, 376–379. On Terradas's early appreciation of general relativity, see "Sesion académica de ciencias, 20 enero 1914," *Boletín de la Real Academia de Ciencias e Artes de Barcelona*, 3 (1909–16), 427.

not exist, and neither does absolute space, inasmuch as we cannot in any way demonstrate its existence by physical means."

Continuing in the same vein, Terradas discusses Dayton Miller's 1904, 1905, and 1906 interferometer experiments to test for aether drift. According to Albert Michelson's hypothesis, inasmuch as the aether is at rest and the earth moves through it, the speed of light on the earth's surface should depend on what direction it is traveling. The interferometer was an instrument that Michelson devised to measure minute phase shifts between two parts of a light beam. It could be used to measure differences in the speed of these two parts if they traveled along different paths in the same medium. The "aether drift," or the effect of earth's motion through the aether on the speed of light, was to be measured by observing the extent of the phase shift between the two beams as the interferometer was rotated through ninety degrees. In the Michelson-Morley experiment *no difference in the velocity of the two parts was detected.* The precise role of this experiment in the elaboration of special relativity is a polemical subject. "The fringes didn't shift because they had no reason to," Terradas (or de Rafael) noted testily. "All that shifted were some aprioristic concepts which had been accepted in science without sufficient examination, with excessive rigidity, and retained stubbornly and futilely by men of science."[35] Terradas's uncompromising posture in favor of relativity was unusual, given a scientific tradition where eclecticism was the norm and where the same eclecticism was frequently a screen for an inability to judge theoretical issues accurately. His advocacy of relativity reached a very wide audience after 1923, when his fifty-page article on the subject appeared in the *Enciclopedia Universal Ilustrada*, together with a biographical sketch of Einstein.[36]

---

[35] Ibid., p. 220. Later in 1921, Terradas gave two additional lectures on relativity in Madrid; see *Ibérica*, 16 (1921), 67.

[36] "Relatividad," *Enciclopedia Ilustrada Universal*, L, 455–512. See Roca, "Incidència del pensament d'Einstein," pp. 172–173. The proofs of the article "Relatividad," dated May 5, 1923, are preserved in the Terradas Archives. In-

Throughout the 1920s Terradas received offprints of Levi-Civita's publications, including those relating to relativity. These publications form part of the Terradas collection, now housed at the Institut d'Estudis Catalans, and include proofs of Levi-Civita's 1920 article on geometric optics and general relativity. This is a striking indication of the speed with which Terradas received information from the research front that contradicts the generalized perceptions of information lag. Terradas also reported to the Spanish mathematical community on Levi-Civita's latest research.[37]

Terradas's activities stand in contrast to clichés regarding the lag between the appearance of physical concepts in the scientific "mainstream" and their reception in Spain. The perception commonly held by Spanish scientists from the late seventeenth century well into the twentieth was that Spaniards were the last to receive new ideas. However, by the second decade of this century the lag was not as great as it appeared. New ideas were often received quickly, but by an extremely limited number of persons. The reason a lag was so generally perceived seems to be the result of the slowness of secondary diffusion of ideas beyond small circles of specialists. The notion or perception of lag in information, which was a keystone in the often articulated defensive view that Spaniards were unable to do science, was developed by scientists and engineers

---

terestingly enough, the article, although perhaps drafted by Terradas, may have been more the work of his assistant Ramon Jardí (1881–1972), according to what Jardí told his own students. See Josep M. Vidal i Llenas, "L'Entrada de la relativitat entre nosaltres," in *Centenari de la naixença d'Albert Einstein*, p. 140. On the other hand, there are passages in this article very similar to others in Rafael's "Apuntes" of Terradas' lectures.

[37] Terradas received proofs of Levi-Civita's article, "L'Ottica geometrica e la relativita generale di Einstein," published in *Revista d'Ottica e Mecanica di Precisione*, 1 (1920), 187–200. At a meeting of the Sociedad Matemática in Madrid in 1929, he presented reports on Levi-Civita's Hamburg seminar on adiabatic invariants and his Bologna lecture on the application of the theory of integral invariants to some problems of astronomy; *Revista Matemática Hispano-Americana*, 4 (1929), 61–62.

who, as students, had to face directly real institutional and cultural barriers to the flow of new ideas.

Blas Cabrera

Cabrera's assimilation of relativity followed a similar course. At the Zaragoza meeting in 1908 he presented a paper on the theory of electrons, explicating Maxwellian and Hertzian electromagnetic conceptions of light. In this presentation, Einstein's theory is again mentioned as a refinement of Lorentz's electron theory, and Cabrera still presupposed the existence of an aether.[38] It was not until 1912 that he was able to state, in discussing special relativity, that there was no experiment to detect the aether.[39]

In a series of lectures on electricity delivered at the Residencia de Estudiantes in January 1917, Cabrera presented a view of the Special Theory in which he was fully cognizant of its revolutionary nature. The failure of attempts to determine absolute motion, he noted, had created the need to "reorganize science" in order to rid it of "such evident contradictions." The confusion which the theory, with its seemingly paradoxical reinterpretation of simultaneity, incites in us is simply the product of "a mental habit, of the supposed independence of space and time."[40]

In November 1921 Cabrera delivered an important address at the Academy of Sciences on the current state of physics.[41] Since he was to become, in the eyes of the Spanish public, the leading spokesman for relativity during Einstein's visit, his

[38] Cabrera, "La teoría de los electrones y la constitución de la materia," cited by Roca, "La incidència del pensament d'Einstein," p. 170.

[39] "Principios fundamentales de análisis vectorial en el espacio de tres dimensiones y en el universo de Minkowski," *Revista de la Real Academia de Ciencias Exactas*, 11 (1912–13), 326–344ff.

[40] Cabrera, *¿Qué es la electricidad?* (Madrid: Residencia de Estudiantes, 1917), pp. 173–176.

[41] *Momento actual de la física* (Madrid: Real Academia de Ciencias Exactas, Físicas y Naturales, 1921).

views on the current state of physics, and relativity's role therein, are of great interest. For Cabrera, quantum theory and the structure of the atom were the dominant issues in 1921. He was an experimental physicist, and his identification of atomic structure as the central issue is hardly surprising. But how did relativity fit into the picture? Although he was not, of course, unmindful of the changes in conceptualization that relativity had introduced into physical research, Cabrera stressed, in this address, the social concomitants of relativity's reception. Referring to resistance to relativity in the name of "immutable principles," he insists on the urgent need to create a favorable environment "to give a greater impulse to the advancement of national science," and notes the difficulty of doing this "in the midst of an absolutely indifferent society, without receiving the heat given off by the favorable or adverse criticism of those who directly surround us."[42] Cabrera, like other Spanish scientists in touch with mainstream European science, understood that both the smallness of Spain's scientific community as well as the lack of scientific discussion in the society at large made it difficult to establish the necessary climate for the critical and informed evaluation of ideas that scientific advance requires.

Turning to current physics, but still with an obvious reflection on the intellectual environment familiar to him, Cabrera observes that a science based on postulates, in the sense of axiomatic truths, is false. Relativity, he notes, had been opposed in the name of the immutability of principles: two of those that it had invalidated were Lavoisier's principle of the conservation of mass and Newton's concept of attraction. But this statement merely serves him as an introduction to quantum theory, which also "contradicts the extrapolation of princi-

[42] Ibid., p. 8. In another context (an observation by Odón de Buen on the origin of the world), Cabrera remarked: "There exists much prejudice and sediment which must be destroyed in order for new theories to take root"; see Mariano Salaverría, "Terminación de la Semana Científica," *El Sol*, September 15, 1921.

ples."[43] The view of physics that Cabrera presented was not so different from his similar presentation in January 1915 at the Madrid Athenaeum, in which he also stressed the structure of the atom in its relation to quantum theory, in particular Sommerfeld's quantum of action.[44] It is clear that Cabrera related the discussion of atomic structure directly to the work that he and Moles were carrying out on magnetons, which Cabrera believed to be "constituent elements of the nucleus."

Cabrera's views are significant because the events of 1923 thrust him into the position of the leading Spanish spokesman for Einstein, a role he might not have played had Spanish physics not been so thinly staffed. In the absence of any voice in theoretical physics, the experimentalist Cabrera shared the defense of relativity with mathematicians.

Among Spanish scientists lecturing on relativity in the 1920s, Cabrera probably had the widest audience—in spite of the fact that his lecturing style was characterized (albeit by a listener skeptical of relativity) by a "diffuse slowness."[45] His success as a spokesman for relativity was due to his ability to interpret Einstein's theories on a number of levels, with equal comprehensibility, whether speaking to a lay or a scientific audience. With regard to his "popular" presentations, he was the only leading Spanish relativist who was interested in the philosophical ramifications of the theory. In a 1921 lecture to philosophers, he endeavored to place relativity in the context of the theory of knowledge. In the investigation of nature, he began, there are but two ways to proceed. Either we travel a path of thought already developed, in order to find a solution, or else there is no path and we must proceed "through tentative probings." Quantum physics, characterized by a mass of experimental data but lacking a logical construction through which to interpret them (because, he believed, of the lack of an

[43] Cabrera, *Momento actual de la física*, p. 12.
[44] "Estado actual, métodos y problemas de la física," in *Estado actual, métodos y problemas de las ciencias* (Madrid: Ateneo de Madrid, 1916), pp 109–143.
[45] Ataulfo Huertas, "La relatividad de Einstein," *Revista Calasancia*, 11 (1923), 246.

adequate mathematical theory), is an example of the latter. Relativity illustrates the former: "It has been constituted in a very short time as a construct of exemplary logic, because appropriate methods of reasoning were known prior to posing the problem which science imposed." All that was necessary was to examine the origin of contradictions in the old theory in order to eliminate false extrapolations from it.[46]

From an analytical, logical standpoint, Cabrera continued, relativity is easy to analyze: "It has to do with a simple change of variables, with the condition that a certain number of analytical expressions . . . be 'invariant' of the transformation." It does not matter which invariants we select to resolve a specific problem: "In all cases we arrive at the same group of Lorentz." When we then look at the physical world, modifying our conceptions in accord with the Lorentz transformation, we note that time and space are no longer independent but now appear to us to be "symmetrically combined, the three dimensions of ordinary space and time, which we now treat as the four dimensions of a hyperspace." Cabrera then notes that a number of the well-known paradoxes associated with the Special Theory—the Lorentz contraction, the variability of mass, and its relation to energy—had attracted resistance from scientists, but these objections were largely of a sentimental nature, he judged.[47]

Turning to general relativity, he observed that Einstein had replaced a rigid system describing the universe in terms of the classical heliocentric coordinates and a single clock determined by the earth's rotation with a flexible system that could be molded to the circumstances of time and space—Einstein's "mollusk of reference." The classical problems of the finitude versus infinitude of the universe and of its origin and end were simply the obligatory consequences of a system of references with indefinite rectilineal axes and a permanently running

[46] Blas Cabrera, "Las fronteras del conocimiento en la filosofía natural," *Verbum*, 14, no. 55; reprinted in *Revista de Filosofía* (Buenos Aires), 14 (1921), 153.

[47] Ibid., pp. 154–155.

clock. In the mollusk of reference, neither time nor space can be infinite.[48]

In the fall of 1921 Cabrera gave a short course of twelve lectures at the University of Madrid, "with incredible attendance" of the scientific community.[49] Cabrera's presentation from the perspective of an experimental physicist provides an illuminating counterpoint to the presentations of Plans, Terradas, and others more attuned to mathematical physics. In Cabrera's view physicists had lost sight of the experimental basis of Galilean-Newtonian relativity (by making it a postulate) and had since striven to make theoretical physics accord as closely as possible to the principles of Rational Mechanics.[50]

Not unexpectedly, Cabrera placed great emphasis on the experimental proofs, not only of the General Theory but of the Special as well:

> But where the relativistic theory has gained its most brilliant proofs has been in the field of electromagnetic phenomena; the sage lecturer, after expounding the variations which relativistic mechanics introduced in the expression of Hamilton's principle and of electromagnetic field equations, referred with deeply felt enthusiasm to predictions of Bohr and Sommerfeld, especially those related to the breaking down of the rays of hydrogen, which experience fully confirmed afterwards. This fact is the one which locates the ideas of Einstein and Planck in the panorama of fully accepted scientific advances.[51]

[48] Ibid., p. 156.

[49] Enrique de Rafael, "Conferencias de relatividad en la Universidad de Madrid," Ibérica, 16 (1921), 306. In September 1921 Cabrera had also given a popular lecture on relativity at the Sociedad de Oceanografía de Guipúzcoa, published as La teoría de la relatividad (San Sebastián, 1921); see review in Ibérica, 19 (1923), 63.

[50] Ibérica, 16 (1921), 324.

[51] Ibid., p. 356. "Sommerfeld was the first to point out and carry through the idea that the variation of mass demanded by the theory of relativity must have an effect on the spectrum. He replaced the classical energy-function by

Sommerfeld's research of the previous decade, which proposed that the fine structure of the spectral lines of hydrogen, revealing relativistic increases in the mass of electrons, was arcane by comparison with the notoriety of the eclipse observations, and surely Cabrera, a few colleagues at his Institute, and Terradas were the only persons in Spain who had followed it.[52] Cabrera also discussed the astronomical evidence supporting general relativity: the perihelion of Mercury in the seventh lecture and the 1919 eclipse results in the tenth. Like Terradas, he was unsparing of those who attempted to explain away the results:

> First they questioned the exactitude of the observations. But then they wanted to seek an *ad hoc* explanation. This is wholly futile; one can explain a single phenomenon in a variety of ways, but the entirety [of phenomena] is more difficult to harmonize. It costs us nothing to suppose a solar atmosphere which refracts rays in a ratio conducive to the production of an observed effect. But if one bears in mind that this is a fact which must be explained *along with many others than come before it*, one sees the inconsistency of such an aprioristic manner of thinking.[53]

Rafael concludes his description of this lecture with this florid characterization of the audience reaction: "The enthusiasm with which today's audience received the luminous observations of the lecturer was depicted so keenly on their faces, that for one moment it seemed that, leaving aside matter and its base demands, the spirits of those present were transported to

the relativistic one . . ."; Max Born, *Problems of Atomic Dynamics* [1926] (Cambridge, Mass.: MIT Press, 1979), p. 41.

[52] On Sommerfeld's research, see Paul Forman and Armin Hermann, "Arnold Sommerfeld," *Dictionary of Scientific Biography*, XII: 529.

[53] Ibid., p. 388. It is interesting to compare Rafael's summary with Cabrera's later volume, *Principio de relatividad* (Madrid: Residencia de Estudiantes, 1923), pp. 254–259. There Cabrera concludes that all the cavils over the eclipse results were based on an unacceptable extrapolation of the Newtonian gravitational hypothesis that had acquired the value of "an imposition of Nature."

the purest realms of ecstatic contemplation of the loftiest ideals of science."

The enthusiasm of Cabrera's audience, composed principally of scientists and engineers (we may presume), was faithfully recorded by Enric de Rafael. But those persons were not so much seeking ecstatic contemplation of pure science as they were celebrating the demise of a discredited science. It is significant that the most positive audience reaction occurred not when Cabrera announced some innovation of the new physics but when he debunked the old. Such *rites de passage* were characteristic of Spanish science in the 1920s—if not in all disciplines, then at least in the most advanced or successful.

Josep Maria Plans

Plans (1878–1934), who won a prize offered in 1919 by the Academy of Exact Sciences for a work explaining "the new concepts of time and space," was the third leading spokesman for relativity in the 1920s. The referee charged by the Academy with evaluating his manuscript was a military engineer and mathematician, Nicolás de Ugarte (d. 1932). Ugarte, although favorable to relativity, had no profound understanding of it, and his report was based on notes provided him by Blas Cabrera.[54] Cabrera makes a number of interesting points. First, the idea for a prize on relativity had been in the air for a number of years before 1919, and that explains why Plans's treatment, later published as *Nociones fundamentales de Mecánica relativista*,[55] was mainly devoted to the Special Theory (only the last two of the volume's nine chapters deal with general rel-

[54] See Ugarte's article, "Las teorías relativistas," *Madrid Científico*, 31 (1924), 178–179, where he notes that relativity "is congenial and attractive owing to its conceptual novelty," but that the destruction of classical mechanics will take a long time and profound modifications of it are unnecessary. Although his enthusiasm for innovation marks his distance from traditional values, Ugarte's closing statement that nothing is absolute except "the Supreme Being" seems designed to allay the fears of conservative Catholics.

[55] Madrid, Real Academia de Ciencias Exactas, 1921.

ativity). But, the report notes, the Academy could not have had in mind an essay on the General Theory, "because at the time when the competition was devised, the theory of relativity was little known." Second, the prize was offered by the exact science section of the Academy, not the physical, and this explains the omission of detail regarding experimental facts. (Still, Plans's ninth chapter is a discussion of the equations of movement in a gravitational field and their application to the deformation of the perihelion of Mercury and the deflection of light rays.) The point is interesting because the conceptualization of both the prize and Plans's response to it must have taken place before the eclipse observations of 1919 had directed public and scientific attention to experimental results. Plans, moreover, always stressed the experimental results in his subsequent popularizations of relativity. In a 1920 article, for example, he noted with respect to the General Theory that one could ask nothing more of a theory than to have its predictions confirmed by observation and, with the hindsight of several more years, he added that "the great success of Einstein's theory and its relativistic mechanics" was the correct prediction of the deformation of the perihelion of Mercury.[56]

According to Tomás Rodríguez Bachiller, Plans was the only person in Madrid in the 1920s who was capable of teaching relativity at an advanced level, and this was undoubtedly because of his ability to introduce original interpretations or formulations into his discussion. Plans himself noted that, while teaching a course on relativistic mechanics in Madrid, it occured to him that, by applying Paul Appell's analogy be-

[56] Plans, "Algunas ideas sobre la relatividad," *Ibérica*, 13 (1920), 380, and "Bosquejo histórico y estado actual de la mecánica celeste," *Ibérica*, 23 (1925), 111. In Cabrera's view, Plans drew his inspiration from Laue's 1912 summary but had introduced a number of original points, mainly having to do with the movement of material points in the universe of Minkowski. The referee's report also notes that, although relativity modifies certain concepts of classical mechanics, "it should not spread panic among partisans [of the latter]"; Nicolás de Ugarte, "Informe de la Real Academia de Ciencias Exactas sobre la Memoria presentada en el concurso de premios del año 1919 . . . ," *Revista de la Real Academia de Ciencias Exactas*, 2nd series, 19 (1920–21), 234–243.

tween the form of equilibrium of a cord and the trajectory of a light ray, it was possible to derive a formula for the deflection of light in a gravitational field with an equation that was simpler, "more natural and logical" than that of Einstein: the deflection was simply the angle of the asymptotes of the hyperbola formed by the ray passing through the field.[57]

Plans's note was one of the few original contributions on relativity produced in Spain in the 1920s. Another by a mathematician was Pere Puig Adam's thesis on four problems in the mechanics of special relativity, carried out in the Mathematical Laboratory under the direction of Plans. It should be noted, in light of Rey Pastor's policy of identifying significant current problems for doctoral candidates, that Puig Adam's avowed motivation in pursuing this topic was not to present "a theoretical work of scientific merit" but only to "handle . . . the fundamental equations of this Mechanics, applying it to specific problems to see the difficulties of calculation that these present and making the necessary comparison with the difficulties of calculation that appear, and the results which are obtained, for the same problems in classical Mechanics."[58] The interest of Spanish mathematicians in relativity was more in the formal representation of specific aspects of the theory than in physical problems.

Plans was well-known for his exaggerated religiosity. Educated by Jesuits before entering the University of Barcelona, he acquired, according to one of his students, a "love for everything jesuitical."[59] For Terradas, who was scarcely less fervent in his piety, Plans, "armed with the divine excellences

[57] Plans, "Nota sobre la forma de los rayos luminosos en el campo de un centro gravitatorio segun la teoría de Einstein," *Anales de la Sociedad Española de Física y Química*, 18 (1920), pt. 1, 367–373. The course was given in 1919–20; see *Revista Matemática Hispano-Americana*, 1 (1919), 226–227.

[58] Pedro Puig Adam, "Resolución de algunos problemas elementales en Mecánica relativista restringida," *Revista de la Real Academia de Ciencias Exactas*, 20 (1922), 161–216. The four cases had to do with the movement of a point on a line or surface, under various circumstances.

[59] Tomás Rodríguez Bachiller, interview, Madrid, April 10, 1980.

of the elect paladins of the Catholic faith," assumed the demeanor of an apostle.[60] Yet, except for one notable lapse, he steered clear of controversial issues involving the conflict of religion and science. The exception had to do with his translation of Sir Arthur Eddington's *Space, Time and Gravitation*, each chapter of which was preceded by a literary quotation. The epigraph in chapter 1 was from Descartes and stated: "In order to reach the Truth, it is necessary, once in one's life, to put everything in doubt—so far as possible." Plans, who found this statement contrary to his firm religious convictions, consulted Enric de Rafael, who assured him that Descartes "excepted religious ideas" although readers may well be unaware of this. Not satisfied, Plans added a footnote registering his disagreement with Descartes: "It is unnecessary to say that the translator does not associate himself with this principle." He apprised Terradas of his decision, asking the latter to advise him if he did not think the note a good idea. Terradas must have agreed with Plans, because the note appeared in print, whereupon Plans was promptly attacked in *El Sol* for having added a gratuitous disclaimer.[61]

Despite Plans's prudence in avoiding socially controversial aspects of the reception of modern science, we might speculate that his choice of Eddington's volume, which was viewed in the English-speaking world as having brought to the public the comforting message that absolute values and standards were still intact and that science and religion were reconcilable,

---

[60] Terradas, in *La Veu de Catalunya*, March 22, 1934, cited by F. Navarro Borras, "Don José María Plans y Freyre," *Anales de la Universidad de Madrid, Ciencias*, 3 (1934), 231.

[61] Arthur Eddington, *Espacio, tiempo, gravitación* (Madrid: Calpe, 1922). Plans to Terradas, January 1, 1922, Terradas Archives, Institut d'Estudis Catalans, Barcelona: "Page 48. Saying of Descartes at the beginning of Chap. I. It really strikes me as very strong, as I told you. Even though de Rafael told me that Descartes did not have religious ideas in mind, people generally don't know that. I am considering this *translator's note* which you can look at. If you don't like it, tell me so." *El Sol*, August 1, 1922. The literary excellence of Plans' translations was noted by Navarro Borras, "José María Plans," p. 243.

may well have figured in a hidden agenda to achieve the same ends in Spain.[62]

RELATIVITY AND SPANISH ASTRONOMY

Unlike their colleagues in mathematics and physics, Spanish astronomers born in the middle decades of the nineteenth century produced original research that merited the approval of foreign scientists. José Joaquín Landerer (1841–1922) produced a valuable series of observations of the satellites of Jupiter. He also concluded that our own moon has no atmosphere, based on his studies of the polarization of sunlight reflected off its surface. Josep Comas Solà (1868–1937) published significant research in the 1890s on the topography of Mars, which Camille Flammarion incorporated into his own work on that planet; in particular, he demonstrated that the Martian "canals" were more apparent than real.[63]

In spite of the high quality of individual research, however, the available facilities, the national observatories in particular, were impoverished. Juan de la Cierva's ministerial inspection visit to the Madrid Observatory around 1900 was notorious:

> On my visit . . . I found: that the most important telescope was not working because the rails on which it turned had to be fixed, and the money had been spent on decorating the offices of the astronomers in the same building; that two other magnificent telescopes acquired for the last solar eclipse visible in Spain were lacking electrical connections of scant cost. They wanted me to see the stars with those telescopes and naturally I could see

[62] On Eddington's message, see Loren R. Graham, "The Reception of Einstein's Ideas: Two Examples from Contrasting Political Cultures," in Gerald Holton and Yehuda Elkana, eds., *Albert Einstein: Historical and Cultural Perspectives* (Princeton, N.J : Princeton University Press, 1982), p. 119.

[63] E. Portela Marco, "José Joaquín Landerer y Climent," and V. Navarro Brotons, "Josep Comas i Solà," in *Diccionario Histórico de la Ciencia Moderna en España*, s.v.; Antoni Roca, "J. Comas Solá, astrónomo de posición," *Mundo Científico*, 6 (1986), 290–303

none! Then the business about the missing light bulb came out. The result was that the three telescopes I mentioned were impossible to use. They showed me, as a curiosity, some magnificent telescope lenses that Godoy . . . had donated in his time. The curious thing was that the boxes containing the lenses were closed and banded; no one had ever opened them.[64]

The Observatory of San Fernando, in Cádiz, was in better condition, and, after the turn of the century, there were two new private observatories of great importance, both in Catalonia: the Fabra Observatory in Barcelona, founded in 1904, and the Ebro Observatory, established the following year in Tortosa. The former, under the direction of Comas, was supplied with excellent instrumentation, including a meridian circle that permitted precise measurements of sidereal time as well as of the right ascensions of stars, and a great telescope with which photographic studies of the Delevan and other comets were made.[65]

As in the sister disciplines discussed above, the quality of Spanish astronomy improved rapidly after 1900. Symbolic of the advance was the great effort mobilized to observe the solar eclipses visible from Spain—those of 1900, 1905, and 1912. These observations were significant not only because they involved the coordination of a large part of the astronomical community but also because they were executed in conjunction with foreign astronomers—hosts of whom, especially in 1905, descended upon the country. Landerer viewed the 1905 eclipse, whose trajectory he had predicted with great accuracy, from Alcoseba, Comas from Vinaroz, and the British and other visiting scientists (including no less than twelve teams of Jesuit astronomers alone) dispersed throughout the peninsula:

[64] Juan de la Cierva, *Notas de mi vida* (Madrid, 1955), pp. 65–66, cited by Juan Vernet, *Historia de la ciencia española* (Madrid: Instituto de España, 1975), p. 223.

[65] Diputació de Barcelona, *Guía de les institucions científiques i d'ensenyança* (Barcelona: Consell de Pedagogia, 1916), p. 55.

John Buchanan in Torreblanca, John Evershed in Pineda de la Sierra, Alfred Fowler and Hugh Callendar in Castellón de la Plana. Coordinately, the eclipse was viewed by Spanish and foreign personnel from hydrogen balloons of the army's Aerostatic Service, directed by Pere Vives i Vich (1868–1938). Vives had first presented his plans for the eclipse in September 1904 at the fourth meeting of the International Commission for Scientific Aerostation in St. Petersburg. Observations were to take place from one captive and two free balloons to obtain readings of solar diffraction, to observe the corona, and to gather meteorological data. In St. Petersburg he consulted with foreign specialists in aerostatic meteorology, notably A. Laurence Rotch of the Blue Hills Observatory in Boston and Arthur Berson of the Prussian Meteorological Observatory. The final plans were devised by Vives and Augusto T. Arcimis (1844–1910), director of the Central Meteorological Institute of Madrid and longtime professor of astronomy at the Institución Libre de Enseñanza. A variety of observations were made from balloons that ascended from Burgos. In one, Emilio Herrera drew the solar corona, from another Arcimis took photographs, and in a third Vives and Berson obtained meteorological data.[66]

The Exposition of Lunar Studies, held in Barcelona in May and June 1912, was nothing less than a celebration of Spanish

---

[66] On the scientific importance of the four "Spanish" solar eclipses, beginning with that of 1860, see M. López Arroyo, "La espectroscopía en el Observatorio Astronómico de Madrid," *Boletín Astronómico del Observatorio de Madrid*, 8, no. 2 (1972), esp. pp. 4–7. On the eclipse of 1905, see Pedro Vives y Vich, "Emploi des ballons pour l'observation de l'éclipse totale de soleil 30 aout 1905," in Quatrième Conférence de la Commission Internationale pour l'Aérostation Scientifique, *Procès-verbaux des Séances et Mémoires* (St. Petersburg: Académie Impériale des Sciences, 1905), pp. 74–75; on the scientific data produced by the various observation points, see Vives, *Avance de los resultados obtenidos en las observaciones del eclipse total de sol de 30 de agosto de 1905* (Madrid. Parque Aerostático de Ingenieros, 1906). The episode is discussed in my introduction to Emilio Herrera, *Flying: Memoirs of a Spanish Aeronaut* (Albuquerque: University of New Mexico Press, 1984), pp. 176–177.

astronomy and meteorology. Stands of European and American observatories and instrument makers alongside those of Spanish exhibitors gave a strong impression of the maturity of Spanish astronomy and the ability of Spanish astronomers to occupy a place in the forefront of research, particularly in the area of photography.[67]

As in other fields, astronomers of the younger generation tended to be more abreast of current developments than their older colleagues. The Madrid Athenaeum's 1915 lecture series on the present state of the science contained two presentations by astronomers of the Madrid Observatory. The first, an overview of problems in astronomy by Antonio Vela (1865–1927), was largely devoted to a history of astronomy, with only a few passing remarks, near the end, on spectral analysis. The second, by Pedro Carrasco (1883–1966), was on the theory of relativity. This presentation, limited to a discussion of the Special Theory, was in part influenced by Cabrera's 1912 article and reveals a cautious, skeptical approach. Relativity changes the way we conceive the universe, he notes at the outset, even though classical science still persists. Rather than overthrow classical science, relativity broadens it, in spite of the long list of classical concepts—including length, mass, action at a distance (he thought), absolute time, and motion—which have been overthrown. Yet it is a simple theory, basically a problem of the transformation of coordinates. Carrasco, without discussing the point, retains an aether concept and notes that to take the velocity of light as a limit "is somewhat repellent to reason." Relativity is not about to be overthrown, nor is it wholly proven; disinterested criticism is called for.[68]

Although he held a chair of mathematical physics, Carrasco

---

[67] Sociedad Astronómica de Barcelona, *Exposición General de Estudios Lunares, Catálogo* (Barcelona, 1912).

[68] "Teoría de la relatividad," in Ateneo de Madrid, *Estado actual, métodos y problemas de las ciencias* (Madrid, 1916), pp. 149–150. Carrasco was not the Ateneo's first choice but a last-minute substitute, probably for Cabrera, who had given the previous two lectures on physics, making no mention of relativity.

was not much of a theoretician. His original work was experimental in nature. For example, he devised a number of novel ways to measure the speed of light.[69] His interest in relativity, therefore, was more on the practical level. To his mind, relativity had made possible great advances in the kinematic study of electrical systems, as well as in establishing the relationship between electrical and material systems (in his conception, it "connects the physics of matter with the physics of aether, and makes presumable a common explanation for both. Whence it may be said that relativity strengthens electrical theories of matter").[70] For Carrasco, the most appealing aspect of relativity was that it opened up new avenues for research in physics.

Carrasco followed the experimental outflow of the General Theory, in particular the eclipse of 1919, with close attention. Indeed, he might well have participated in some of the observations himself except that, "during the first years of the European war, the astronomers of Madrid, lacking adequate means, lost the opportunity to participate in the resolution of very interesting scientific problems stimulated by Einstein's theories."[71]

Evaluating the experimental results stimulated by the General Theory, Carrasco arrived at the following balance sheet in 1920:

> *Positive*, for the movement of Mercury;
> *Probably affirmative*, for the deflection of stellar images in the vicinity of the sun;

[69] See Carrasco, "Método para determinar la velocidad de la luz," *Anales de la Sociedad Española de Física y Química*, 17 (1919), 296–306, 316–330, which describes various ways to measure the speed of light with a diapason (tuning fork), and "Nuevo método para medir la velocidad de la luz. Determinación de algunas constantes físicas, que dependen de la medida de pequeños intervalos de tiempo," *Revista de la Real Academia de Ciencias Exactas*, 17 (1918–19), 201–216, 340–357, which discusses the problem of measuring small intervals of time on photographic film.

[70] "Teoría de la relatividad," p. 163.

[71] "Conferencia de don Pedro Carrasco. Lo que es y debiera ser el Observatorio Astronómica de Madrid," *El Sol*, April 19, 1924.

Negative, for the rays of the solar spectrum; and we could even add, on the basis of an article by Evershed on the spectrum of Venus:

 Probably negative, for the deflection of the spectral line of Venus.[72]

Carrasco clearly recognized the ambiguous nature of the English results that, taken together, yielded a value for deflection somewhere between Einstein's value and that obtained in Newton's theory, where one assumes that light rays are subjected to gravitational attraction. The results were received as a confirmation of the General Theory largely because of the prestige of Eddington and Frank Dyson, whose readings of the conflicting values of the Sobral 4-inch instrument (which gave a higher value than Einstein's) and the Sobral and Principe astrographs (which gave lower values) were weighted in Einstein's favor.[73] Carrasco, who had worked with Dyson, must have been swayed by the latter's interpretation that was favorable to relativity, and accordingly he presented the results, though not definitive, as "a brilliant triumph of the English expeditions."

Einstein's formula for the gravitational red shift was not verified experimentally until 1962. From Carrasco's perspective, the mixed results of initial observations, following Einstein's 1911 paper, up to 1920 suggested that there was an internal contradiction in the General Theory, "the same as was found before, in classical science, in the contradiction between the Michelson experiment and the phenomenon of the aberration of light." It is possible that Augusto Righi and others may have uncovered inconsistencies in the way the Michelson-Morley experiment had been interpreted. More likely, there were

---

[72] "Estado presente de la teoría de la relatividad," *Anales de la Sociedad Española de Física y Química*, 18 (1920), pt. 2, p. 94.

[73] See John Earman and Clark Glymour, "Relativity and Eclipses: The British Eclipse Expeditions of 1919 and Their Predecessors," *Historical Studies in the Physical Sciences*, 11 (1980), 76, who point out that Dyson simply omitted any mention of the Principe results, those least favorable to Einstein, in his report to the Royal Astronomical Society.

problems of interpreting the results of the deflection of light rays by the sun. It is naturally repugnant to an astronomer, he noted, to consider the sun as "a trimmed-down geometrical entity," when in fact it is a body characterized by extreme physico-chemical complexity. "But right away astronomers and physicists must see a greater complexity in the phenomenon, since for them the light ray is an abstraction and the physical reality is the electromagnetic wave which traverses a very complex medium, including the electromagnetic field of the sun." To Carrasco, both the Einsteinian solution and that of his opponents who thought the deflection was due to simple refraction were simplistic. "It is the contrast between a physical mechanism and a mathematical symbol, whose value and beauties are undeniable, but which, as a real representation of our physical world meets with enormous difficulties." Physicists, he noted, prefer to deal with concrete entities, even if these are hypothetical, rather than with abstract ones. In this attitude Carrasco seems close in spirit to the position of those English physicists who were unable to dispense with electromagnetic models. Yet he retained an open mind: "If the astronomical results of the relativistic theory were to be fully confirmed, and, in connection with them, we were to have a group of non-gravitational physical phenomena which would have, as well, a relativistic explanation, we would have to contemplate a revision of theoretical physics according to a relativistic criterion, and we would have taken a giant step towards the yearned-for unity of science."[74]

In his 1915 presentation Carrasco had noted that once infinite velocity had been abandoned, instantaneous action at a

[74] Carrasco, "Teoría de la relatividad," pp. 96–97, 99. In 1928 Carrasco published a small volume entitled *Filosofía de la mecánica*, in which his reluctance either to abandon the aether or to admit the speed of light as the limit of velocities is conjoined with acceptance of other elements of the theory. For example, he accepted the equivalence of accelerated motion and gravity, using Einstein's famous elevator example (pp. 115–116). Elsewhere in the book, though, he seems unable to abandon pre-Einsteinian views typically held by anti-relativist British physicists. See discussion in Thomas F. Glick, *Einstein y los españoles* (Madrid: Alianza, 1986), chap. 6.

distance was ruled out. Carrasco was here repeating a commonly held misconception. Since special relativity is a kinematic, not a dynamic, theory, it does not rule out the possibility of action at a distance. Since Carrasco was confused on this point, he could not bring himself to break completely with the necessity for an aether—"the most necessary substance in physics . . . in spite of its hypothetical character."[75]

Carrasco, in spite of his conceptual difficulties with special relativity, was virtually the only Spanish astronomer who could handle the theory at all. In general, those astronomers who knew mathematical physics (Carrasco held the chair at Madrid) were able to follow the theory. In particular, most of the British astronomers with whom Carrasco had studied had, as students, passed through the Cambridge mathematics curriculum and were able to interpret relativity in the context of a continuing discussion in British physics of a possible linkage between electromagnetism and gravitation. Carrasco's association with this tradition adequately equipped him to discuss the theoretical ramifications of the observational results of general relativity. Astronomers who lacked commensurate mathematical background and who were mainly oriented to observation (most American astronomers, as well as Comas Solà) were unable to follow relativity theory.[76]

TOWARD EINSTEIN'S TRIP: SPAIN AND EUROPE

By 1923 Spanish mathematics, physics, and astronomy were represented in the international science community by a small number of researchers who were able to participate in research

[75] "Teoría de la relatividad," p. 163; Filosofía de la mecánica, p. 164. On the confusion over action at a distance, see the exhaustive discussion by J. M. Sánchez Ron, "Studies of Relativistic Action-at-a-Distance Theories," doctoral dissertation, University of London, 1978, and "Einstein y Lorentz: Significado de la relatividad especial y la inconmensurabilidad entre paradigmas," Pensamiento, 38 (1982), 425–440.

[76] Jeffrey Crelinsten, "William Wallace Campbell and the 'Einstein Problem': An Observational Astronomer Confronts the Theory of Relativity," Historical Studies in the Physical Sciences, 14 (1983), 2.

problems defined by that community as having significance. In order to achieve such participation, peripheral scientists had to adopt the language of the center; visit, study, or perform research at central institutions; and correspond with scientists there. The research could then be continued at home institutions, such as Cabrera's Institute or the Mathematical Seminar, modeled after those of the center. Because of the smallness of the pool of national scientists, first-rate research was confined to a very few subdisciplines, with one or two professors and their immediate disciples working on problems originally defined by "central" figures and carried on under conditions of close communication with them.[77]

Science and Language

In the 1920s it was patently clear that contact with mainstream science was dependent upon the linguistic competence of Spanish scientists. The issue was raised constantly, frequently in a defensive context because doubts about the ability to communicate effectively reminded the scientific community of its tenuous position on the periphery of European science. The preceding generation had been autodidactic in languages. José Gómez Ocaña noted that "Necessity makes its own laws and on me it imposed autodidacticism. When I had to read a book in French, I plunged ahead through its pages with no other aid than that of a dictionary. Nor did I have an English teacher, not counting a friend who taught me how to pronounce a few phrases."[78] Although the foreign fellowship program of the Junta para Ampliación de Estudios presupposed mastery of a foreign language, the linguistic competence of scientists was still a sensitive issue in the 1920s. Weakness in language skills implied that the persons or disciplines so accused (as in a stri-

---

[77] See Thomas Schott, "Fundamental Research in a Small Country: Mathematics in Denmark, 1928–1977," *Minerva*, 18 (1980), 243–283, especially pp. 246–247, 280–281.

[78] Alberto Ruiz de Galarreta, "El doctor José Gómez Ocaña: Su vida y su obra," *Archivos Iberoamericanos de Historia de la Medicina*, 10 (1958), 401.

dent 1920 polemic over the linguistic competence of the Madrid Faculty of Medicine) were unable to keep abreast of current research.[79]

In the exact sciences, French, the language of the foreign literature most read at the turn of century, became increasingly less important. It is significant, in this regard, that Echegaray did not participate in the early discussion of relativity, of whose significance he was but dimly aware, because he subscribed only to French journals and was poorly informed on developments elsewhere. The early Spanish discussants of Einstein's theory knew German well. Terradas had attended elementary school in Germany and was proficient in German. Cabrera, Rafael Campalans, and other scientists were able to communicate with Einstein in German during his visit.

Discussing the linguistic competencies necessary for the study of chemistry, Enric Moles, who had himself studied in Geneva and Zurich, did not even mention French:

> Nowadays in order to learn chemistry one is not completely helpless knowing English. Nevertheless, for advanced study, of rigor and seriousness, the German language is an unavoidable instrument. I want my students to acquire it. I know too well that the law gives them the right to choose between English and German and that the professor cannot formally impose one or the other. Still, I don't think it is too difficult to obey the law and teach what must be taught at the same time.[80]

In physics the shift away from French was marked. In a study of bibliographical references in articles published in the *Anales de la Sociedad Española de Física y Química*, broken down into three periods, Manuel Valera Candel has documented the shift in linguistic focus in Spanish physics over the first third of the century. In the first period (1903–15), 32.5 percent

[79] See Sebastián Recasens, "En defensa de la Facultad de Medicina," *El Sol*, February 3, 1920.
[80] "Una conversación con el químico Sr. Moles."

of all references are to materials in French, 30 percent in German, 25 percent in English, and the rest in Spanish. In the second period (1916–30), French references drop to third place (16.4 percent of references), behind German (37.2 percent), and English (30.3 percent). Spanish references (15.5 percent) occupied fourth place. In the last period (1931–37), English references led (39.8 percent), followed by German (34.2 percent), and Spanish (15 percent). French now occupied last place, with less than 10 percent of all references.[81] Of course, a similar shift could be documented in other national literatures of the same period. What makes the Spanish case interesting is that the educational infrastructure required to educate scientists in German and English was still extremely weak, and we must assume, even in the 1930s, a high incidence of autodidacticism. In all probability, most holders of foreign grants learned the requisite language, or substantially perfected their knowledge of it, on the site and not in Spain.

Spaniards Abroad

At the heart of the renewal of Spanish science in the early twentieth century was the grant program of the Junta para Ampliación de Estudios. Under the Junta's aegis, most of the leaders of Spanish mathematics, physics, and astronomy studied abroad between 1910 and 1920. Such persons not only studied abroad, but in most cases returned with research programs they continued in Spain, frequently in conjunction with the foreign center. Among early grants in mathematics, Rey Pastor studied analysis and higher geometry in Berlin in 1910–11; the engineer Rafael Campalans studied differential and integral calculus at the Sorbonne in 1911; and Pedro Pineda Gutiérrez studied the theory of groups with Hermann Weyl in Zurich in 1916–17. In physics, Cabrera studied magnetochemistry with Pierre Weiss, also in Zurich, from 1910 to 1912, developing the lines of research that were to occupy the rest of his

[81] Valera, *Producción española en física.*

58

career. Virtually all of his experimental research over the next decade cited Weiss's magneton theory. Martínez Risco studied physical optics in Zeeman's laboratory in Holland in 1911. In the same year, Pedro Carrasco received a grant to study astrophysics at the Astrophysical Laboratory at South Kensington, as well as with Alfred Fowler at Imperial College, London, Hugh Newall at Cambridge, and Frank Dyson at Greenwich. Julio Palacios worked on low-temperature physics in Holland and Germany in 1918. By 1919, Olegario Fernández Baños, then studying mathematics on a Junta grant in Italy, could allude to the founding of the Junta as having signaled "the hour of renovation" of Spanish science.[82]

In 1920–21 the physical chemist Miguel Catalán (1894–1957) was awarded a Junta grant to study atomic spectroscopy in Fowler's laboratory, leading to his famous research on multiplets in the manganese spectrum. Arnold Sommerfeld learned of Catalán's research on a trip to Madrid in April 1922, and this led to a Rockefeller (International Education Board) grant, which brought Catalán to Sommerfeld's laboratory in Munich in 1923–25; when Catalán returned to Madrid, he brought with him Karl Bechert, also on an I.E.B. grant, to continue research on theoretical spectroscopy.[83]

A number of conclusions are suggested by the experiences of those Spaniards who studied or performed research abroad. There was a strong correlation between study abroad and high scholarly productivity. Seventy-five percent of Spanish physicists who studied abroad between 1911 and 1937 were ranked by Valera in the most productive category of scholarly output.[84] Nevertheless, the research profiles of other Spanish scientists show profound discontinuities since many were unable

[82] [Olegario] Fernández Bagnos (Baños), "Lo spirito scientifico in Spagna," *Intesa intellettuale*, 2 (1919), 155.

[83] Junta para Ampliación de Estudios, *Memorias, 1910–11* (Madrid, 1912), pp. 43, 59, 85; *Memorias, 1916–17* (Madrid, 1918), p. 42; Valera, *Producción española en física*, p. 1253; Varadaraja V. Raman, "Miguel Antonio Catalán," *Dictionary of Scientific Biography*, III, 124–125.

[84] Valera, *Producción española en física*, p. 1218.

to continue projects begun abroad, lacking the necessary equipment in Spain. Examples are Martínez Risco whose work on physical (including relativistic) optics was performed almost wholly abroad—with Zeeman in 1911 and again in 1928, and finally in France where he was exiled in the 1940s; and Julio Palacios, whose research on low-temperature physics in Holland with Heike Kamerlingh Onnes during World War I was suspended when he returned to Madrid "because he lacked the necessary equipment to continue the same kind of experiments in which he was trained at Leiden."[85]

The high productivity of returnees, however, reflected their ability to participate in research-front science by virtue of their continuing relationships with foreign centers. Again, Catalán provides the prototypical example. In Spanish spectral research, a high number of studies were written, according to Valera, either abroad or in conjunction with some foreign center: Sommerfeld's Institute of Theoretical Physics in Munich or the South Kensington Astrophysical Laboratory.[86] The tight relationship of the Mathematical Seminar with Italian mathematics is another example. This process, which the Junta made possible, of inserting Spanish scientists in international disciplinary or subdisciplinary networks was a characteristic mode in which basic research of high quality was made possible in a small country. As Rey Pastor understood it, "To train [scientists] we need centers of research, and to found centers we need capable men. This is the vicious circle in which

---

[85] Ibid., p. 1253; Blas Cabrera, in Academia de Ciencias Exactas, *Discurso leído en el acto de su recepción por don Julio Palacios Martínez* (Toledo, 1932), p. 68. Martínez Risco (1888–1954), professor of optics in Madrid, was the only experimental physicist active in Spain in the 1920s who pursued relativistic research interests. He performed interferometer experiments in the 1920s but appears not to have published any results on relativistic optics until the later 1940s, when he was in exile in France. See, for example, "Concept interférential des images optiques dans la théorie de la relativité," *Comptes Rendus de l'Académie des Sciences*, 228 (1949), 2014–2016, and other articles published in the *Journal de Physique* and reproduced in Martínez Risco, *Oeuvres scientifiques* (Paris: Presses Universitaires de France, 1976).

[86] Valera, *Producción española en física*, p. 1243.

Spain is locked, and to break out of it, there is no other solution than to seek spiritual communion in other countries of brilliant scientific culture and perfect organization, to send young men to associate with great teachers, in the very well-springs where science rises." Such men, whom he styled "modern argonauts," returned to Spain not only with new theories and methods, but with something even more important: "a new spirit of research."[87]

Subsidiary but related phenomena were extended tours of foreign research centers and attendance by Spaniards at international scientific meetings. The first had the aim of unearthing models for the establishment of new scientific and technical institutions or facilities or the upgrading of inadequate existing ones; the second was for purposes of establishing or maintaining lines of communication as well as presenting research results.

Foreign technical tours were popular in the late nineteenth century and by the twentieth had lent themselves to a distinctive genre of scientific literature. Casimir Lana Serrate, who held a prewar Junta grant to study physical chemistry in Berlin, made extensive technical tours in the postwar period when he was professor of metallurgy at the Industrial School in Barcelona. His first reports were on technical education (the Industrial School had been founded in 1904 to train industrial engineers for the textile industry; as such, it had an important chemistry laboratory). In 1918, having already visited technical institutions in eight countries, Lana visited Cambridge, Massachusetts, reporting on technical education at the Massachusetts Institute of Technology (where, he noted sadly, not a single Spaniard was currently enrolled).[88] The following decade, when Lana was director of the Hispano-Suiza metallurgical laboratory, his travel interests turned to the steel industries and the laboratories that served them.[89]

[87] Julio Rey Pastor, "Aspecto social de la vocación científica," *La Nación*, November 4, 1923.
[88] See Lana's reports in *Ibérica*, 10 (1918), 201, 252–255.
[89] See, for example, "La significación de los laboratorios en las modernas in-

Another tireless traveler of the same period was the Jesuit astronomer Lluis Rodés Campdera (1881–1939). Rodés studied physics at the University of Barcelona, learning astronomy at the Observatory of Valkenburg, Holland, under Theodor Wulf. In 1914 he was in Sweden to view the solar eclipse of August 21, and in 1916 he traveled to the United States to continue his astronomical education at various American observatories. In 1918 he wrote popular reports on current research at the Yerkes and Harvard observatories. At the former, he noted research on the Doppler-Fizeau effect, while at the latter he was impressed by the archive of photographs and visual observations, undoubtedly thinking of the relevance of such a collection to the Ebro Observatory, his home institution.[90] Another peripatetic astronomer was José Tinoco of the Madrid Observatory, who was in Paris in 1924 investigating procedures for studying radiotelegraphic signals.[91]

Because of World War I and its aftermath, Spanish participation at scientific meetings did not become highly visible until the 1920s. Enric de Rafael was a delegate to the mathematical congress at Stockholm in 1920. Eduard Fontserè attended international meteorological congresses throughout the decade. Rodés attended the Rome Congress of Astronomy and Geophysics in May 1922, and read a paper on the action of the sun on magnets at the Los Angeles meeting of the American Association for the Advancement of Science in 1923.[92] The pace of Spanish participation at international meetings picked up dramatically in the middle of the decade. José G. Alvarez Ude and Antoni Torroja Miret represented Spain at the International Congress of Mathematicians at Toronto in August

---

dustrias siderúrgicas y de construcción de máquinas," *Ingeniería y Construcción*, August 1923, pp. 366–369, and "La pista de pruebas en la nueva fábrica de automóviles 'Fiat' en Turin," *Ibérica*, 11, pt. 1 (1924), 217–219.

[90] Luis Rodés, "Una visita al Observatorio de Yerkes," *Ibérica*, 10 (1918), 232–236, 296–300; "El Observatorio de Harvard College," *ibid.*, pp. 377–381. Rodés became director of the Observatorio del Ebro in 1920.

[91] *Ibérica*, 11 (1924), 211.

[92] *Ibérica*, 17 (1922), 382–383; 20 (1923), 261.

1924, and Obdulio Fernández and Enric Moles were delegates to the International Congress of Chemistry at Bucharest in June 1925.[93] The Bologna meeting of the International Congress of Mathematicians in September 1928 provided an opportunity for Spanish mathematicians to strengthen and, in many cases, renew contacts with their Italian counterparts. The Spanish delegation of fourteen included not only Plans and Terradas, but a half dozen other figures who are identifiable as having played roles in the reception of relativity in Spain: Alvarez Ude, Teófilo Martín Escobar, Pedro González Quijano, Carlos Mataix Aracil, José Antonio Pérez del Pulgar, and Rey Pastor.[94]

EUROPEANS IN SPAIN: LEVI-CIVITA, PLANS, AND GENERAL RELATIVITY

The countermovement, involving lecturing or research by European scientists in Spain, was to a great extent dependent upon previous contacts made by Spanish scientists abroad. In the late teens, Spain had benefited from the dislocation of the war by attracting a number of foreign scientists, particularly physicists, to its laboratories. Jakob Laub, Einstein's early collaborator who had begun the decade at the University of La Plata in Argentina, worked in Blas Cabrera's laboratory in 1915. Another central European, B. Szilard, collaborated in the Institute of Radioactivity in Madrid in 1918–19. Szilard's favorite research topic was the determination of radium and thorium content in minerals containing them, for which he

[93] *Ibérica*, 11, pt. 2 (1924), 85; 12, pt. 1 (1925), 361. Enric de Rafael was a delegate to the Mathematical Congress at Stockholm in 1920.

[94] *Atti del Congresso Internazionale dei Matematici* (Bologna: Nicola Zanichelli, 1928), vol. 1. On p. 32 are listed official delegates of Spanish institutions, most of which produced relativists: Ministerio de Instrucción Pública (Terradas), Academia de Ciencias Exactas (Plans, Alvarez Ude, Terradas), Escuela Central de Ingenieros Industriales, Madrid (Mataix Aracil), and Escuela de Ingenieros de Caminos (González Quijano). La Confederación Sindical Hidrográfica of Zaragoza was represented by Manuel Lorenzo Pardo, who was prominent in the reception of Einstein in that city in 1923.

constructed a number of measuring devices. In 1918 he built a new one in Torres Quevedo's laboratory, which he described in a pamphlet published by the Institute.[95]

A more public phenomenon was the steady stream, beginning in 1921, of foreign scientists—including the major figures in the discussion of relativity—who came to lecture in Spain. The first of the series was Tullio Levi-Civita's course on classical and relativistic mechanics given in Barcelona and Madrid in January-February 1921. The invitation had come from Terradas, who had met the Italian mathematician at the 1912 meeting of the International Congress of Mathematicians at Cambridge, England.[96]

Levi-Civita was at this time a leading figure in general relativity research, having been drawn into contact with Einstein because of the latter's adoption of absolute differential calculus in the fomulation of the theory. The absolute differential calculus, or tensor analysis, is an extension to n-dimensional manifold of the idea and method of calculus. It was invented by Gregorio Ricci-Curbastro (1853–1925) between 1887 and 1896. In the 1890s Levi-Civita (1873–1941) expanded the scope of Ricci's method from its original applications to differential geometry to embrace a wide variety of mathematical and physical problems, including non-Euclidean spaces. When Einstein was struggling with the formulation of the General Theory in Zurich, his friend Marcel Grossmann drew his attention to Ricci and Levi-Civita's work. In 1915 Einstein and the latter began a brief, although intense, correspondence on the mathematical properties of the gravitational field equations that Einstein was developing. Levi-Civita published a major mathematical contribution to relativity, the theory of affinity

---

[95] Lewis Pyenson, *The Young Einstein* (Bristol: Adam Hilger, 1985), pp. 231–234. Laub was again in Madrid briefly in 1920. B. Szilard, *Nuevo electrómetro para la medida de la radioactividad* (Madrid: Instituto de Radioactividad, 1918); see *Ibérica*, 11 (1919), 35. Szilard worked on radioactive fertilizers, among other things. The Instituto was located on Calle Amaniel 2, next to the Faculty of Sciences, and was directed by José Muñoz de Castillo.

[96] See *La Veu de Catalunya*, January 20, 1921.

connected spaces, in 1917.[97] The last two lectures of the Spain course (in Madrid these were given on February 1 and 2, 1921) were on general relativity.[98]

Levi-Civita's visit to Spain provided an opportunity to solidify contacts between Spanish and Italian mathematics. The intimate nature of that bond was exhibited by the extraordinary interest that students showed in his course, particularly in the portions dealing with relativity. In Barcelona a fourth session, not included in the original schedule, was added, "owing to the insistence of the audience. . . . It was, for this reason, an intimate session in which there was established that contact between professor and students which creates dialogue." In this additional talk he clarified certain concepts raised in the first lecture on the stability of movement.[99]

This lecture tour took place just as relativity was beginning to awaken the interests of the Spanish scientific community, beyond the few specialists like Cabrera and Terradas who had followed the theory before the war. Levi-Civita's visit, therefore, was the occasion for the emergence of a distinct circle of scientists who would diffuse relativity in Spain. Table 1 is a list of those invited to a dinner for the Italian mathematician, held in Madrid on February 1, 1921. Of the sixteen participants, nine were mathematicians, including virtually the entire leadership of Spanish mathematics, all of whom were favorable to relativity. Levi-Civita, as previously noted, had already received a number of Rey Pastor's students in Rome, including one—Fernando Lorente de Nó (1896–1955)—with research interests in relativity. The large representation of mathematicians at the dinner explains a number of the idiosyncracies of the reception of relativity in Spain. First, mathematics was the

[97] On Levi-Civita and Einstein, see Judith R. Goodstein, "Levi-Civita, Albert Einstein and Relativity in Italy," in *Tullio Levi-Civita*, Convegno Internazionale Celebrativo del Centenario della Nascità (Rome: Accademia Nazionale dei Lincei, 1975), pp. 43–51.

[98] *Questions de Mecànica clàssica i relativista* (Barcelona: Institut d'Estudis Catalans, 1922).

[99] "Curso Levi-Civita," *Ibérica*, 15 (1921), 98–99.

TABLE I
GUESTS AT A DINNER FOR LEVI-CIVITA (MADRID, 1921)

Primary Disciplines	Scientific Middle Class
Mathematicians	*Engineers*
Julio Rey Pastor	Emilio Herrera
Josep Maria Plans	Mariano Moreno-
Luis Octavio de Toledo	Carraciolo
José Sánchez Pérez	Gregorio Uriarte
Cecilio Jiménez Rueda	Joaquín La Llave
Sixto Cámara	
José Gabriel Alvarez Ude	
Ignacio Suárez Somonte	
Ruperto Fontanilla	
Physicists	
Blas Cabrera	
Julio Palacios	
Astronomer	
Pedro Carrasco	

SOURCE: *El Sol*, February 2, 1921.

strongest and largest discipline in the exact or physical sciences at this time; identification with relativity enhanced its prestige. Second, its members had close connections with Italian mathematicians, also relativists, which explains their interest in the General Theory. Third, mathematicians were needed by physicists and others interested in understanding the new ideas to explain to them the distinctive language—absolute differential calculus—in which these were expressed. Spanish mathematicians consistently emphasized, in their public utterances, the services lent by their discipline to Einstein's theory. Fourth, and consequently, scientific interest in relativity was preponderantly in the General Theory, partly because of the period in which relativity's major impact was felt (1921–24), and partly because the predominance of mathematicians in its reception ensured a focus on what was agreed to be the mathematically most interesting aspect of the theory.

Josep M. Plans stressed exactly this point later in the same year at the Oporto congress of the Spanish Association, in his inaugural lecture of the mathematics session entitled "The Historical Development of Absolute Differential Calculus and Its Current Importance."[100] In the history of the interpenetration of mathematics and the physical sciences, Plans began, there have been cases of two different kinds. In the first, a physical problem stimulates a useful abstraction: vibrating chords and the properties of heat gave rise to the Fourier series. In the second, the opposite is true. Here, an abstraction lends itself, at some time after its initial conceptualization, to the explication of some physical problem. Plans's examples are the non-Euclidean geometry of Lobatchevsky, Bolyai, and Riemann in special relativity, and absolute differential calculus in general relativity. Relativity, Plans asserted, was "the scientific event of greatest consequence at the present time" and stressed "The great services lent, in the hands of the Italian school of Ricci and Levi-Civita, to Einstein's theory of relativity and gravitation, by absolute differential calculus, which, as our colleague Sr. Terradas has so wisely said, has become the language appropriate to the study of Reimannian space-time of four dimensions, just as ordinary vectorial calculus serves for the Euclidean space of three dimensions."[101]

It is interesting to contrast this programmatic statement by Plans with Cabrera's statement of the same year, cited earlier in this chapter. For the experimental physicist, relativity was epistemologically important, but the structure of the atom was an issue of greater current significance. For the mathematician, general relativity was the central scientific issue because of the signal role of mathematics in its conceptualization. In 1924

---

[100] Asociación Española para el Progreso de las Ciencias, Congreso de Oporto [1920] (Madrid, 1921), I, 23–43.
[101] Ibid., pp. 24, 41.

Plans wrote a manual of absolute differential calculus, obviously designed to make general relativity accessible to physical scientists and engineers. Like his earlier volume on relativistic mechanics, this book had its origin in a prize offered by the Academy of Exact Sciences, "desiring a book wherein one might acquire the primordial ideas of this powerful mathematical resource developed some years ago by the mathematical school of the University of Padua."[102] The last two chapters of the volume consider Einstein's gravitational equations and the field theories of Weyl and Eddington, respectively.[103]

The intent and value of Plans's volume were not lost on the Spanish mathematical community. The prize, which Plans won, was offered in anticipation of Einstein's visit, and there was no doubt about its purpose:

> The rationale for the competition and [Plans's] book must be sought in the rapid development in Spain of interest in Einstein's theory of relativity; said interest met not only with new and profound conceptual difficulties, but also with others of a formal nature derived from the use which all the commentators from [Marcel] Grossmann (Einstein's first collaborator) on made of absolute differential calculus, the creation of Professor Ricci of the University of Padua, subsequently developed and enriched with algorithms by his disciple and colleague Professor Levi-Civita, whose personal contribution to relativistic theory is well known and was decisive for its principles.[104]

This review, by Fernando Lorente de Nó, appeared in the *Revista Matemática Hispano-Americana* and further served to emphasize the point that mathematics had provided the means for

---

[102] *Nociónes de cálculo diferencial absoluto y sus aplicaciones* (Madrid: Real Academia de Ciencias Exactas, 1924), p. 5.

[103] In the last chapter, in particular, Plans was aided by Fernando Peña; see Peña's article in *Revista Matemática Hispano-Americana*, 6.

[104] Fernando Lorente de Nó, review of Plans, *Nociones de cálculo diferencial absoluto*, in *Revista Matemática Hispano-Americana*, 7 (1925), 206.

resolving the conceptual problems of General Relativity. In addition, Plans's volume had a further impact that transcended the original purpose of satisfying the public's urge to learn something about relativity. José María Orts stated it well:

> Once the moment had elapsed when, as a result of the rapid expansions of the ideas of Einstein, everyone (including those who, lacking the requisite mathematical background, could not pass beyond the frontiers of the land in which the new theory settled) was talking about the subject . . . when the flood of more or less well-informed popularizing lectures and pamphlets ceased . . . when this first period, which might be styled that of *relativistic fever* had gone by, a cooling set in among the general mass of pseudo-scholars, and the new doctrines condensed around a smaller circle composed of those few who were able to handle the elements necessary to penetrate to their depths. And from this small nucleus of *conscientious relativists*, among whom the works of Ricci, Levi-Civita, Einstein, Weyl, Eddington, Laue, etc., were reflected, we can say, although we knowingly wound his innate modesty, that the most enthusiastic, he who delves into the theory to the greatest depth, he who has contributed with tangible proofs to the study of Relativity in Spain, is the author of the book under review .[105]

It is obvious that for Orts, a mathematician who had studied in Italy, the *relativistas conscientes* were mathematicians, whether relativity occupied them centrally or not (Orts's own specialty was probability theory).

There can be no doubt of the effective role played by Plans's volume in Spanish scientific education. According to one of his students, Plans's manual had, by 1934, placed absolute differential calculus "within the reach of persons of middling scientific education. . . . Even the most recent graduating classes

---

[105] José María Orts, review of Plans, *Nociones de cálculo diferencial absoluto*, in *Ibérica*, 24 (1925), 335.

learned from this book how to handle this potent instrument of calculation, which two years ago I included in my syllabus of Mechanics as an auxiliary discipline."[106]

When speaking of relativity, Spanish mathematicians continually stressed their discipline's service to it and to science in general. Consider Rey Pastor's response (in 1928) to Ramiro Ledesma's question of whether it was true that Einstein's problems in formulating a unified field theory arose from his lack of mathematical skills. Nineteenth-century mathematics, regarded until recently as abstract or useless, Rey Pastor explained, was the basis of Einstein's theory, just as Italian physicists of the Renaissance grounded their concepts in ancient Greek mathematics. The same was true of Copernicus and the Pythagoreans. Newton had to create his own mathematical instrument (infinitesimal calculus), but Einstein had the good fortune to find his already made. "It is not, therefore, the lack of mathematical knowledge which impedes reaching the clarity desired in the theory of relativity," Rey Pastor asserted. "What happens is that we are dealing with very complex calculations."[107] The intent to demystify relativity (contrasting, for example, with Sir Arthur Eddington's delight in confounding the public with paradoxes and boasting that only a few could understand the General Theory) complemented Rey Pastor's program of presenting mathematics to the Spanish public (the intelligentsia, in this case) just as he presented it to the engineering schools (see chapter 6): as a practical instrument for the development of practical knowledge, including the theory of relativity.

The overwhelming support given to relativity by Spanish mathematicians confirms a central presupposition of this book: that by the supercharged nature of its reception, relativity obliged scientists to confront both themselves and their disciplines. As Jeffrey Crelinsten observed, "To a great extent,

[106] Navarro Borras, "José María Plans," p. 242.
[107] Ramiro Ledesma Ramos, "El matemático Rey Pastor," *La Gaceta Literaria*, 2, no. 30 (March 15, 1928), 1.

the appearance of Einstein's theory of relativity crystallized many facets of the national character of the scientific enterprise, and focused the attention of disciplinary leaders upon the strengths and weaknesses of their professional community."[108] What was true for the American astronomers studied by Crelinsten was true for Spanish mathematicians. Plans and Rey Pastor, with specific reference to relativity, focused on the strengths of Spanish mathematics, while Terradas was perhaps more concerned for its weaknesses. In either case, relativity was a touchstone by which the program of the discipline could be assessed.

In March–April 1922, Weyl, Sommerfeld, Otto Honigschmid, and Kasimir Fajans all lectured in Madrid and Barcelona. Weyl spoke mainly on non-Euclidean geometry, with one lecture devoted to the relativistic view of space.[109] While in Madrid, Weyl attended a meeting of the Mathematical Society, where he gave his "pertinent observations" on remarks by Pérez del Pulgar on uniform velocity and by Emilio Herrera on a "difficulty" suggested by the theory of relativity.[110] Sommerfeld lectured on X-ray spectra and Bohr's research on the spectrum of hydrogen, but embodied in his remarks was a philosophical note, linking quantum mechanics' overthrow of classical notions of causality with the theory of relativity. When speaking of the quantification of energy and its application to special relativity, Sommerfeld remarked that "classical or rational mechanics is the least rational of all." It was precisely this kind of paradox that delighted the popular readership that followed developments in the new physics.

---

[108] Crelinsten, "William Wallace Campbell," p. 88.

[109] This lecture series was published in German as *Mathematische Analyse des Raumproblems* (1923) and was dedicated to Terradas. On Weyl's visit, see *Revista Matemática Hispano-Americana*, 4 (1922), 50–54, 59. In a letter dated January 15, 1922, Plans informed Terradas of the Madrid faculty's appropriation of funds for the visits of Weyl and Sommerfeld (Biblioteca Esteve Terradas, Institut d'Estudis Catalans, Barcelona).

[110] *Revista Matemática Hispano-Americana*, 4 (1922), 101 (meeting of April 1, 1922). Herrera's "difficulty" is discussed in chapter 7.

Honigschmid and Fajans, both colleagues of Sommerfeld's at Munich, closed the cycle with lectures on atomic weights and radioactivity, respectively.[111]

The following year, Cabrera reciprocated with a short course in Munich on the structure of matter and magnetic properties. Sommerfeld, Fajans, Max Wien, and Karl Herzfeld participated in the discussion. Cabrera's international reputation had been rising rapidly. In spring 1925 he lectured on magnetism and the structure of the atom at the Société de Physique in Paris. Then, while Spanish delegate to the International Research Council meeting in Brussels that October, he was invited to Berlin to talk on magnetochemistry. According to one report, these scientific interchanges had begun with the Madrid visit of the four Germans, which had been reciprocated by the lectures of Cabrera, the biochemist Antonio de Gregorio Rocasolano (who had spoken at Göttingen on catalytic chemistry in June 1924), and Moles in Germany.[112] With the exception of Cabrera, there would appear to have been no direct connection between the Germans' visit and the Spaniards' lectures in Germany; but the perception of a quickening of interchange was accurate nonetheless.

By the mid-1920s, therefore, the Spanish scientific community had become accustomed to relatively frequent exposure to foreign scientists of the highest caliber, particularly Germans and Italians. After Einstein's trip, subsequent lecture tours by Henrick Lorentz and Wolfgang Ostwald (in 1925), Eddington and Vito Volterra (in 1932), and Francesco Severi (1928 and 1935) simply added to the newly acquired patina of international scientific respectability, whereby Madrid and Barcelona had become stops on the international lecture circuit. Lorentz was presented to the public not only as the "creator of the electron theory" but also as "the main precursor of Einstein." Blas Cabrera, in introducing the visitor, noted that Einstein had called Lorentz "the father of physicists." Lorentz

[111] *Ibérica*, 17 (1922), 340–341.
[112] *Ibérica*, 11 (1924), 66; 12 (1925), 290.

gave two lectures, one on the theory of magnetism and another on Bohr's theory of the structure of the atom. Lorentz was awarded the Echegaray Medal of the Spanish Academy of Sciences, before which Josep M. Plans again alluded to the invention by Lorentz of "benchmarks for the development of Einstein's theories, with his famous transformation of electromagnetic equations and the notion of local time."[113]

According to Olegario Fernández Baños, the fact that European scientists, already in 1919, were passing through Spain, "collaborating directly with the men of the new Spanish scientific renaissance," was an evident sign of "a new scientific life."[114] Einstein's visit, however, was the crucial episode that converted scientific prestige into a broad popular awareness of, and support for, pure science.

[113] *Revista Matemática Hispano-Americana*, 7 (1925), 168.
[114] Fernández Baños, "Spirito scientifico in Spagna," p. 158.

TWO

The Einstein Phenomenon

THE MAKING OF THE MYTH

Einstein's leap into world renown began on November 6, 1919, when English astronomers announced to a joint extraordinary meeting of the Royal Astronomical and Royal Society that observations of the total eclipse of the sun on May 29 had confirmed Einstein's prediction, in the General Theory of relativity, that light rays passing in the vicinity of a great mass would be deflected by the force of its gravity. The news was quickly seized upon by the British press, particularly the London *Times*, which, the very next day, proclaimed that a scientific revolution had occurred and Newton had been overthrown. In the House of Commons, the physicist Joseph Larmor, who was the member for Cambridge University, "said he had been besieged by inquiries as to whether Newton had been cast down and Cambridge 'done in'."[1]

Clearly, the *Times* was in close contact with the English scientific establishment, which was rocked by the news and received it in greater intensity than would have been the case had the observations been carried out by non-English scientists. In the first place, the fact that an English team had corroborated the theory of a German scientist was bound to arouse strong feelings in the aftermath of the war, whether in response to the renewal of scientific internationalism or to the concerns of patriots that German science had tarnished the image of Newton and, therefore, England. Second, English physicists who were

[1] *London Times*, November 8 and 9, 1919, cited by Donald Franklin Moyer, "Revolution in Science: The 1919 Eclipse Test of General Relativity," in Arnold Perlmutter and Linda F. Scott, eds., *On the Path of Albert Einstein* (New York: Plenum, 1979), pp. 55, 80.

74

habituated to a mechanical model of the aether and who therefore had had great difficulty in coming to terms with the Special Theory of relativity, one of whose two postulates, that of
relativity, requires that there be no privileged systems of reference (thus implying that the aether is superfluous) now
girded themselves for another battle to save Newtonian principles. Although Sir Oliver Lodge, the leading defender of the
aether concept, announced in advance his willingness to accept
the General Theory if the deflection observed should turn out
to be 1.75 seconds (as Einstein predicted), he still admonished
that he was not prepared, thereby, to abandon the aether and
also offered a number of rival explanations to explain the
eclipse results with modified Newtonian principles.[2]

The situation in England, both scientific and political, therefore, was peculiarly suited to enhance the dramatic nature of
the eclipse results, and the spirited public debate that followed
the announcement also launched the Einstein myth. The pub-
licity Einstein received in the first few months after the English
announcement set the tone for much of the characteristic popular discussion of relativity over the following decade. Perhaps
most significantly, both the Special Theory (whose popularization had been delayed by the war) and the General Theory
were discussed at the same time.[3]

The public was won to Einstein's cause by the observational
proofs of the General Theory (the anomalous motion of the
perihelion of Mercury and the deflection of light rays by gravity) and cared little for the underlying theory. Some of the paradoxical results of the Special Theory were mixed into relativity popularizations at this time in order to pique public interest
even more. As early as December 1919, Sir Arthur Eddington,
the astronomer who had led the observation party on the is-

[2] On Lodge's opposition to relativity, see ibid., pp. 71, 85, and Stanley
Goldberg, "In Defense of Ether: The British Response to Einstein's Special
Theory of Relativity, 1905–1911," *Historical Studies in the Physical Sciences*, 2
(1970), 102–104.
[3] Philipp Frank, *Einstein: His Life and Times* (New York: Alfred A. Knopf,
1947), p. 61.

land of Principe, was already stimulating the public's imagination by noting that, should he move vertically at 161,000 miles per second, he would shrink in height from six to three feet. Eddington, who delighted in emphasizing the paradoxical nature of relativity, set a standard mode for its popularization. Confusion in the press between the Special and General Theories led to the widespread belief that the eclipse results had confirmed not only the General Theory but also the paradoxical effects of special relativity, such as the slowing of time.[4]

By early 1920, Einstein had gained a high degree of notoriety (on February 2, he took note of the "flood of newspaper articles" that had already appeared); however, the geographical patterning of his fame was not altogether uniform. French newspapers, for example, scarcely mentioned Einstein in the fall of 1919, and neither did those in Spain.[5] By the time Einstein lectured on relativity in Prague and Vienna in early 1921, his fame had preceded him, and when he toured the United States with Chaim Weizmann in April and May of that year he had already become an international celebrity.

Michel Biezunski has identified three elements of the Einstein myth, as it was manifested in France in the 1920s. First, Einstein was a genius because he was German. Second, his personality made him the ideal incarnation of the message he bore. Third, he was incomprehensible.[6] The first element had

---

[4] Ronald Clark, *Einstein: The Life and Times* (New York: World, 1971), p. 242. On popular and journalistic confusion between the General and Special Theories, see Jeffrey Crelinsten, "Physicists Receive Relativity: Revolution and Reaction," *Physics Teacher*, 18 (1980), 190.

[5] Ibid., p. 246; Michel Biezunski, "Einstein à Paris," *La Recherche*, 13, no. 132 (April 1982), 503. It is important to determine the chronology of Einstein's rise to fame because, although the Einstein myth had the same or similar dimensions in a wide variety of settings by the mid-1920s, the initial burst of popularization was triggered by different stimuli in different countries. Such problems can be resolved by comparative research of newspaper materials, which can also be used to appraise the geographical and social limits of the myth's diffusion.

[6] Michel Biezunski, "La diffusion de la théorie de la relativité en France," doctoral dissertation, University of Paris, 1981, p. 114.

to do, of course, with the overbearing presence of German science in the international scientific community. Particularly in France the image of German science was highly colored by political antagonisms, which came amply to light during Einstein's 1922 visit to Paris. Nevertheless, it was a routinely held opinion among European scientists of the early twentieth century that "contemporary physico-chemical doctrines carried the mark of the countries in which they were born."[7] Spanish scientists in general subscribed to the French view of German science and even extended it to apply to the science of developed Western countries generally.

Einstein's personality, or rather his persona, seemed to connote, if not inscrutability, then at least incongruence. The press featured minute descriptions of his physique and countenance, contrasting, for example, the sensuality of his mouth with the dispassion of his eyes, whose frequent twinkles only served to perplex his observers more. His famous unkempt hair became a kind of visual token of his audacious ideas that had apparently brought an element of disorder into the Newtonian system. He was said to have been inept at mathematics (at least as a student), and the omnipresent violin he carried beneath his arm merely underscored the notion that Einstein was more the model of the artist than the scientist. The fact that Einstein did not fit the stereotype of the man of science "explained" why his ideas did not fit that of a scientific theory. As Biezunski points out, Einstein was German without being German (he was a naturalized Swiss), Jewish without being Jewish (a nonbelieving Zionist), a scholar lacking scholarly airs, and famous without wishing to be.[8] In addition, the contrast between the simplicity of his personality and the com-

[7] Harry W. Paul, *The Sorcerer's Apprentice· The French Scientist's Image of German Science, 1840–1919* (Gainesville: University of Florida Press, 1972), p. 13.

[8] Biezunski, "La diffusion de la relativité," pp. 137–138. To Biezunski, Einstein presented a familiar public image that was *almost* understandable, at the same time as he presented intellectual paradoxes that defied comprehension. This insurmountable contradiction, or tension, lies at the core of the Einstein myth.

plexity of his ideas was continually stressed. Such paradoxes made Einstein an incomprehensible figure.

The separate elements that made up the Einstein myth had different origins and different functions. Some, perhaps most, were created, or at least encouraged, by Einstein for personal motives. His penchant for wearing old clothes, for example, helped create an offbeat image that kept people at a desired personal distance from him.[9] His public persona was well constructed for the purpose of satisfying the public's curiosity without revealing much of his personal life. On his trips, accordingly, he was adept at telling a wide variety of interlocutors exactly what they wanted to hear, and he was always ready to deliver printable, pithy (although frequently not very original) pronouncements on any topic, from politics to folk dancing. His second wife, Elsa, had the corresponding ability to reply to questions on the intimate side of Einstein's life without revealing anything intimate. Still other elements of the myth, such as his having been a poor mathematics student, had both a personal function (to enhance Einstein's view of himself as a simple man with simple ideas) and a social one (to defeat those opponents of relativity who claimed that Einstein had made physical problems unreasonably abstract by veiling them with complex mathematical formulas). An example of Einstein feeding his own mythology is an anecdote recounted by the Spanish humorist and cinemast Tono (Antonio de Lara Gavilán) who met Einstein at a party in California in the early 1930s. Tono, whose English was notoriously bad, apparently conducted an extended colloquy with the physicist, after which his astounded friends asked him:

"What did he tell you?"

"Things about life: we reached the conclusion that, in it, everything is relative."[10]

[9] Lewis Pyenson, following Georg Simmel, argues that, by rejecting new clothes, Einstein assumed the role of a stranger and rejected the conventions of majoritarian culture; *The Young Einstein* (Bristol-Boston: Adam Hilger, 1985), p. 71

[10] Fernando Vizcaíno Casas, *Personajes de entonces* (Barcelona: Planeta, 1984), pp. 235–236.

Other elements, those having to do specifically with the theory of relativity, originated among scientists, both those who favored it and those who opposed it. This aspect of the myth had to do mainly with the supposed incomprehensibility of the theory and had different functions, depending on what group was involved. Although scientists and intellectuals both said similar things about Einstein's theories, they frequently had different motives for doing so.

The inherent difficulty of distinguishing the subjective from the objective elements of the Einstein myth, as well as of determining the origins of such elements, is illustrated by the question of whether Einstein, when playing music, could "count" or not. According to Boris Schwarz, "There is an apocryphal story that Einstein, while playing with Artur Schnabel, lost his place in the music, whereupon Schnabel said mockingly, 'Professor, can't you count?' " Schwarz, who played often with Einstein, suggests that this anecdote is "pure fiction"; when Einstein played with him, he never miscounted.[11] The anecdote, however, is more than likely true, because the many musicians who witnessed Schnabel and Einstein performing together in Berlin synagogues after the advent of Hitler noted such incidents, which Einstein acknowledged. On one occasion in the late 1930s, the Stradivarius Quartet visited Einstein in Princeton, New Jersey, for an evening of chamber music. The violist, Marcel Dick, recalled: "I do not remember the repertory we played but in a Haydn quartet—he played first violin, of course—after a rest of three measures he failed to re-enter. We stopped, and with an unforgettable angelic smile he said, 'I never knew how to count.' "[12]

---

[11] Boris Schwarz, "Musical and Personal Reminiscences of Albert Einstein," in Holton and Elkana, eds., *Albert Einstein: Historical and Cultural Perspectives* (Princeton, N.J.: Princeton University Press, 1982), p. 410. Compare the music counting anecdotes in Jamie Sayen, *Einstein in America* (New York: Crown, 1985), p. 137, where a musician both denies that Einstein had counting problems and also reports his willingness to go along with this aspect of the myth. Whether Einstein could really "count" or not is less significant than what the myth held in this regard and Einstein's fueling of it.

[12] Marcel Dick, personal communication, Cleveland, June 27, 1982. Dick,

Clearly, counting rests did not hold Einstein's attention, and musicians did in fact chide him. It is also true that Einstein participated in the joke, suggesting that this element of the myth was, in part at least, autogenerated. The story also demonstrates the interrelatedness of different elements of the myth. That Einstein "couldn't count" would have been unremarkable had it not been for the widespread, although erroneous, belief that he had difficulty with mathematics, particularly simple math. In this sense, we are dealing with an artifact of the Einstein myth; musicians would not have chided Einstein if this fabrication had not fit in with other elements of the myth.[13]

The notion that relativity was incomprehensible in part originated among scientists opposed to the theory. Since Einstein questioned the basis of classical physics, to claim that his theory was incomprehensible was a way to avoid coming to grips with its implications. His theory, moreover, did not seem properly experimental, and therefore did not appear, to many experimentalists, to constitute *real* physics. The magical effect of mathematical abstractions actually matching observations, which dazzled Einstein's supporters, was also the source of his opponents' distress.[14] Among nonscientists, however, incomprehensibility played a different role. A characteristic position of intellectuals was to express undisguised resentment at being forced to admit their inability to understand a revolutionary scientific idea. In most European countries, the literary intelligentsia adopted an ambivalent position.

---

who was later first violist of the Cleveland Orchestra, heard of Einstein's synagogue performances from Victor and Vitya Babin.

[13] Cf. Clark, *Einstein*, p. 250, for anecdotes in which Einstein stresses his lack of education, inability to "count," and so on. Einstein, although endowed with fine mathematical intuition, was "not especially skilled in formal manipulation" or in calculation, according to a former assistant at Princeton; see Sayen, *Einstein in America*, p. 87. The fact that Einstein underestimated his own mathematical abilities would also suggest that this aspect of the Einstein myth was autogenerated.

[14] See Jean Eisenstaedt, "La relativité générale à l'étiage, 1925–1955," *Archive for History of Exact Sciences*, 35 (1986), 176.

On the one hand, there was a snide undercurrent of resentment against scientists for having displaced the traditional arbiters of culture. In this vein, it was common to find reiterations of complaints of anti-relativist scientists that only a handful of people could understand relativity. On the other hand, many of the same commentators attempted to acquire at least a patina of familiarity with the theory. Lacking scientific background, they adopted a mode a of discourse characterized by what Biezunski calls "semantic slide."[15] In discussing relativity, nonscientists used the terminology of the theory to formulate interpretations that fit meanings already known: to use "relativity" to convey the sense of "relativism," for example. In this way, intellectuals generated a dialogue that ostensibly discussed the philosophical meaning of Einstein's theory, but that in fact was devoid of physical content.

Intellectual legitimation and justification are powerful motivations informing the attitudes of specific groups toward specific ideas. I will argue that the abstract quality of the theory and its mathematical language were perceived by some intellectuals as delegitimizing their role as public arbiters of ideas (chapter 7). I have already noted the opposite process: that Spanish mathematicians attempted to demystify relativity, for if it were truly incomprehensible, then their commitment to it would cast doubts on the legitimacy of their discipline.

POPULARIZATION AND THE PROCESS OF DIFFUSION

This raises the issue of how much scientific information one can reasonably expect to be transmitted. A number of recent commentators have attempted to build a case for the structural impossibility of transmitting any but a minute fraction of scientific knowledge to nonscientists. In one view, only one percent of what a scientists knows can be communicated, and of that one percent, after allowances are made for various kinds

[15] *Glissements sémantiques*, Biezunski, "La diffusion de la relativité," p. 129.

of formal and nonformal screening and for the inevitable distortions caused by "translating" *scientific concepts into non-scientific language*, only one percent of that is really assimilated.[16] Even if only a tiny fraction of scientific knowledge is assimilatable by nonscientists, do the latter really want to learn? According to Beaudouin Jurdant, there exists *no* demand for scientific popularization among the public at large. If such demand exists, it has nothing to do with the quest for scientific knowledge but simply reflects supply.[17]

The question of demand for popularization raises the issue of the nature of the diffusion of scientific ideas. In regard to the related problems of the reception of relativity and the diffusion of the Einstein myth, there would appear to be two separate processes involved. Information about relativity diffused in a hierarchical fashion, downward through the urban and educational hierarchies (from higher to lower order cities and educational centers; from high-brow to low-brow publications). As information percolated through the hierarchy, the screen or barrier effect caused by lack of preparation in mathematics and/or physics grew increasingly severe.

Superficially, at least, the diffusion of the Einstein myth also exhibits hierarchical characteristics, inasmuch as it appears to have diffused, via the mass media, from larger centers to smaller ones (or, in the Spanish case, outward from the three cities visited by Einstein). But at the local level, within each city, such information, newsworthy and of great popular interest, spreads by simple contagion, or word of mouth. Since relativity was difficult to understand, it is hard to distinguish between a situation where some knowledge of it, in debased form, has reached a person through the educational hierarchy, as in a *tertulia* (Spanish literary salon, typically held in a café), where a physicist "explains" relativity to a bullfighter, or by simple word of mouth—the bullfighter hears in the neighborhood bar that the inventor of relativity is in town. At the low-

[16] M. W. Thistle, "Popularizing Science," *Science*, 27 (1958), 951-955.
[17] Jurdant, as discussed by Biezunski, "'La diffusion de la relativité," p. 124.

est levels, the hierarchy virtually disintegrates (although in chapter 6 I provide evidence of a very low-level perception of relativity which has clearly diffused through the educational hierarchy). When, in the process of hierarchical diffusion, information about relativity is stripped down to a few clichés (for example, "Light has weight"), we can no longer discern a structured pattern of flow.

A statistical comparison of the receptions of Darwinism, relativity, and psychoanalysis in Spain reveals a common pattern. Figure 1 records production of articles and books through six five-year periods, beginning with the period of first citation (1861–65 for Darwinism; 1906–10 for relativity and psychoanalysis).[18] The process of reception, in each case, begins with an extended period of gestation or incubation lasting up to ten years (periods 1 and 2), in which the level of production is low. The reasons for the long incubation period are complex and may include political and social factors as well as cognitive and linguistic ones. In each case, discussion in the initial two periods was confined mainly to those who read the requisite foreign language and, in the reception of relativity, had the necessary mathematical background. The period of gestation yields to a third period, by which time the initial small groups of discussants have diffused the idea widely enough to create a secondary demand for translation of pri-

[18] The three sets of figures are of unequal value and the graph is meant only to illustrate the general pattern of reception. The figures for psychoanalysis are most accurate and are derived from Francisco Carles Egea, "La introducción del psicoanálisis en España (1893–1922)," doctoral dissertation, University of Murcia, 1983; and Isabel Muñoz González, "Evolución de los conceptos psicoanalíticos en España (1923–1936)," licenciate thesis, Univesity of Murcia, 1983. I take the period 1906–1910 as the first in which psychoanalysis was discussed, although two translations of Freud and Breuer's "Preliminary Communication on Hysteria" had been published in the 1890s, with no repercussion. The figures on relativity are derived from sources mentioned in this book, omitting newspaper articles. The Darwinism figures are the least exhaustive; I have derived them from "Bibliografía hispánica sobre Darwin y el darwinismo," Anthropos, 16–17 (October 1982), 15–54, in which the criteria for selection of titles are not always apparent.

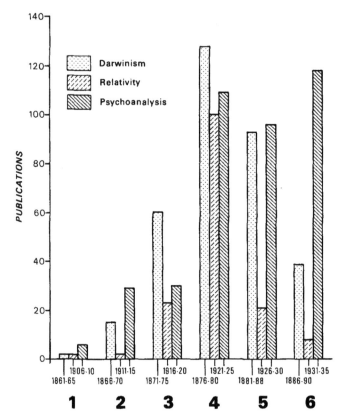

1. Receptions of Darwinism, Relativity, and
Psychoanalysis in Spain.

mary texts and production of "high-brow" treatises of popularization. This resulted, in the case of Darwin, in the translation of the *Descent of Man* (1870) and *The Origin of Species* (1877); in that of Freud, of the *Obras completas* (beginning in 1922, but contracted for during World War I); and in that of Einstein, in the translation of major synthetic accounts of relativity beginning in 1920. The fourth period is marked by the logarithmic growth of titles, peaking after just two or three years, as the phase of genuine popularization occurs, with all of the attributes associated with intellectual fads (public lectures, newspaper articles, and so forth). The last two periods (5 and 6) are marked by decline (in the cases of relativity and Darwinism) or leveling off (psychoanalysis), as popular interest falls off and the scientific community assimilates the ideas. This period has interesting consequences because, during it, one can mark the end of the period of primary diffusion, as Darwin was discussed in small Spanish towns around 1881, and Freud around 1931. Subsequently, there are waves of secondary diffusion, more complex and harder to describe, as the ideas are reinterpreted and reworked, frequently (on popular levels) without attribution of author (most obvious in the case of Freudian psychology). This process is seen as global, encompassing both scientific and nonscientific receptions that are viewed as a continuum whose specific patterns of growth and diffusion involve interactions among various levels and domains of discourse.

This curve is a good predictor of events in the diffusion of relativity and popularization of Einstein's image in Spain. At the end of the gestation period, which extended from 1905 to 1915, scientists had heard of relativity, and information about it had reached substantial numbers of engineers. Thus demand for high-level popularization and translation occurs toward the end of the third period. The invitation tendered to Einstein in 1920 to visit Spain also fits the logic of the curve; such a trip would only make sense given a large enough audience for his appearance. But when Einstein actually arrived in Spain, the process sped up, due to a media blitz, suggesting a permutation in the curve, with the explosive phase of diffusion occurring at a steeper pitch than would have been the case had there been no visit. The visit also caused secondary disturbances, inasmuch as less attention was focused on Einstein beyond the range of the three cities he visited. Belated invitations from Valencia and Bilbao, two cities farther down in the urban hierarchy than Madrid and Barcelona and lower than Zaragoza in the educational hierarchy, follow a pattern of distance decay.[19]

To return now to Jurdant and demand for popularization, none of the processes just described requires demand. They simply represent the automatic and predictable flow of information, taking into consideration the cognitive screen stipulated. Clearly, however, this kind of formulation is insufficient to explain the interaction of specific groups in the course of the diffusion process, nor the kind, quality, or focus of the popularization. In Spain, high-brow popularization was typically produced by mathematicians and physicists at the behest of engineers and to serve their requirements. At this level, therefore, there was considerable transmission of physical (and mathematical) ideas. Most of the "popular" writing on relativity was, in turn, produced by engineers, a professional group that had a great deal to gain by associating itself with a prestigious scientific theory. Successful popularizers of relativity gained the prestige accruing to those who possessed the magical power of being able to render the incomprehensible comprehensible. The material presented in chapter 6 shows that there was a genuine demand for popularization: at the highest level, of scientists by engineers; then, of engineers by certain sectors of the general public.[20]

At the heart of the difficulty in popularizing relativity was the inextricable linkage in Einstein's conceptualization between the physical phenomena treated in the Special and Gen-

---

[19] On these abortive invitations to Einstein, see *La Voz Valenciana*, March 3, 1923; *Las Provincias*, March 4; *ABC*, March 9, and *Heraldo de Aragón*, March 15. These are discussed in my *Einstein y los españoles* (Madrid: Alianza, 1986), chap. 5, sec. 3.

[20] Biezunski's working through of Jurdant's cynical view of popularization, while stimulating, nevertheless suffers from an overly brutal dichotomization of society into scientists and *grand publique*. In chapter 6, I will argue that the popularization of relativity in Spain was not a function of supply, but of demand by engineers to acquire specific information regarding the new physics. At the next lowest level, I can more readily admit that popularization by engineers was in part a function of supply, because the acquisition of prestige was sufficient reason for engineers to promote the popularization of relativity.

eral theories (electromagnetic phenomena of moving bodies, gravitation) and the problems of observation and measurement attendant upon describing them. It is clear why this should have been so, because the heart of Einstein's critique of Newtonian physics was that there were no privileged observers. Einstein reiterated this point in his Madrid talks, but popular commentators did not pick up the theme with any consistency. At an elementary level, the concept of no privileged reference systems was seemingly perceived by intuition (riding on a moving train next to a stationary one) and hence not really questioned. Whether this common experience was an Einsteinian intuition or simply a reaffirmation of Galilean relativity was a distinction not made in the popular press.

The Special Theory was said to have been intelligible to those who knew college algebra and the General Theory not to be intelligible at all without tensorial calculus. Yet, on the level of popular reception, there were greater problems by far in explaining the Special Theory, with all its paradoxes, than there were with the General Theory, and a typical pattern was for an individual to vault over the Special Theory, placing his doubts in abeyance, and to accept the General Theory, after the eclipse of 1919. Popular confusion between the two theories provided an ideal context for such a pattern. Problems with the Special Theory had to do, first, with the inability to abandon aether and, second, with philosophical problems caused by the necessity of accepting the speed of light as the limit of velocities, and, even more, by the incapacity to comprehend the kinematic (as opposed to mechanical) sense of special relativity.

The acceptance of the Special Theory was further complicated by the examples Einstein used to illustrate it—for example, the various problems based on the distinction between a train moving at a very high velocity and another system (a station, an embankment) at rest. Such examples could be grasped by persons with no knowledge of higher mathematics at all, or at least by persons with an informed or intuitive grasp of descriptive geometry. Many readers, however, including mathematically gifted ones, lacked this talent completely. Simultaneity created confusion as to whether the events so described were real or simply artifacts of observation. Adding to the confusion was the Lorentz contraction, also an observational phenomenon, which was crucial in the elaboration of the Special Theory and appeared implausible to the general reader, who was never quite sure whether it was a "real" phenomenon or not. The railroad, the station, and the embankment constituted what Nigel Calder calls "the boondocks of Special Relativity."[21] Today's popularizations are more effective because they concentrate more on physical, particularly astronomical, phenomena (in general relativity, and in the Special Theory the properties of elementary particles) and less on the paradoxes of measurement. Everyone, however, was able to grasp the experimental proofs of the General Theory, and the average relativity-watcher was willing to accept them as demonstrating the correctness of the theory even if he was unable to grasp most, or any, of it.

The Scientist as Hero

Einstein's trip to Spain was a repetition of a pattern already established on his previous sojourns in the United States, France, and Japan, and one that would be repeated in 1925 on his voyage to Argentina, Uruguay, and Brazil. Although the Jewish dimension was lacking in Spain (as in Japan), the typicality of the Spanish excursion is noteworthy. Einstein's reception in Spain in 1923 falls so close to the general pattern that one must conclude that the presence or absence of a Jewish community, or its relative size or influence, had little to do with the Einstein phenomenon.[22]

When Einstein arrived in the United States in April 1921 the myth was already fully developed. Received as the unconven-

[21] Nigel Calder, *Einstein's Universe* (New York: Viking Press, 1979), p. 2

[22] A different case is made by Marshall Missner, who detects socially based differences in Jewish and non-Jewish attitudes toward Einstein; see "Why Einstein Became Famous in America," *Social Studies of Science*, 15 (1985), 267–291.

tional bearer of an incomprehensible theory, Einstein elicited some responses with a strongly American coloration. His unconventionality appeared to have won him respect and admiration in a society that placed a high value on individuality. On the other hand, the fact that only a few could comprehend his message clashed with democratic values. As the *New York Times* editorialized, "The Declaration of Independence itself is outraged by the assertion that there is anything on earth, or in interstellar space, that can be understood by the chosen few." That document had espoused truths held to be self-evident, and the newspaper implied that the same standard should be applied to scientific truths. Paul Carter, commenting on this passage, noted that "In a democratic age, common sense was felt to be the property of common men."[23] In the United States, the popularization of science was deemed socially meritorious, with science presented as the key to social betterment. It followed that, if science was not intelligible to the general public, it was socially invalid.[24]

Einstein drew large crowds to his lectures on relativity, first at Columbia University and City College of New York in April, and then at Princeton in May. He and Chaim Weizmann, whom he accompanied on a Zionist fund-raising tour, then visited Chicago—where Einstein met Robert Millikan—and Cleveland. Einstein's experiences in the latter city were revealing. Although it was, by far, the most provincial city he had yet visited as a public figure, he had attracted such notoriety that the surging crowds awaiting him at the railroad station actually placed him in danger of physical harm:

A little man in a wide brimmed, high crowned black felt hat, with long hair, looking the traditional musician, appeared on the back platform and smiled at the reception committee. . . . As he emerged from the runway into

[23] Paul A. Carter, *Another Part of the Twenties* (New York: Columbia University Press, 1977), p. 68.

[24] Peter Morris Dixon, "Popular Criticisms of Relativity around 1920," Honors thesis, Harvard College, 1982, p. 6.

Union Station the crowd closed in behind him and grew denser as the people pushed together up the stairway. They shouted hysterically and rushed forward to see him, following in his wake.

The veterans [accompanying Einstein's party] began pushing people right and left and hurrying until they reached the pavement. . . . At this point the safety of the scientist for a moment seemed threatened by the eagerness of the crowd to see the visitor. With the aid of traffic policemen and mounted men who rode back and forth trying to break up the jam, the veterans managed to push Prof. Einstein into the automobile with Prof. Weizmann and Cleveland reception committeemen.

The crowd closed around the car. Those in the rear pushed forward, but the automobile honked threateningly until a path was made. People attempted to cling to cars in the escorting line and kept policemen busy driving them off. [The motorcade proceeded to City Hall pursued by people on foot.]

The scene was repeated, though less violently, from City Hall to The Hollenden [Hotel].

"Who is it?" asked one man who was running full speed from the station after the line of automobiles.

"Prof. Einstein; he invented the theory of relativity," was the answer.

"What's that?"

"I don't know."[25]

One way in which popular interest in relativity was awakened, as this incredible scene attested, was through crowd reactions, whereby passers-by were initiated into this arcane world by those whose curiosity had already been aroused. But Einstein, according to a headline in the *Plain Dealer*, stood "ready to show theory is so simple." Relativity theory can be put into "simple and brief language," he assured a reporter through an interpreter, "but before that can be done so that it

[25] *Cleveland Plain Dealer*, May 26, 1921.

would be intelligible, it is necessary that the person should give several weeks of study to the underlying principles on which it is based." He did not specify the principles in question, but when the reporter inquired whether the theory did away "with the old philosophical and scientific idea of an infinite time and an infinite space," Einstein replied that "time and space have heretofore been regarded as something separate and apart from the earth, as being absolute and all reaching. But by the theory of relativity time and space are seen to be interwoven with all the other laws, so that they cannot be separated." It is interesting that Einstein, who frequently gave pithy popularizing statements of this kind, did not specifically address the issue of finitude versus infinity, which was the cosmological issue that most caught the public's imagination. His response comes across as somewhat evasive, if not cryptic. Later, it became current in the popular press to do away with modifiers (such as "absolute") completely and to assert that Einstein had done away with time and space. The general public was as unaware of Newtonian concepts of time and space as they were unable to grasp Einstein's, or to comprehend the nature of the revolution in physics. Einstein concluded this interview abruptly, without replying to a question concerning the fourth dimension.[26]

In Cleveland, Einstein found time to visit Dayton Miller (1866–1941), the Case Institute physicist whose continued attempts to confirm the existence of the aether and therefore disprove the Special Theory aroused the hopes of Einstein's opponents throughout the 1920s. That the *Plain Dealer* did not comment on this visit was simply symptomatic of the fact that the vast majority of the media assumed that relativity was already proven, so incontrovertible did the 1919 eclipse observations appear.

The "Curiosity Files" in the Einstein Archive preserve, in the form of unsolicited mail received by Einstein from the gen-

---

[26] *Cleveland Plain Dealer*, May 27, 1921. On the public's unawareness of Newtonian physics, see Biezunski, "Einstein à Paris," p. 507.

eral public, a wealth of evidence relating to the social diffusion of the myth in the decade after his first visit to America. A 1925 letter from a high school chemistry teacher explains that "We Americans are beginning to see a few things we could not see eleven years ago"—possibly a reference to the writer's own introduction, in his student days, to special relativity, which was not much appreciated in the United States before World War I. By 1930 the image of Einstein the worker of wonders was set in popular culture. Arriving in New York, the physicist received an anonymous postcard, sent to his ship, which read: "Welcome to our City!! The Master of Space!"[27]

In March and April 1922 Einstein made a celebrated appearance in Paris. There, the scientific community accurately reflected a climate of public opinion still remarkably hostile toward Germany and Germans even though the war had been over for three years. The French, therefore, made more of Einstein's association with Germany than was the case in other allied countries. In the United States, Einstein had attracted virtually no anti-German flack, although in New York the anti-Semitic councilman Bruce Falconer did suggest that Einstein should be regarded as an enemy alien. No American scientists criticized Einstein's science as German. A sizable group of French physicists, however, was openly opposed to Einstein and his theories for ostensibly political reasons. The French Society of Physics, whose membership included a large number of fiercely nationalistic engineers, took no part in Einstein's reception, and thirty members of the French Academy of Sciences announced that they would walk out of a session planned to honor Einstein. Visceral anti-relativists were able to dismiss him simply on the grounds of his having been a *boche*.[28] This was not the case elsewhere. In other countries, like-minded persons who were neither anti-German nor anti-Semitic, argued against relativity without reference to Einstein's nationality.

[27] David W. Rial to Einstein, November 23, 1925, Einstein Duplicate Archive, Seeley Mudd Library, Princeton University.
[28] Biezunski, "La diffusion de la relativité," pp. 60–61, 152.

The high points of Einstein's Paris tour were his lecture, arranged by his leading French supporter Paul Langevin, at the Collège de France on March 31, followed by three special sessions to allow physicists to bring forward difficult or controversial points of relativity theory for discussion (April 3, 5, and 7); and a presentation—debate, really—on April 6 at the Société Française de Philosophie. In the March 31 lecture, Einstein summarized the findings of both special and general relativity, building from the contradiction between the constancy of the speed of light for all observers and Newtonian laws of mechanics, concluding with a description of the four-dimensional universe. He spoke in French (he had used German in the United States)—"His language is very clear. The very *gaucherie* of his vocabulary creates an image," one observer noted. Langevin sat behind him, to supply difficult words, and Einstein complemented the spoken words with gestures, those of a sculptor "whereby his hand caresses the forms intended, however unreal."[29] The dynamics of the discussion sessions, where die-hard antirelativists such as Georges Sagnac, whose opposition to relativity led him to persist in attempting to confirm the existence of an aether wind, interrupted the discussion to interpose his own theories, were determined by Paul Painlevé's "conversion" to relativity. Painlevé, the leading figure in French mathematics, had been skeptical of relativity and queried Einstein regarding what was known as "Langevin's paradox," which had to do with a very fast train, bearing a clock initially in time with that of the station of departure. Observers at rest will see the train's clock moving less rapidly than the one at the station. If the same train returns to its point of departure at a similarly high speed, its clock will be behind that of the station. To Painlevé, this formulation appeared to violate the laws of symmetry, but in a personal interchange between the first and second sessions Einstein was able to satisfy Painlevé's doubt by explaining that two distinct reference sys-

[29] Raymond Lulle, in *Oeuvre* (Paris), April 4, 1922, cited by Biezunski, "La diffusion de la relativité," p. 43.

tems were involved, and not just one.[30] The second session began with an explanation by Langevin of how Einstein had resolved Painlevé's doubt.

At the April 6 session at the Société de Philosophie, Einstein exchanged views with Henri Bergson, Emile Meyerson, and other philosophers, although Langevin and the mathematicians Hadamard, Painlevé, and Elie Cartan also participated. The crucial interchange was between Einstein and Bergson over the distinction between philosophical and physical time. For Einstein, there was only a psychological, subjective time and a physical time. Simultaneity had been perceived in accord with the former with no contradiction because of the extraordinary velocity of light. But simultaneity is a mental construct, with no physical reality.[31]

Spaniards were well informed of Einstein's presentations in Paris. Corpus Barga, *El Sol*'s man in Paris, gave a full report on the Collège de France and Société de Philosophie lectures, making clear that Einstein had been able to overcome Painlevé's mathematical purism by interposing some physical realism.[32] The same newspaper had registered its editorial indignation over the exclusion of women from the sessions at the Collège de France. Painlevé himself was stationed at the door, *El Sol* noted, in a "lack of gallantry, inexplicable in a Frenchman." To the Madrid daily the only explanation possible was a desire not to popularize Einstein's theory; the motive seemed clear in the light of the historic role that French women had played in the diffusion of "general ideas."[33]

The French press gave ample emphasis to the incomprehen-

---

[30] Biezunski, "Einstein à Paris," p. 508. The more usual example of Langevin's paradox was the space traveler.

[31] Einstein, *Bulletin de la Société Française de Philosophie*, session of April 6, 1922, p. 107, in Jean Langevin and Michel Paty, eds., *Le séjour d'Einstein en France en 1922*. Cahiers Fundamenta Scientiae, no. 23. (Strasbourg: Université Louis Pasteur, 1979).

[32] Corpus Barga, "El 'colloquium' de Einstein con los sabios franceses," *El Sol*, April 14, 1922.

[33] "Einstein y las damas," *El Sol*, April 11, 1922.

sibility of relativity and its social role. Snobbism was a theme much discussed in this regard. Attendance at Einstein's lectures became a badge of honor to elements of the literary class who wished to preserve their title to intellectual leadership by associating themselves with relativity. Snobbism, according to one French commentator, consisted in seizing upon and promoting ideas that were incomprehensible to the common man: symbolism in literature, cubism in art, relativity in science.[34] The preponderance of French press commentary on Einstein's visit, though, was favorable to the man and his cause, and virtually all the pro-Einstein scientists (including Langevin, Charles Nordman, and Painlevé himself) also wrote in the popular press. Einstein convinced the French public during his visit but not the majority of scientists, who were unable to shed their commitment to classical physics.[35]

In November 1922 Einstein undertook a long trip that was to keep him away from Germany for four months. Ever since the murder of the Jewish foreign minister Walther Rathenau in June, Einstein had felt increasingly uncomfortable in a climate of open anti-Semitism. He was thus able to take advantage of a number of attractive invitations from abroad to remove himself from Germany in a difficult moment. After brief stops in Ceylon and China, Einstein arrived in Japan in mid-November to begin a lecture tour sponsored by the Kaizosha publishing house. Relativity had long been known to Japanese scientists: Ayao Kuwaki had written on the Special Theory (which he regarded as a modification of the electron theory) as early as 1907, and Jun Ishiwara (1881–1947), who had studied with Einstein in Zurich in 1913, wrote extensively on both the Special and General Theories throughout the decade preceding Einstein's trip. Ishiwara accompanied Einstein on his lecture tour, serving as interpreter. Among Einstein's addresses was one entitled "How I created Relativity Theory," delivered at

[34] G. de la Fouchardière, in *Oeuvre*, April 2, 1922, cited by Biezunski, "La diffusion de la relativité," p. 101.
[35] Biezunski, "Einstein à Paris," p. 510.

Kyoto University. This lecture provides an interesting example of the mine of information generated by Einstein's foreign trips which has remained almost wholly untapped by historians of physics. In this lecture, of which only a Japanese translation of the German original survives, Einstein stated quite clearly that he knew of the Michelson-Morley experiment as a student, and that he was stimulated by it to question the current explanations of the motion of the earth relative to the aether. This was the view held popularly in the 1920s but which recent revisionist historiography has sought to discredit.[36]

Beyond the bounds of the Japanese scientific community, the situation was similar to the experience of other countries that had received visits from Einstein. It was not clear whether more than a few of the many persons who heard him lecture on relativity actually understood anything, in spite of Ishiwara's stentorian efforts at translation, which extended the duration of his first lecture to six hours (with an intermission after three). As in the United States, politicians wrangled over whether or not the theory was comprehensible. In a full-fledged cabinet debate, the Ministers divided over the issue: the Minister of Education claimed that ordinary people could understand the theory, the Minister of Agriculture said they would understand only "vaguely," and the Minister of Justice, who himself had been defeated by the mathematics of the theory, "insisted that there could be no midway between understanding and not understanding. If they understood, they understood clearly. If they did not understand, they did not understand at all."[37]

The German ambassador's report to the Foreign Office gives a vivid indication of the character of Einstein's public reception:

[36] Sigeko Nisio, "The Transmission of Einstein's Work to Japan," *Japanese Studies in the History of Science*, 18 (1979), 1–8; Tsuyoshi Ogawa, "Japanese Evidence for Einstein's Knowledge of the Michelson-Morley Experiment," ibid., pp. 73–81.

[37] *Japan Weekly Chronicle*, cited by Clark, *Einstein*, p. 300.

Professor Einstein arrived in Japan on November 17 and departed again on December 29. His trip through Japan resembled a triumphal procession. Whereas the visits of the Prince of Wales and Field Marshall Joffre were attended by royal and military pomp, advanced detailed planning and officious play-up in the press, there was none of this during Einstein's reception; instead the participation of the entire Japanese people was in evidence, from the highest dignitary to the rickshaw coolie; spontaneous, without preparation and window dressing! When Einstein arrived in Tokyo, there were such crowds at the station that the police could do nothing to control this life-and-limb-threatening crush of people. The Tokyo reception was repeated again and again in the other cities where Einstein held lectures or where he simply recuperated from the strain of his trip while viewing both the land and its people. Since there is no reason to assume that the thousands upon thousands of Japanese who streamed into Einstein's lectures—at 3 yen per head—had any interest in the theory of relativity (which is incomprehensible to the layman), many Germans here view all this as an attempt to achieve a kind of parity; after visits from England (Prince of Wales), America (Denby) and France (Joffre), it was only fitting to honor a German! . . . The climax of the distinctions awarded this famous man was this year's Chrysanthemum Festival! . . . All eyes were riveted on Einstein; everyone wanted at least to shake the hand of the most famous man of present times. An admiral in full uniform forced his way through the rows of people, went right up to Einstein and said: "I admire you" and, thereupon, departed.

The newspapers were replete with stories about Einstein—both true and false. Timidly, one or another individual even dared to defend Newton or Galileo. One professor even had the courage to debate the nature of the absolute with Einstein, but finally had to admit, to the joy of the audience, that he had been blinded by error. There

97

were also caricatures of Einstein in which his short pipe and his thick comb-resistant hair played a major role; these caricatures also alluded to the fact that the clothing Einstein selected to wear was not always appropriate to the occasion at hand.[38]

Einstein was impressed by what struck him as the unity of Japanese culture with its natural environment and commented on Japanese music, art, and architecture in this context. He was widely recognized, at least by the intelligentsia, as the hero of peace, the flower of German science, the artistic physicist. In spite of the vast cleavage separating European from Japanese culture, the major elements of the Einstein myth were all present during his Japanese tour.[39]

Departing Japan, the Einsteins continued around the world by ship and arrived in Palestine from Port Said on February 2, 1923. In Palestine, as in the United States and Japan, he was pursued by great crowds who pressed upon him at close quarters. Besides his public support for Zionism, Einstein had been keenly interested in developing the scientific capability of the Jewish settlement, having previously expressed the view that scientific institutes related to the practical problems of settlement should be established at Hebrew University: those of agriculture, chemistry, and microbiology should come first, in his opinion. Einstein gave the inaugural lecture at Hebrew University, visited the Technion and the Migdal agricultural station, where he stooped down to examine some plants at close range, and continued to promote scientific development, although the primary focus of his tour was political. At a reception in Jerusalem on February 6, after a motorcade along streets lined by cheering children, Einstein, obviously moved,

[38] Christa Kırsten and Hans-Jürgen Treder, eds., *Albert Einstein in Berlin, 1913–1933*, 2 vols. (Berlin: Akademie-Verlag, 1979), I, 230–231.

[39] On Einstein's encounter with Japan, see Tsutomu Kaneko, "Einstein's View of Japanese Culture," *Historia Scientiarum*, 27 (1984), 51–76, and "Einstein's Impact on Japanese Intellectuals," in Glick, *The Comparative Reception of Relativity* (Dordrecht: D. Reidel, 1987), pp. 351–379.

said: "I consider this the greatest day of my life."[40] In Palestine, Einstein seemed to identify wholly with the rebirth of a Jewish national culture and to have inspired the settlers with his enthusiasm and support. The depth of feeling on both sides had an electric quality. Was there any emotion left for the physicist to experience on his world tour? On February 23, 1923, Einstein arrived in Barcelona.

[40] See catalog of the exhibition, *Einstein, 1879–1979* (Jerusalem: Jewish National and University Library, 1979), esp. pp. 38 and 41. Clark, *Einstein*, p. 393.

THREE

Barcelona: Einstein and Catalan Nationalism

IN HIS two weeks in Spain, Einstein interacted with a great many persons, both scientists and other intellectuals. Typical of his personality, these interactions were mainly one-way, affecting the Spanish interlocutor far more than the visiting physicist. Our understanding of such interactions must, of necessity, be superficial, inasmuch as the press reports, although copious, were for the most part written by nonscientists, and few accounts were left by those with more intimate exchanges with Einstein. Einstein granted few formal interviews, and, as a result, the press coverage was largely superficial, concentrating on mood and gestures rather than substance. Since these impressions, which can also be viewed as accretions to the Einstein myth, were transmitted to the public, however, their substance acquires social significance.

Terradas traveled to Germany frequently and must have met Einstein there sometime between 1918 and 1920. The invitation to visit Spain was his idea, but the first round of negotiations in the spring of 1920 were carried out by Julio Rey Pastor, who happened to be in Leipzig. The heart of the negotiation had to do with what Einstein referred to as his "language phobia." Rey Pastor replied that this posed no problems for Spaniards. "Our French is rather deficient," he explained, "and you can make mistakes without fear." If necessary, the audience could be limited to German speakers.[1] Before leaving for

---

[1] Rey Pastor to Einstein, Leipzig, April 22 and 28, 1920. Einstein Duplicate Archive, Seeley Mudd Library, Princeton University. In spite of Rey Pastor's role in the negotiations, it was well known that the invitation had first been proposed by Terradas on one of his frequent visits to Germany (Francisco

Spain, Rey Pastor, writing from Frankfurt, pressed Einstein for a commitment, adding that the Junta para Ampliación de Estudios had set aside 2,000 pesetas, a sum that could be increased if Einstein "should wish to stay in Madrid beyond one month. . . . It is our wish that you remain in Madrid as long as possible so that we may derive the greatest possible benefit from your valuable teaching."[2] Einstein's reply was short and again addressed the language issue:

> I will accept your invitation on the condition that I limit my lectures to the area of science and that I can avail myself of drawings and mathematical formulae. Given my total inability to speak Spanish and my deficient knowledge of French, I would be unable to present my lectures if I were to simply use words alone. German is the only language in which I can speak intelligibly on my theory. For the time being, I would propose next fall (October) as a date for the lectures.
>
> Be advised that I look forward to seeing you again and acquainting myself with your beautiful country.[3]

It is interesting to note that Einstein, at this early date subsequent to his rise to international fame, had not yet worked out the method of popularization that he would later perfect, first in Paris, then in Spain, of lecturing in French with the use of blackboards for complementary explanation.

Einstein, of course, did not travel to Spain in 1920. He explained to Rey Pastor that, besides his exhaustion and the demands of his position in Berlin, "there are some individuals here who have taken it amiss that I have not applied as much energy to my duties here as of recent days."[4] In fact, Einstein's

---

Vera, "El doctorado 'honoris causa' y otras grandes menudencias," *El Liberal*, March 16, 1923). Einstein's actual stipend was 3,500 pesetas (*El Correo Catalán*, February 25, 1923).

[2] Rey Pastor to Einstein, Frankfurt, May 11, 1920. Einstein Duplicate Archive.

[3] Einstein to Rey Pastor, Berlin, June 3, 1920. Einstein Duplicate Archive.

[4] Einstein to Rey Pastor, July 14, 1920; Einstein Duplicate Archive.

refusal to travel at this time was undoubtedly due to his acceptance of an extraordinary professorship at Leiden, where he began in October.

There were no further negotiations regarding the Spain trip until July 1921, when Terradas wrote to reiterate the invitation. Einstein replied that it was impossible for him to get away from Berlin that summer but that he hoped to be able to do so during the academic year 1921–22.[5] It is difficult to say whether at this time Einstein was really planning a trip to Spain, or whether he was just being cordial to his Spanish colleagues. The journalist Ricardo Baeza, in an interview held in London in June 1921, asked Einstein if he would visit Spain. Einstein replied that he would "when invited" and then added: "Do you really believe there is any interest [there] in the theory of relativity?"[6] The visit, of course, was again delayed, but by May 1922 plans were already quite advanced for the following year, inasmuch as the faculty of sciences at the University of Zaragoza was discussing its own invitation to Einstein. The final arrangements were made in Germany that summer by Lana Serrate who informed Terradas, upon the latter's arrival in Germany, of Einstein's definitive acceptance.[7]

In spite of the fact that we have less information regarding Einstein's days in Barcelona than we have for the Madrid stay, and that much of what Einstein did in Barcelona went unrecorded, some details that seem to have scant significance be-

[5] Terradas to Einstein, July 1, 1921, asking if Einstein can come to Spain the following winter or spring, and adding that Weyl is coming in winter, Sommerfeld and Fayans in April. Einstein to Terradas, July 16, 1921. Einstein Duplicate Archive; original of second letter preserved in Terradas Collection, Institut d'Estudis Catalans, Barcelona (reproduced in *Ciència*, no. 20 [October 1982], p. 43).

[6] Ricardo Baeza, "Delante del profesor Einstein," *El Sol*, July 3, 1921.

[7] Mariano Torres Lacrué, *Biografía científica de la Universidad de Zaragoza* (Zaragoza, 1967), p. 142, n. 393; *La Veu de Catalunya*, February 24, 1923. On May 4, 1922, Lana wrote to Einstein to ask whether he might translate a popularization of relativity into Spanish in advance of his trip to Spain. "You have become so popular that everyone wants to know about you and your theory," he explained (Lana Serrate to Einstein, Einstein Duplicate Archive).

came part of Catalan mythology. The story of Einstein's arrival in Barcelona was the one most frequently repeated in subsequent years. According to *La Veu de Catalunya*, he arrived by train from Toulon and, before proceeding to the Hotel Colón, "he presented himself at the house of Sr. Terradas."[8] Most of the other reports, however, state that he and his wife arrived without being met by anyone and that they proceeded to a "humble pension" on a street near the Rambla de Santa Monica (the Cuatro Naciones, tradition has it). When the proprietor learned, from the newspapers, the identity of his guest, he went to Einstein's room and found the physicist seated on the bed, playing the violin. "I am a modest citizen," he explained to a municipal functionary, "and I chose a room which corresponds to my category." The hotelier insisted that Einstein go to the Ritz, as had been planned for him.[9]

Einstein's methods of popularization differed from place to place and over time, as he strove both to refine his presentations and to tailor them to the specific audience. In Barcelona he first gave a series of three lectures on relativity (Special Theory, General Theory, and current problems) at the Diputació (provincial government building), sponsored by the Mancomunitat, the Catalan regional authority. The first two of these lectures were reasonably easy for the scientifically educated to follow; the third, which dealt with Einstein's problems with unified field theory, was not. (See the summaries of the Madrid lectures in Appendix 3.) Afterward, at the Academy of Sciences, he lectured to a less specialized audience on the phil-

[8] *La Veu de Catalunya*, February 24, 1923.

[9] *El Debate*, February 25, 1923; Joaquim Maria de Nadal, introduction to Ramon Muntanyola, *Vidal i Barraquer, cardenal de la pau*, new ed. (Barcelona: Montserrat, 1976), p. 14. Nadal, as president of the Municipal Cultural Commission, accompanied Einstein to the Casas Consistoriales at 12:30 P.M. on February 27 (*Las Noticias*, February 28, 1923). For recent evocations of this incident, see Joaquín Edwards Bello, "Einstein en Barcelona," *La Vanguardia*, April 29, 1955, and A. Coll Gilabert, "Einstein: El desconocido era un gran genio," *Diario de Barcelona*, March 11, 1979. The Terradas family preserved a handwritten note that Einstein had dispatched from the pension, announcing his arrival.

osophical consequences of relativity, concluding with remarks on the cosmological implications of a finite universe. In spite of the fact that admission to the first three lectures was restricted, by invitation, to those with scientific education ("knowledge of physics and mathematics is presupposed"), there was not enough room in the hall to accommodate all who wished to hear him: "They piled together at the doors, anxious . . . to hear the magic words which, when obeyed, would reform old systems and concepts, opening new horizons for science through the fourth dimension: time."[10]

Clearly there were numerous nonscientists in attendance at the scientific sessions. Some were there as representatives of official entities. Carles Pi Sunyer, for example, attended as director of the agricultural division of the Industrial School.[11] Besides reporters and other writers who had been granted admission to the first session, the press noted the presence of "the first two rows of grave men, with white beards and bald heads predominating," as well as "the mathematicians, weighed down with spectacles."[12] Einstein spoke slowly. "He observes before him his listeners's faces with brows knitted, even more wrinkled by the torture of incomprehension and through the difficulty of following him on his disconcerting flight." A fat man sleeps. Einstein picks up a piece of chalk and writes, explaining, "This is easy to see." The atmosphere was tense, in part because Josep Comas Solà was a domineering figure and his opposition to relativity was notorious: "From the other end [of the room] the astronomer Comas Solà lies in wait, more than observes, with an expression of surprise and suggestion proper to one of those Chaldean shepherds who studied the marvelous netting of the constellations."[13] Indeed, ges-

---

[10] *La Veu de Catalunya*, February 20, 1923 (according to this report, the first three lectures were organized as a course costing 25 pesetas); *La Vanguardia*, February 28, 1923.

[11] Oriol Pi-Sunyer, personal communication, July 26, 1982.

[12] J[oaquín] Arraras, "Una lección de Einstein," *El Debate*, March 2, 1923; *L'Esquella de la Torratxa*, 46 (1923), 139 (March 3).

[13] Arraras, "Una lección de Einstein."

tures were all that were allowed Comas during these sessions; he promptly complained in an article in *La Vanguardia* that no public discussion had been permitted.[14]

The press reports were divided over the degree of understanding displayed by the audience. The mathematician Ferran Tallada (1881–1937) and the dramatist Josep Maria Sagarra (1894–1961), writing in *La Vanguardia* and *La Publicitat*, respectively, were both of the opinion that very few had understood the lectures. On the other hand, the philosopher Joaquim Xirau (1895–1946), who prepared short summaries for *La Publicitat*, spoke of the "marvelous clarity" and "simplicity" of Einstein's exposition.[15] But Xirau was primarily interested in the final lecture, that at the Academy of Sciences on philosophical consequences, where Einstein contrasted relativity with Kantian notions of experience. According to Kant, Einstein said, all knowledge has an a priori base. Relativity is not contrary to this general line of thought but imposes some rectifications upon it. Simultaneity, therefore, loses its a priori characterization and, under the General Theory, a priori geometrical space also loses this status. There can be no geometry apart from physics.

The Catalan press's reaction to relativity, from the most frivolous positions to others more serious, was similar to that in other countries where Einstein spoke. The principal problem confronting editors was whom to send to cover the lectures. *El Noticiero* explained that it was difficult for a "press boy" to take intelligible notes; therefore, in this newspaper only short précis of the lectures were given.[16] Most of the re-

[14] José Comas Solá, "Las conferencias del profesor Einstein," *La Vanguardia*, March 14, 1923. In the same article Comas states that he was able, during Einstein's visit, to confront him directly.

[15] Fernando Tallada, "Einstein en Barcelona, I," *La Vanguardia*, March 4, 1923; Josep María de Sagarra, "Einstein," *La Publicitat*, March 4, 1923; and J.X.P. [whom I assume to be Joaquim Xirau Palau], "Les conferències del professor Einstein," *La Publicitat*, March 4, 1923.

[16] "El doctor Einstein en Barcelona," *El Noticiero Universal*, February 28, 1923.

porting was done either by "boys" or by columnists—intellectuals like Sagarra or Carles Soldevila (1892–1967) who had no scientific background. Xirau was an exception. Therefore, the best one could hope for was an elegant evocation of the general atmosphere of the event, such as Sagarra's piece discussed below. Soldevila's comments were more typical of the literary approach, repeating a cliché that followed Einstein wherever he went: "Einstein is famous," he wrote in *La Publicitat*, "because a few hundred mathematicians have believed him worthy to be so."[17] The rest of us, he wrote, must take relativity's worth purely on faith. Indeed, some scientists themselves doubted the capacity of most of the public to understand the problem. Comas asserted that those in attendance felt defrauded because they were unable to experience the revelations they had anticipated.[18] Tallada, in a popularizing series written for *La Vanguardia*, also noted "spirits in suspense and filled with confusion and dismay."[19] Here I point out only that such perceptions were general and probably psychologically accurate to the extent that there was a pervasive sense of let-down after the lectures. The reasons for this form part of our discussion of Einstein's reception by the intelligentsia (see chapter 7).

Catalan humor magazines had a field day with Einstein. *L'Esquella de la Torratxa* published a parody of relativistic reasoning, replete with nonsensical formulas aping those of mathematical physics, and *En Patufet* presented a satirical poem on the relativity of time (see chapter 8). *La Campana de Gracia* contented itself with a philosophical reflection: "We have always believed it better to declare oneself ignorant before a scholar, than a scholar before an ignoramus."[20]

One of the journalists who covered Einstein's visit, Miguel-

[17] Carles Soldevila, "La popularitat d'Einstein," *La Publicitat*, February 25, 1923.

[18] Comas, "Las conferencias del profesor Einstein."

[19] Tallada, "Einstein en Barcelona."

[20] "Relativitzant," *L'Esquella de la Torratxa*, 46 (1923), 176 (March 16); "Pel·licula de la setmana," *En Patufet* (1923), 192; "Einstein a casa," *La Campana de Gracia*, 53, March 3, 1923.

Emilio Durán, found himself by the side of the composer Jaume Pahissa i Jo (1880–1969) at a reception for Einstein at City Hall. Pahissa had written a popularization of relativity entitled *Idea de la teoría de la relatividad de Einstein.*[21] Durán, who did not have a scientific background, was more than happy to stay with Pahissa through the official activities, hearing his layman's explanation of Einstein's theory. Pahissa, a typical figure of the "scientific middle class," had studied exact sciences at the University of Barcelona. He gave Durán (sotto voce, one assumes) a short and quite effective synopsis of the theory, stressing the importance of mathematical methods in relativistic physics and enumerating those physical phenomena, such as the anomalous advance of the perihelion of Mercury, that the General Theory explained. For Pahissa, it was immaterial whether or not the general public could understand how the theory had been deduced. It was sufficient to direct its attention to the observations that confirmed the theory.[22] Indeed, stressing the experimental confirmations of Einstein's theory was one of the hallmarks of the Catalan reception of relativity. Although the disorientation and incomprehension (and attendant defensive reactions) were also present in Catalonia, popular opinion there laid greater stress than in Madrid or Zaragoza on the concrete, easy-to-grasp evidence that confirmed the elusive theory. Pahissa noted in conclusion that Einstein played the violin better than many professionals, and that if the physicist could encroach upon his field, he had the right to comment on "that marvelous theory of relativity."

EINSTEIN AND THE ANARCHO-SYNDICALISTS

Another episode drawn from the book of Einsteinian mythology, the one relating to the words of encouragement he directed to an anarchist labor meeting in Barcelona, present the

[21] Barcelona, *La Publicitat*, 1921. On Pahissa, see *Gran Enciclopedia Catalana*, XI, 69.

[22] Miguel-Emilio Durán, "Einstein en Barcelona: La teoría de la relatividad y la música," *Las Provincias*, March 6, 1923.

historian with an intractable problem, given the divergence of reports in the press coverage of the episode. Einstein had become an authentic hero of the working classes, not only because of his pacifism but also because of his refusal to sign the infamous "Manifesto of the 93." In anarchist publications, this aspect of his public persona was stressed as often as were his scientific accomplishments. For example, the *Noticiari de l'Ateneu Enciclopèdic Popular*, whose readership was primarily anarcho-syndicalist, pointed out that "Einstein has struggled for liberty, justice and understanding among individuals and nations."[23]

According to the press reports, on Tuesday, February 27, a syndicalist delegation called upon Einstein at the Ritz and accompanied him to union headquarters in the Baixa de Sant Pere. (The members of the delegation are not enumerated but included at least two of the most important leaders of the anarchist syndical confederation, the CNT: Angel Pestaña and Joaquín Maurín.[24] As to what transpired next, all accounts agree on a number of points: that Pestaña introduced Einstein; that the latter was surprised by the massive illiteracy rampant in Spain (mentioned by Pestaña); that he replied to Pestaña's allusion to repression by opining that it was more due to stupidity than to evil; and that he encouraged the workers to read Spinoza, "a source of many good ideas and timely advice."

Most of the reports, including those of the press agencies, added that Einstein had remarked to Pestaña, "I too am a rev-

[23] *Noticiari de l'Ateneu Popular*, IV, no. 35 (March 1923), 17–18.

[24] Joaquín Maurín (1896–1973) wrote to Einstein from New York (May 9, 1950): "May I . . . evoke the name of a common acquaintance, Rafael Campalans? An old friend of his, I was one of the group which invited you to address a Workers' Union in Barcelona. You were kind enough to come and, though you did not lecture, you spoke with us. I still remember one bit of advice you gave: read the *Ethics* of Spinoza." Copy of letter in the Hoover Institution Archives, Stanford, California, shown to me by John Stachel. Maurin had enclosed a list of philosophical questions that he wished Einstein to answer. No reply is preserved. According to the German consul (see Appendix 2), Einstein accepted the syndicalist invitation at the behest of Campalans. Soon after the visit, Maurín left the CNT and joined the Communist party.

olutionary, but in the area of science. I am concerned with social questions, as are other scientists, because they constitute
one of the most interesting aspects of human life."[25] These
words attributed to Einstein circulated throughout Spain, not
only in the daily press, but also in anarchist publications. For
example, *Redencion*, an anarchist weekly published in Alcoy,
accompanied the last of a three-part series by Charles Nordmann popularizing relativity with an article by Francisco Pellicer entitled "Scientific Revolution and Economic Revolution." Noting Einstein's supposed self-identification as a
revolutionary in Barcelona, Pellicer added that anarchists
might well reply: We too are revolutionaries, "if economic
ones." He continued, by way of explanation:

> The law of physical relativity is without doubt the harbin-
> ger of the law of moral relativity, which we as anarchists
> support in order to defeat those who, as Einstein ob-
> served, are more stupid [than evil]. . . . They believe that
> one must not proceed against an institution because . . .
> they regard it as immutable and not relative, as we would
> classify it in the economic interpretation of history, as
> Einstein [did] in the scientific interpretation of natural
> laws.

Pellicer concludes by identifying Einstein as one who had pro-
tested the manifesto of the ninety-three "lackey/scholars of the
German Empire."[26]

The story was widely reproduced (in the London *Times*, for
example),[27] because of the notoriety of the Spanish anarchists
who, since 1919, had been implicated in violent street fighting
and a series of political assassinations. Einstein, however,
firmly denied ever having uttered the controversial phrase. On

[25] *El Diluvio* and *El Noticiero Universal*, February 28, 1923.

[26] Francisco Pellicer, "Revolución científica y revolución económica," *Redención*, March 22, 1923.

[27] London *Times*, March 2, 1923: The syndicalists had "laid their griev-
ances" before Einstein, who replied: "I also am a revolutionary, though only
a scientific one."

the train that carried him from Barcelona to Madrid on March 1, he told Andrés Révéscz, reporter for *ABC*, that he was not a revolutionary, not even in science, and that he did not believe in a socialist society, "nor in the Communists' program of production."[28] It is unlikely that Einstein had ever uttered the controversial phrase. It represented neither his social-democratic political opinions, nor his conception of his place in science. Indeed, in his lecture on the General Theory, Einstein explicitly presented relativity not as a revolution but as a translation of the physics of Galileo and Newton. "The language varies," he said, "but at base the idea is the same." In the same frame of mind, he explained to Révécsz that he was no revolutionary, not even in science, since he wanted to preserve whatever could be saved of classical physics and to eliminate only that which hindered the progress of science.[29]

Conservative comment on this episode was not only harsh, but linked up with the incomprehensibility issue through the perspective of social class. Wenceslao Fernández Flores, for example, noted that the syndicalists adulated Einstein for his refusal to sign the Manifesto, an act of minuscule importance compared to what he had really accomplished, which "escapes the comprehension of those syndicalists," none of whom had sufficient scientific background. The syndicalists wanted to turn Einstein into a "promotion poster for Inquisitorial

[28] *ABC*, March 2, 1923; also quoted in *La Vanguardia*, March 3.

[29] The full text of this lecture is reproduced in Appendix 3. For Einstein's views on scientific revolutions, see I. Bernard Cohen, *Revolution in Science* (Cambridge, Mass.: Harvard University Press, 1985), pp. 435–445. In all probability, reporters, unable to follow the verbal interchange between Einstein and Pestaña, attributed to the former the latter's words. In a report of the exchange published in Madrid (*El Liberal*, February 28, 1923), Pestaña remarks in French that "just as the German scholar has revitalized [heard as "revolutionized"?] science with his theories, the workers aspire to revitalize the existing economic order by means of the union." In this version, seemingly unique among the reports published in the Spanish press, Einstein "replied modestly that he had done nothing more than to deduce consequences from scientific principles and he counseled the workers moderation in destruction or renovation because not everything old, he said, is bad."

Spain." Einstein, in the columnist's opinion, *should* have said that the slayings in Barcelona were little as compared to those in Russia, Hungary, or fascist Italy. Of what importance is all this beside the overthrow of Euclidean theory, Flores concludes, putting the rhetorical question in Einstein's mouth.[30] Ramiro de Maeztu, while castigating the syndicalists for blind faith in what they could not understand, suspected they had invited Einstein because he represented change in what was once believed.[31]

EINSTEIN AND CATALAN NATIONALISM

The political movement that sought cultural, administrative, and economic autonomy for Catalonia within the Spanish state had been gathering force since the turn of the century under the leadership of, first, Enric Prat de la Riba (1870–1917), and then of his successor as leader of the conservative Lliga Regionalista, Francesc Cambó (1876–1947), as well as a number of liberal groups such as the Esquerra Republicana of Francesc Macià (1859–1933). A major step toward the realization of the Catalanist political program had been achieved in 1914 with the foundation of the Mancomunitat, an official entity charged with implementing a limited measure of cultural and administrative autonomy. Part of the Catalanist cultural program, as Prat de la Riba formulated it, was the creation of institutions to propagate not only Catalan science, but science in the Catalan language. On the other hand, the industrial bourgeoisie that Prat de la Riba represented had been incapable of identifying itself fully with modern science. Physics and chemistry were highly favored, of course, because of their recognized contribution to industrial modernization, but Catholic ideology was a powerful impediment to other fields. A prime example was Prat's pronounced hostility to Darwinism; in this he was faith-

---

[30] Wenceslao Fernández Flores, "Einstein y los comunistas," *El Diario Español*, April 7, 1923 [dated March 6].

[31] Ramiro de Maeztu, "Fuera de la cultura," *El Sol*, March 6, 1923.

ful to the late nineteenth-century orthodox Catholic position that proclaimed its support of *true* science, but not to false-hoods such as evolution. In particular, he opposed the educational philosophy of Francesc Ferrer's Modern School, which was overtly and explicitly Darwinian and whose pedagogical mission was the education of the working classes who populated the factories of the Catalan bourgeoisie.[32]

It was a commonplace in the popular reception of relativity to hail Einstein as a new Darwin and to contrast the nearly universal acclaim accorded the physicist with the religious and political polarization that had characterized the nineteenth-century polemics over evolution.[33] This historical background provides the immediate political context of Einstein's visit to Barcelona, during which exponents of Catalan nationalism attempted to associate Einstein with their cause. His visit provided an opportunity for conservative Catalanists to associate their program with the most modern and revisionist scientific ideas without having to admit, at the same time, that traditional values were under attack. The favorable disposition of the Catalan bourgeoisie toward Einstein was in part a compensation for its previous inability to accept modern science in its Darwinian guise.

In the Sala d'Actes of the Diputació, where Einstein delivered his three lectures on relativity, symbols of the nationalist cause, in particular the distinctive shield with red and yellow bars, were omnipresent, and behind the dais was a large Catalan flag with the four bars, a detail noted by all the newspapers. Directly in front of the flag was the blackboard made famous by Josep Maria de Sagarra in an article in *La Publicitat*. While Einstein spoke, Sagarra, unable to understand a single word, pondered the historical and cultural significance of the event. He noted people staring fixedly at Einstein and lamented that the forms drawn by the physicist on the blackboard said so lit-

[32] Joan Senent-Josa, *Misería y dependencia científica en España* (Barcelona: Laia, 1977), p. 69.

[33] For a comparison of Einstein and Darwin, see, for example, Maríano Poto, "Einstein y su teoría," *El Liberal*, March 1, 1923.

tle to his brain: "But all my attention settled on the hand of that man, his manner of writing and gesturing with his arm. The little steps he took, his hesitant words, the insinuating sweetness of his voice. That fine eye of a carnivore, the disproportioned nose, its end lightly reddened; the hair, the suit, the invisible aureole." All of this had tremendous symbolic value for Sagarra, and the mere observation of Einstein speaking satisfied the curiosity that had brought him to the lecture. When Einstein wrote on the blackboard, it occurred to him that the words of great men should always be saved: "When Professor Einstein erased the white inscriptions in front of the decorated black curtain, my heart impelled me to say to him: 'Do us a favor! Don't erase it! They can bring out another blackboard!!' " Not only the blackboard, but also the *flag* behind it should be preserved, in order to explain Einstein's theories to future Catalan men of science, he said.[34]

On February 27, when Einstein was received at City Hall (Fig. 2), the acting mayor, Enric Maynés (1883–1951), welcomed the visitor in Catalan. In all previous functions, French and German had been used. "If your fatherland is proud of you," the mayor said, "this pride is shared by everyone because of science's power of solidarity. You are not a stranger to us for science's fatherland is the world." Einstein replied that he wished for Barcelona a new human community that would overcome every kind of political and personal rancor.[35] The following day, *La Veu de Catalunya*, the organ of the Lliga, reproduced the mayor's words and commmented editorially that Einstein "was able to speak the European language of science, with an absolute and cordial identification between the scholar and the city."[36] If science was international, the Catalan language could be too. Nevertheless, the mayor's choice of language initiated a brief polemic. Regina Lamo, writing in the radical republican daily *El Diluvio*, attacked Maynés for

[34] Sagarra, "Einstein."
[35] *La Publicitat* and *Diario de Barcelona*, February 28, 1923.
[36] "La visita de Einstein" (editorial), *La Veu de Catalunya*, February 28, 1923.

2. Reception for the Einsteins at the City Hall of Barcelona.

having spoken in a tongue unintelligible to Einstein. Still, the same paper failed to back its columnist editorially, attacking in turn those who were opposed to the use of Catalan on the grounds of the official status of other languages. The newspaper concluded that Barcelona, rather than extending a cold and formal welcome to Einstein, had expressed itself in the warmth of its own language.[37]

Reflecting on the general atmosphere of Einstein's visit, *L'Esquella de la Torratxa* noted wryly: "In Barcelona the theory of relativity has served us for political ends, and we present Einstein as a perfect regionalist, as a kind of Cambó of mathematics." The concept of relativity provided the humorists

[37] Regina Lamo, "Interpretaciónes sentimentales. Einstein el precursor," *El Diluvio*, March 2, 1923; "Comentarios optimistas. Sobre el discurso de Maynés" (editorial), ibid., March 3, 1923.

with a ready-made context for a sardonic comment on the current political situation in Catalonia and the Lliga's proclivity for compromising the Catalanist cultural program:

> Indeed, wasn't the Lliga the precursor of inexistence of the straight line and, therefore, of the predominance of the curved line? Don't you know that all the parallels—the Monarchist Federation, the regionalists, Maura's faction—intersect in an electoral point? Doesn't it act as if neither time nor space existed, without granting greater importance to Montserrat than to Covadonga? . . . Catalunya is the land of relativity. We Catalans favor relativity, except for Macià's men, who are for all or nothing.[38]

The claim that a certain cultural group could be characterized as especially congenial to the relativistic perspective, as the Catalans are in their capacity for political compromise or in having some innate understanding of "relativity," was a recurrent theme in its popular reception.

Political satire aside, Einstein directly influenced the political philosophy of at least one Catalan politician, Rafael Campalans (1887–1933), an engineer who at the time of Einstein's visit was director of the Industrial School. At one point during the visit, Campalans was explaining his philosophy of a nationalistic socialism when Einstein, laughing, rejoined: "Das passt nicht zusammen!" ("The two don't go together!") Later, he told Campalans that he had finally understood this seemingly paradoxical concept, but that the politician would do better to omit reference to the word "nationalism" in his program. According to Einstein, that term was not applicable to the struggle of oppressed national minorities to win recognition, since it was impossible to avoid the then predominant connotation of the word: that of conservative and imperialist nationalism that had characterized German politics. In a speech in the Cortes of the Second Republic nine years later, Campalans recalled that because of his discussions with both

---

[38] "Einsteiniana," *L'Esquella de la Torratxa*, 46 (1923), 139.

Einstein and Jacques Hadamard, the French mathematician who had visited Barcelona in 1921, he had decided to omit this word from his political platform, for its habitual connotation in Europe was that given the word by Léon Daudet and, in Spain, by the stridently nationalistic conservative politician Antonio Royo-Villanova.[39]

There is one more curious footnote to Einstein's connection with Catalan nationalism. Soon after the conclusion of his Spanish tour, Einstein resigned from the League of Nations Committee of Intellectual Cooperation. To a German pacifist periodical he indicated that the League's inability to counter the will of the great powers had convinced him to resign, and he intimated to Madame Curie that the French occupation of the Ruhr was the incident that precipitated his resignation. Other commentators, however, have asserted that Einstein resigned because of the "Catalan Question" (*katalonische Frage*). The story first appears in Carl Seelig's biography of Einstein and has been repeated since, with no direct evidence that this was the case. However, there is some indirect evidence that might well be taken as an allusion to the Catalan problem. Einstein's letter of resignation contains the following lines: "The Commission has given its blessing to the oppression of the cultural minorities in all countries by causing a National Commission to be set up in each of them, which is to form the only channel of communication between the intellectuals of a country and the Commission. It has thereby deliberately abandoned its function of giving moral support to the national minorities in their struggle against cultural oppression."[40]

---

[39] Santiago Riera i Tuebols, "Rafael Campalans, enginyer i polític," *L'Avenç*, no. 16 (May 1979), 8; Campalans, "Discurs del 27 de juliol de 1932 sobre l'ensenyament a Catalunya," in Albert Balcells, ed., *Ideari de Rafael Campalans* (Barcelona: Pòrtic, 1973), pp. 122–123. See also Campalans to Einstein, May 17, 1924 (Einstein Duplicate Archive) in which the Catalan politician recounts the repression carried out in Catalan cultural institutions by the dictatorship of Primo de Rivera.

[40] Carl Seelig, *Albert Einstein: A Documentary Biography* (London: Staples Press, 1956), p. 175: "His definite secession from the League ultimately took

EINSTEIN'S diary reference to his days in Barcelona is disappointingly brief. It reads simply: "Stay in Barcelona. A great deal of strain but kind people (Terradas, Campalans, Lana, Tirpitz's daughter), folk songs, dances, Refectorium. It was lovely!"[41] A note of interest, of course, is the reference to the daughter of Admiral von Tirpitz, a German nationalist politician whom Einstein particularly loathed. (She may be one of the women pictured at the station in Fig. 5).[42] The Refectorium was a restaurant on the Rambla del Centre and was much frequented during the 1920s by Catalan nationalist politicians. Einstein sipped a cup of café au lait there, probably in the company of Campalans. There is a sentimental evocation of the moment in *La Campana de Gracia*, whose writer noted Einstein's disregard for formal etiquette in appearing in a "cabaret" not frequented by intellectuals. The writer (along with other diners, we can assume) rose to his feet in respect when Einstein passed his table.[43]

While in Catalonia, Einstein made two trips outside Barcelona. On Sunday, February 26, 1923, he visited the Romanesque monastery at Poblet, accompanied by Bernat Lassaleta, professor of chemistry at the Industrial School. There is no record of the visit except for two photographs and Einstein's signature in the monastery's guest book.[44] On Monday he visited Tarrassa, site of a famous Roman basilica. In other free moments he toured Barcelona without attracting the attention

place on the Catalan question." The letter of resignation is reproduced in Einstein, *The World As I See It* (New York: Philosophical Library, 1949), p. 54.

[41] Einstein Duplicate Archive. The complete text of the portion of Einstein's travel diary relating to Spain is reproduced in Appendix 1.

[42] Ronald Clark, *Einstein: The Life and Times* (New York: World, 1971), p. 186.

[43] "Einstein a casa," *La Campana de Gracia*, March 3, 1923 (no. 2808); on the Refectorium and Catalan nationalism, see the cartoon in *L'Esquella de la Torratxa*, 39 (1917), 84.

[44] See the photograph of Einstein at Poblet, *Mundo Gráfico*, February 7, 1923, reproduced in *Ciència*, 3 (October 1980), p. 152.; Agustí Altisent, *Història de Poblet* (L'Espluga de Francolí, Abadia de Poblet, 1974), p. 641.

of passers-by.[45] Virtually all of his sightseeing was in the company of three German-speaking friends, Terradas, Lana, and Campalans. With Lana he visited the Escola del Mar, an experimental outdoor school established by the city for handicapped children in 1922, and the port. On a formal visit to the rector of the university, Einstein was accompanied by Terradas.[46]

It is clear that Einstein held Terradas in high esteem. Einstein met Julio Rey Pastor in Buenos Aires several years after the Spanish trip and told him, "I met an extraordinary man: Terradas. His head is one of the six best in the world." Similarly, in 1930, Einstein remarked to Antonio Fabra Ribas, Catalan socialist and German correspondent of *El Sol*, "Terradas has a great mind, very original above all. I have known many men in the course of my life and I do not hesitate to affirm that the Spanish professor is one of those who has interested me most."[47] In spite of these strong words, we know virtually nothing of Einstein's personal relationship with Terradas, but once during the Barcelona visit the two men were observed in a long conversation on the subject of relativity. At one point Einstein interrupted Terradas to say: "I see, Sr. Terradas, that you know more about it than I do!"[48]

[45] *El Diluvio*, March 1, 1923: "Einstein has passed through Barcelona without the people's (*pueblo*) taking notice." Perhaps this is not an accurate reflection of reality, since the intent of this editorial comment was to distinguish (speciously, I think) between the common people who had not been aware of Einstein's presence and the "citizens" who had been.

[46] *La Vanguardia*, February 27 and 28, 1923, and *El Noticiero Universal*, February 28.

[47] José Gallego-Díaz, "Alberto Einstein, símbolo de nuestro tiempo," *ABC*, April 19, 1955; A. Fabra Rivas, "Una visita a Einstein," *El Sol*, March 27, 1930. While in Madrid, Einstein had said of Terradas: "C'est une singularité"; Vera, "El doctorado." Another version is given by Theodore von Karman, *The Wind and Beyond* (Boston: Little, Brown, 1967), p. 340: "Albert Einstein once said of [Terradas]: 'I was in Spain, looked at Spanish science, and discovered Terradas.'" Beginning in 1924 Rey Pastor was professor of mathematics at the University of Buenos Aires, where he spent half of each year, and in Madrid, where he spent the other half.

[48] Joan Sales, *Cartes a Marius Torres* (Barcelona: Club Editor, 1976), p. 458.

On Wednesday, February 28, Einstein's last day in Barcelona, Campalans was his host. At the Industrial School (Fig. 3), Campalans organized a performance of the sardana, the Catalan national dance, by a troup called La Penya de la Dansa. As always, Einstein was ready to offer an extemporaneous critique: "It is a very distinguished dance which reveals what the Catalan people are and it ought to be known among the other

3. Einstein at the Industrial School
of Barcelona.

According to Sales, this anecdote was common knowledge in Barcelona (personal communication, June 9, 1981).

nations: it is a work of art closely related to sport." Einstein was presented with records, probably of sardanas, and continued to play them.[49]

That same evening, Campalans gave an elegant farewell supper for the Einsteins, attended by Terradas (whose copy of the menu is reproduced in Fig. 4) and the German-speaking politician Miquel Vidal i Guardiola, among others. According to *La Veu de Catalunya*, "The dinner was served according to a

4. Menu in "Relativistic Latin" presented to guests at Rafael Campalans's banquet for the Einsteins.

[49] *La Veu de Catalunya*, March 1, 1923; J. M. G., "Albert Einstein i la Sardana," *La Publicitat*, February 28; "Albert Einstein," *La Sardana*, 3 (1923), 105. In May 1934 Einstein exchanged letters with Adolfo Marx, evidently a German Jew residing in Barcelona. Marx conveyed to Einstein an invitation from the Catalan Generalitat to visit Barcelona and receive honorary Catalan citizenship. In his reply Einstein mentioned that he had been given records in 1923 and still played them (Marx to Einstein, May 3, 1934; Einstein to Marx, May 21; Einstein Duplicate Archive).

menu, beautifully printed in Gothic characters, in two colors, and written in relativistic Latin, to preserve the character, more or less, of the theory of relativity.' "[50] Thursday morning, March 1, Einstein boarded the train for Madrid (Fig. 5).

On March 6, Einstein was nominated as a corresponding member of the physical sciences section of the Academy of Sciences and Arts of Barcelona by Lassaleta, Tallada, Ramon

5. Einstein departing Barcelona.

[50] *La Veu de Catalunya*, March 1, 1923.

Jardí, and Tomàs Escriche i Mieg. In a vote taken on May 22, thirty academicians voted in favor, with three opposed.[51] Comas doubtless took this opportunity to seal his protest, although his antirelativist articles continued into the next decade.

Reflecting on Einstein's days in Barcelona several days later, the editor of the *Diario de Barcelona*, Joan Burgada i Julià, noted the contrasting reception accorded the visitor in the two cities. In contrast to an "unrestrained and somewhat impulsive enthusiasm" observed in Madrid, Einstein had been received in Barcelona effusively, "but with an austerity contrary to any hint of showiness." The initiated had reflected on his words "and that was all," whereas in Madrid the rector of the University himself, according to Burgada, declared "his plan to undertake an arduous campaign so that we might all become relativists."[52] This interesting commentary combines respect for Einstein with the Catalan bourgeoisie's characteristic reticence toward modern science, manifested in the desire to restrict the philosophical implications of the new theory to the narrowest possible limits, and Burgada cites Einstein to that effect.

Burgada was correct. In Madrid, the "Einstein phenomenon" showed itself in full strength, a fact made possible by, or at least related to, pervasive and explicit acceptance of the philosophical ramifications of the Einsteinian universe as different groups, both professional and lay, understood it.

[51] Acadèmia de Cièncias of Barcelona, Expediente, March 6, 1923. The Academy's archives contain a folder devoted to Einstein, containing the *expediente*, the ballot, the Academy's notification of Einstein on June 5, Einstein's reply of June 20, as well as many newspaper clippings related to Einstein's visit. The ballot is reproduced in *Ciència*, no. 3 (October 1980), 149. Tomas Escriche (b. 1844) was a physicist; see *Gran Enciclopedia Catalana*, VI, 779, where an erroneous date of death (1916) is given. The two other negative votes may have been cast by Eduard Alcobé and Josep Tous i Biaggi.

[52] J. Burgada y Juliá, "Einstein in España," *Diario de Barcelona*, March 10, 1923.

FOUR

Madrid: The Two Aristocracies

EINSTEIN was to spend ten days in the capital. He was met at
the station on March 2, 1923 by Cabrera (Fig. 6), whom Einstein recognized "at once." Cabrera then introduced the members of the welcoming party: Pedro Carrasco, Francisco Vera
(mathematician and popular science writer), Josep Maria
Plans, and other delegates from the Faculty of Sciences, as well
as the anatomist Julián Calleja, who headed a separate delegation of physicians from the College of Physicians. Einstein
held in his hands two photograph albums and Fernando Lorente de Nó's translation of his short volume on the theory of

6. Einstein arriving in Madrid (Blas Cabrera is at the extreme right).

relativity. Einstein uttered a few words "of courtesy" in French, and then departed with Elsa's cousins Lina and Julio Kocherthaler (who resided in Madrid) for the Palace Hotel, where plans for the week were made.[1] There a reporter observed him surrounded by people arranging his lectures, all speaking French:

> These men are in great part unknown to the public at large. Modest, quiet, reserved men who discharge a meritorious task in the shadows. Paradoxical men in this land of theatrical southern temperament.
>
> They are the colleagues of the German scholar, those who will be able to penetrate the substance of his explanations. Einstein's arrival draws them from their solitary and obscure work for an instant and brings their names and appearances before the public eye. Their names are Cabrera, Carrasco, Plans. . . . They are the men who tend the little light of mathematical research among us. Cabrera, effusive and cordial, stands out from the group.[2]

At 11 A.M. that day the Kocherthalers accompanied the Einsteins on a drive through the city, where they were recognized by "many passers-by." Einstein spent the rest of the day with Cabrera; a photograph taken in the latter's laboratory at the Institute of Physical Research, at the Hippodrome, appeared in the press the next day (Fig. 7).[3] In the evening, Cabrera had planned to attend a concert. However, Einstein's wish to see something more typically Spanish was honored, and they attended, instead, a musical review called "Tierra de Carmen" at the Apolo.[4]

Saturday evening Einstein delivered his first Madrid lecture at the University, on special relativity. In the audience were "men of science, mathematicians, physicists and philosophers," and also politicians: the former prime minister Anto-

---

[1] *El Debate* and *El Sol*, March 2, 1923.
[2] *El Heraldo de Madrid*, March 3, 1923.
[3] *La Voz* and *La Vanguardia*, March 3, 1923.
[4] *El Debate*, March 3, 1923.

7. Einstein in Cabrera's laboratory (with Cabrera at Einstein's right, Julio Palacios at his left).

nio Maura, the minister of public instruction Joaquim Salvatella, and Amalio Gimeno, politician and physician who had taken Darwin's side in the debates over evolution of the 1870s.[5] Naturally, the scientific presence was predominant, as Francisco Vera's description in *El Liberal* makes clear: "There is Plans, a great expert in this material . . . tall, thin, nervous, a contrast to the equanimity of Cabrera, Einstein's good friend . . . and Luis Octavio de Toledo, worthy of El Greco's brush, with eyes searching of mathematics; and Commander Emilio Herrera, the bold pilot and illustrious scientist."[6]

Einstein was introduced by Pedro Carrasco. Then he spoke, illustrating his remarks on a blackboard, as in Barcelona: "He seizes the chalk to trace some figures on the blackboard, and with one hand in his pants pocket he strolls around for an in-

[5] On the atmosphere at the first lecture, see *El Debate* and *El Liberal*, March 4, 1923.

[6] *El Liberal*, March 4, 1923. I have supplied given names.

stant, his eyes fixed on the ceiling, deep in thought."[7] Directly
following the lecture the College of Physicians hosted a banquet at the Palace, organized by Drs. Ignacio Bauer and Toribio Zúñiga. (Zúñiga had founded the College to enhance the
scientific prestige of the medical community.) In attendance
was a veritable "Who's Who" of the Madrid medical world,
including Gustavo Pittaluga, Florestán Aguilar, Julián Calleja,
and many more. On Einstein's left were seated the chemist
José Carracido, president of the Academy of Sciences, and Angel Pulido, a doctor famed for his successful campaign to gain
official recognition of the rights of Sephardic Jews to Spanish
citizenship.

After the banquet Einstein exchanged views with an editor
of *El Imparcial* on the problem of representing his theory in the
mass media the journalist wrote:

> He is the first to recognize that his theories meet with
> great difficulties when one attempts to bring them to the
> comprehension of the great public. But, in spite of that,
> he found in Spanish journalists an admirable predisposi-
> tion towards the assimilation of his theory.
>
> And above all—he added—one notes a yearning for
> knowledge, in all social classes, indicative of a spiritual
> state from which very satisfying results might be ex-
> pected.[8]

On Sunday morning, March 4, the Kocherthalers took the
Einsteins on another drive through the city, after which Einstein prepared a response to the address Cabrera would make
at the Academy of Sciences later in the day. The session opened
with King Alfonso XIII in the chair and, of course, the cream
of Spanish society in attendance: *ABC* noted Leonardo Torres
Quevedo, Ignacio Bauer, Nicolás de Ugarte (military engineer), Cecilio Jiménez Rueda and Eduard Torroja (mathematicians), Eduardo Hernández Pacheco (geologist), Ignacio Bo-

[7] *El Liberal*, March 4, 1923.
[8] *El Imparcial*, March 4, 1923.

lívar (zoologist), and others.[9] Carracido, as president of the Academy, made a short address (which Einstein characterized in his diary as wonderful—*wunderbar*) concerning the three-tiered structure of science: at the lowest level, the discovery of fact; then, that stage at which fact is converted into an instrument of experimentation (he gives the spectral theory as an example); and finally, the highest stage, where the brain commands—the rule of pure theory, of ideas rather than apparatus. The highest stage was exemplified by the theory of relativity.[10]

The main interest of the Academy session was an exchange of remarks between Cabrera and Einstein. Cabrera first asserted that the postulate of the constancy of the speed of light which underlay the Special Theory was both experimentally proven and logically irrefutable. The old physics was a dead letter that the prejudice of its supporters was powerless to save. For Cabrera the experimentalist, there was no doubt of the Special Theory's veracity, because "the confirmation of the change in the mass of electrons [as a function of] velocity, which leads to the identification of matter with energy, has consecrated the Special Theory, as of this date." We have noted the significance of experimental confirmations in the acceptance of general relativity in the aftermath of the 1919 eclipse observations; but Cabrera was one of the few who stressed the experimental confirmation of the Special Theory. For Cabrera, relativity was proven and needed no further support.[11] He therefore turned to Einstein's other contributions to science. After reviewing his research on Brownian movement and on the photoelectric effect, Cabrera noted that Einstein's

[9] *ABC, El Sol, El Imparcial*, March 6, 1923. Einstein had been an honorary member of the Sociedad Española de Física y Química since 1920.

[10] Based on a report in *El Imparcial*, March 6, 1923. In the officially published version the three-stage model is not nearly so explicit; Real Academia de Ciencias Exactas, Físicas y Naturales, *Discursos pronunciados en la sesión solemne que se digno presidir S. M. el Rey el día 4 de marzo de 1923, celebrada para hacer entrega del diploma de académico corresponsal al profesor Alberto Einstein* (Madrid, 1923), pp. 23–25.

[11] *Discursos*, p. 9.

"efforts to find a direct proof for the quanta of light, something very recent, have not achieved the results desired." He concluded by directing to Einstein the following words, in the name of Spanish scientists:

> We recognize our debt to Humanity and our desire is to settle it soon. I assure you of it, in the name of the generations present here and those of the immediate future. You are still young. I hope that at the end of your life, which will also be that of our generation, scientific Spain, which today is scarcely an embryo, might have reached that position which it has the inexcusable obligation to occupy. That, at least, is the way we, for whom optimism is the driving virtue of progress, think. [12]

Einstein's reply, translated into Spanish by the chemist José Casares Gil, began by indicating his appreciation of Cabrera's words, "because they demonstrate the deliberate and affectionate way in which you have studied my life's work." Einstein then turned to the weakness in his current approach to quantum, just mentioned by Cabrera:

> You have also taken into consideration the weak point of the theory of light quanta, a difficult subject for our generation of physicists. I believe that those difficulties can only be overcome by means of a theory which not only modifies basically the principle of energy, but which perhaps also expands that of causality. Just a while ago [Hugo] Tetrode noted such possibilities exactly. Even though the principles for the solution of this basic problem have, up to now, acquired little solidity, the new impulse towards the unification of all the forces of nature, born in the theory of relativity, nevertheless promises satisfactory results. The method employed in this is purely [that of] speculative mathematics, typified by the names of Levi-Civita, Weyl [and] Eddington. In this way one can really achieve the total liberation of the bases of phys-

[12] Ibid., pp. 14–15.

ics from the disturbing dualism, summed up in two words, gravitation and electricity.[13]

The king himself then awarded Einstein with the diploma naming him a corresponding member of the Academy (Fig. 8).

"Afterwards," Einstein's diary records laconically, "tea with an aristocractic lady companion." This overcondensed entry was a mammoth understatement, since it referred to a "tea of

8. Reception at the Academy of Sciences, March 4, 1923.
Left to right: Joaquín Salvatella (Minister of Public Instruction),
the German Ambassador, King Alfonso XIII, Einstein,
José R. Carracido, and Blas Cabrera.

[13] Ibid., pp. 19–20.

honor" offered by the Marqueses de Villavieja,[14] attended by many of the most outstanding members of the Madrid intelligentsia. Einstein's aversion to "social life" was a feature of his personality which he always made clear to reporters, perhaps as part of his routine to keep them away. He told Andrés Révéscz, for example, "Social life bothers me," to which his wife added that now that she knew this (as if Einstein had been carrying through on a well-known joke between them), they would not be so active socially.[15] The guests at this reception included the Kocherthalers, Salvatella, Carracido, Cabrera, Gregorio Marañón, Gonzalo R. Lafora and José M. Sacristán (neurologists), Teófilo Hernando (internist), Pittaluga, Aguilar, Hugo Obermaier (a German paleontologist attached to the Madrid Museum of Natural History), Alberto Jiménez Fraud (director of the Residencia de Estudiantes), Manuel B. Cossío (art historian), the philosophers José Ortega y Gasset and Manuel García Morente, and, among writers, José Maria Salaverría, Miguel Asúa, and the novelist Ramón Gómez de la Serna, who, admiring Einstein's "disordered" hair, was heard to characterize the physicist as an "inspired Italian violinist." Among the many aristocrats present was the Vizconde de Eza, a trustee of the Junta para Ampliacion de Estudios.[16] It is interesting to note how many of those present were to write, or had already written, about Einstein: Cabrera, Ortega, García Morente, Maeztu, Salaverría, and Gómez de la Serna.

The object of this "little party" at the Marqueses' palace, according to ABC's society writer Gil de Escalante, was to establish a relationship between two aristocracies, "that of blood and that of intelligence." In a similar vein, Salaverría noted that there were no bankers, industrialists, or politicians present, only "*nobles* (either of blood or intelligence)," meaning, I

[14] The newspaper reports say Villavieja or Torrevieja. The German consul says Torrevieja. The lady in question was the current holder of the title, the Marquesa de Villavieja, Doña Petronilla de Salamanca y Hurtado de Zaldívar.

[15] *ABC*, March 2, 1923.

[16] J. M. Salaverría, "Las originalidades einsteinianas," *ABC*, March 10, 1923.

think, that the elite was present as a class, not in representation of particular interests. Révéscz had already discovered that Einstein had no taste for social life, his colleague noted wryly, yet here he was smiling while "seated on an ample sofa, surrounded by cushions and women."[17] During the evening, Einstein and the violinist Antonio Fernández Bordas improvised "an intimate concert."[18] This episode is extremely interesting in the light of the notion of civil discourse among a disunified elite. I do not believe that the decision to undertake such a dialogue is necessarily a deliberate one. But ultimately such a decision reaches the level of consciousness, as it did in Spain in the 1920s. This particular document shows that in 1923 such a discourse was an explicit objective of the political Right, as evidenced in the society pages of the staunchly conservative, monarchist newspaper *ABC*.

Einstein's activities on Monday morning, March 5, are not recorded. He had lunch with Kuno Kocherthaler. The afternoon was devoted to a special meeting of the Mathematical Society. There is no doubt that the Society was the center of relativistic thought in Madrid (along with the Mathematical Laboratory). At its meeting of February 3, when the membership had heard Enric de Rafael lecture on "the movement of a solid around an axis in relativistic mechanics," Emilio Herrera "proposed an exchange of impressions about the theory of relativity before Professor Einstein's arrival, in order to be able to direct suitable questions to him." The members agreed and Cabrera suggested that they should submit questions in writing for the next session. In fact there were two special sessions wholly devoted to the theory of relativity, both characterized by "lively discussion," held on February 20 and 22. Present, among others unnamed, were Herrera, Cabrera, Plans, Palacios, Manuel Lucini, Vicente Burgaleta, Fernando Peña, Juan López Soler, and Pedro M. González Quijano. It seems clear

[17] *ABC*, March 6 and 10, 1923. Both Salaverría and Escalante wrote for this monarchist newspaper.

[18] *El Imparcial*, March 6, 1923. Fernández Bordas (b. 1870) had performed in Germany; see *Enciclopedia Universal Ilustrada*, XXIII, 769.

that the purpose of these meetings was for Madrid mathematicians to clear up their own doubts about relativity in advance of Einstein's visit, although as Tomás Rodríguez Bachiller later recalled, an equivalent objective was "the preparation of the student public."[19]

Both Lucini (an engineer who had met Einstein while studying in Zurich before the War) and Rafael published accounts of the meeting with Einstein.[20] According to Lucini, two main points, one relating to special relativity and the other to the General Theory, were discussed. The former, raised by Burgaleta, had to do with the impossibility of transmitting signals at velocities greater than the speed of light. Burgaleta introduced a spherical solution to D'Alembert's equation which suggested that it "might establish a system of signals whose speed of propagation would be greater than that of light." This was a problem of great interest to Burgaleta, who had written: "In my view, the existence of velocities greater than that of light is undoubted, without that constituting a serious objection to the theory of relativity, but rather to certain forms of expression of which relativists are very fond, doubtless because of their habit of shocking the world with their conclusions."[21] Einstein replied (according to a paraphrase) "that there can be no system of signals in which there are no interruptions in propagation; and even though Sr. Burgaleta insisted on the example of [John Henry] Poynting, to which the gradient mentioned could be ascribed, the German scholar remained unconvinced and the discussion ended."

The problem relating to the General Theory was raised by Plans and had to do with the apparent impossibility of reconciling a kinematic relativity for rotation with the limitation

[19] *Revista Matemática Hispano-Americana*, 5 (1923), 51, 76; Glick, "Tomás Rodríguez Bachiller," *Dynamis*, 2 (1982), 403–409.

[20] Manuel Lucini, "El profesor Einstein," *Madrid Científico*, 30 (1923), 65–66; Enrique de Rafael, "El profesor Albert Einstein en Madrid," *Anales del Instituto Católico de Artes e Industrias*, 2 (1923), 160–164.

[21] Vicente Burgaleta, "Una paradoja relativista," *Madrid Científico*, 30 (1923).

that the speed of light imposes on possible velocities. What could one understand by the term "absolute rotation," Plans asked, "especially in reference to the idea expressed in the fifth edition of Weyl's work, *Raum, Zeit, Materie*, where the *limits of the universe* are defined as the convergence of all the normal directions at a point on the same world line, and which relative to the said limit, and along an infinitesimal parallel trajectory, it is possible to define an absolute rotation." This is an interesting question because, as Lucini pointed out, in a movement of rotation it is difficult to do without a notion of absolute space. Admitting the kinematic and dynamic *relativity* of movements of rotation, centrifugal force would develop in a body that turns in relation to exterior masses just as it would if these masses rotated around the body (Mach's principle). There is no way to tell which of the two systems, the body or the exterior masses, turns and which is in repose.

Einstein replied (in de Rafael's paraphrase) that

> for the time being, in his theory (contrary to that of [Willem de] Sitter, which admits the finitude and curvature of space, but not of time), there is nothing wrong with speaking of an absolute direction, that of time, and, therefore, with respect to it, absolute rotations. Therefore the universe, because of the presence of matter, is curved, but not in all directions, because one of them, time, is rectilineal and, when you take these two things together [the finitude of space and the infinitude of time] the universe therefore turns out to be cylindrical. Inasmuch as the differentials of the world line are of proper time, one sees the possibility of defining rotations which can be called absolute with respect to the directions normal to [the world line].

Enric de Rafael then explained his preoccupation with how problems of classical mechanics might be posed mathematically in relativistic mechanics. Specifically, he referred to "Poinsot's motion," a geometrical representation of rotary motion "by the rolling of the ellipsoid of inertia of a body on a fixed plane."[22] Einstein replied that

> since the concept of a rigid solid does not exist in relativistic theories, no such formulation can be made based on it. It can be said that a solid is a system in respect to which its points have rectilineal world lines, equal and parallel to each other and to the axis of proper time; but which on going to other axes or coordinate systems, the transformation formulas cannot be established *a priori*, as in classical kinematics and mechanics; solids are converted into elastic bodies and their kinematics and dynamics are those of continuous and elastic media. Only experiment can determine the elastic constants and in conformity with it we must establish the equations of motion and deformation with respect to any coordinate system, the physical interpretation of which coordinates cannot be established *a priori* either.[23]

The session concluded, Einstein, wearied by the formality of the interchange, found a convenient means of escape. Rodríguez Bachiller raised some problems that he was having in interpreting Lorentz's theory of electrons. Einstein accompanied him to a small room where, for half an hour, he explained the theory completely to the young mathematician, "with extraordinary clarity." Bachiller retained this memory for the rest of his life, noting that the great man seemed to prefer the company of students to that of their professors.[24]

At 8:30 in the evening, Einstein was taken to see Santiago Ramon y Cajal whom, he had told Andrés Révész, he had known by reputation for twenty years.[25] "Visit with Cajal,

[22] René Taton, "Louis Poinsot," *Dictionary of Scientific Biography*, XI, 62.

[23] Anales del Instituto Católico de Artes e Industrias, 2 (1923), 163. De Rafael adds that "This lack of aprioricity is indicated in the books of Eddington and Plans, where they try to establish the basic tensor of [Karl] Schwarzshild for the case of a gravitational field created by a single material point."

[24] Glick, "Tomás Rodríguez Bachiller."

[25] *ABC*, March 2, 1923.

wonderful old man (*wunderbarer alter Kopf*)," Einstein's diary records.

It must have been a short visit because the second lecture, on general relativity, followed. Francisco Vera was surprised to note that the greater part of those who had heard the first lecture were also present at the sequel. Another account noted: "A numerous public—the professorial, scientific and curious of Madrid—fills the large hall. In this atmosphere, there is a certain new, religious anxiety in anticipation of the man who comes to revolutionize what is established and catalogued in books and minds, to show us infinite, unknown horizons."[26] Before speaking, Einstein "exchanges a few words with Carracido . . . [and] smiles an enigmatic smile." Vera continually interrupted his account of the lecture to record "feminine eyes" following the speaker. Both at this lecture and the third, Einstein had to search for words, would utter the German word, and then waited for someone in the audience to supply the French translation: "Occasionally he lacks the precise word and then, with a smiling gesture, he utters, in consultation, the German word which two, three, ten voices translate immediately. . . . In this way, the teacher is instantly converted into a disciple of his students."[27] Or, he would know the French word but was unsure of its pronunciation. In one instance, it was *similitude* whose pronunciation eluded him. Vera observed: "Several voices, joyous, note: *similitude*, and the reporter thinks he perceives a puerile vanity in these ingenuous listeners, for whom the identification of a word—only one— has raised to the category of *collaborators* of Einstein."[28]

It is interesting to note that two different observers drew the same conclusion regarding the listeners's eagerness to associate themselves with Einstein's intellect. Also significant, in view of the habitual problems that the Spanish scientific community had in communicating with mainstream science, was

[26] *El Liberal*, March 8, 1923; *El Imparcial*, March 6.
[27] *El Imparcial*, March 6, 1923.
[28] *El Liberal*, March 8, 1923.

the high number of German-speakers in the audience and their evident ease in using the language in an extemporaneous and very public situation.

The activity on Tuesday, March 6, was a sightseeing trip to Toledo, with the Kocherthaler brothers and their wives, Ortega, and Manuel Cossío, the art historian. Among the historic and artistic sites visited were the Hospital de Santa Cruz, the Plaza de Zocodover (where Einstein was recognized), the cathedral, the medieval synagogues (the Tránsito and Santa María la Blanca), and the church of Santo Tomé where the

9. Einstein and Ortega in Toledo, March 6, 1923. Left to right: Manuel B. Cossío (art historian), Einstein, Lina Kocherthaler, Julio Kocherthaler, María Luisa Cazurla (art historian, married to Kuno), Kuno Kocherthaler, Ortega y Gasset.

party viewed El Greco's *Burial of the Count of Orgaz*. Mrs. Einstein noted that she had not seen her husband so "enthused" for a long time.[29] Einstein's diary entry is eloquent in its simplicity:

> Trip to Toledo camouflaged by many lies. One of the most beautiful days of my life. Radiant sky. Toledo was like a fairy tale. An enthusiastic old man, who has supposedly produced some significant writing on Greco, guides us. The streets and market place, view of the city, Tajo with stone bridges; stone-covered hills, lovely plain, cathedral, synagogue. Sunset with glowing colors on our way home. A little garden with a view near the synagogue. A magnificent picture of Greco in a small church (burial of a nobleman) among the most profound things that I saw. A marvelous day.

Ortega later recounted to Pío Baroja that Einstein had not wanted to visit the cathedral, but only Santa María la Blanca. The thought of his ancestors worshipping there "moved him. It was the force of tradition," the anti-Semitic Baroja concluded darkly.[30]

Virtually nothing is known of what Einstein discussed with Ortega. Ortega's own account, a counterpoint between Einstein's universality and a *costumbrista* description of Toledo, is uninformative. Einstein's popularity was one topic of discussion. As they passed through the medieval ambience of the Plaza de Zocodover, where the admiring crowd swirled around the physicist, Ortega joked that Einstein was "already well-known in the thirteenth century." Einstein smiled, but rejoined: "I have no historical sensibility. I am only really interested in the present." Then he added that the typical concentration of scholars on only "a small niche of problems" was a characteristic product of specialized education in Germany which, in that country, had become "a true curse. In human

[29] *ABC*, March 7, 1923.
[30] Pío Baroja, *Memorias* (Madrid: Minotauro, 1955), p. 739.

terms, it is monstrous to serve science and not to serve for anything other than it."[31] In Toledo also, Ortega apparently introduced the physicist to the philosophy of Franz Brentano, who
had spent his declining years in Zurich in virtual isolation from
the academic community, Einstein included. In Madrid the
two scholars discussed relativity, Ortega commenting to Einstein that "you will end up making physics into geometry," referring to the tendency of relativistic mechanics to absorb dynamics into kinematics.[32]

At noon on Wednesday, March 7, Einstein was received at
the royal palace, accompanied by Carracido. Einstein recorded
the event in his diary: "Audience with the King and Queen
Mother. The latter reveals her knowledge of science. One sees
that no one tells her what he is thinking. The King, simple and
dignified. I admired him in his way." Some time during the
day, before his lecture, a delegation of engineering students
visited Einstein and invited him to their Association of Engineers and Architects. Einstein promised to address them the
following day about the relationship between relativity and the
applied sciences.[33]

At the conclusion of the second lecture, Einstein had advised
the audience that it would be difficult for them to appreciate
the contents of the third, on problems raised by the theory
(and in particular his quest for a unified field theory), without
having knowledge of absolute differential calculus. This was
worrisome to Vera who, although he had a mathematical education, arrived unsure of his ability to understand. The hall
was filled once again: "A similar audience and even greater

[31] Ortega, "Con Einstein in Toledo," *La Nación*, April 15, 1923; reprinted
in *El tema del nuestro tiempo*, 18th ed. (Madrid: Revista de Occidente, 1976), pp.
195–202.

[32] José Ortega y Gasset, "La metafísica y Leibniz [1926]," *Obras completas*, 11
vols. (Madrid: Revista de Occidente, 1963–1969), III, 433, and "Conversión
de la física en geometría [1937]," ibid., V, 286. In Ortega's office at the Librería
Calpe, there was a signed photograph of Einstein that dominated the famous
tertulia. Ortega's remark about the geometrization of physics was a much-re-
peated theme; cf. Eugeni d'Ors' remark cited in chapter 5.

[33] *ABC*, March 8, 1923.

anticipation, if that is possible, than on the previous days," *El Imparcial* reported, and yet the number of those "seriously prepared," in the journalist's calculation, did not encompass one-fifth of the audience. Vera observed "the military, among whom the artillery officers and engineers predominate, lined up in one row," noting in particular the conspicuous presence of Herrera and Joaquín de La Llave. The audience seemed not to understand the lecture, according to both the press—"the very lack of understanding perhaps stimulates applause"—and Einstein himself, who noted in his diary: "Attentive audience, which surely understood almost nothing because of the difficult problems being treated." The third lecture turned on the dualism created in physics by "the existence of magnitudes representable by symmetrical tensors (gravitation) and anti-symmetrical ones (electromagnetism), and Weyl's proposed solution to this dualism, which is a generalization from Einstein's principle (the invariance of the length of the differential of the arc of the universe)." Einstein proposed his own solution by creating a second-order tensor with both symmetrical and antisymmetrical characteristics.[34]

Following the lecture a reception was held at the German embassy. As at the banquet at the Palace Hotel and the Marqueses' tea, Einstein was again surrounded by physicians, whose omnipresence at such events was a hallmark of the social contour of Spanish science during the 1920s. There was Sebastià Recasens, dean of the Faculty of Medicine, and many doctors whom Einstein had already met: Pittaluga, Hernando, Calleja, Aguilar. Also present were Carracido, Cabrera, García Morente, and María de Maeztu.[35] Einstein was not impressed. "Ambassador and family, splendid, straightforward people. Party, painful as usual," he wrote in his diary.

At eleven on Thursday morning, March 8, Einstein was awarded an honorary degree at the University of Madrid (Figs.

[34] *El Liberal, El Imparcial,* and *El Debate,* March 8, 1923; *Ibérica,* 19 (1923), 293.

[35] *ABC,* March 8, 1923.

10 and 11). Plans read a biography of the laureate, who then read a short speech stressing that he belonged to a group of extreme exponents of the unification of the sciences. "Isolated facts," Einstein asserted, "only interest me in relation to the basic system of ideas."[36] A number of students then made pre-

10. Rector José R. Carracido and Einstein (in his academic hood) at the University of Madrid, March 8, 1923.

[36] *ABC*, March 9, 1923.

11. Einstein and members of the Faculty of Sciences, University of Madrid. Left to right, standing: Edmundo Lozano Rey (Zoology), Josep M. Plans (Celestial Mechanics), José Madrid Moreno (Natural Sciences), E. Lozano Ponce de León (Optics), Ignacio González Martín (General Physics), Julio Palacios (Thermology), Angel del Campo (Chemical Analysis). Seated: Miguel Vegas (Analytic Geometry), José Rodríguez Carracido (Pharmacy, Rector), Einstein, Luis Octavio de Toledo (Mathematical Analysis, Dean of Science), Blas Cabrera (Electricity).

sentations. The rhetoric, especially that of the students, must have been intense, for Einstein recorded in his diary: "Honorary Doctor. Real Spanish speeches with accompanying Bengal fire. The German ambassador spoke on the subject of German–Spanish relations—long speech but the content was good, German through and through. Nothing rhetorical."

Einstein then kept his date with the engineering students, arriving at the Association of Engineers, the alumni association of the Catholic Institute of Arts and Industries, at 12:30 P.M. There, Einstein gave a short talk on the finite nature of the universe in which he illustrated the geometry of three-dimensional space with an analogy from a space of two dimensions. The Minister of Economic Development was also present and followed Einstein with a nationalistic invocation: "The rebuilding of Spain requires the natural development of scientific technology, conjoined to political technology, because all our efforts will be useless without the art of governance."[37] Einstein's diary records the episode: "Then a visit with technical students. Talking and nothing but talking, but well intentioned."

In the evening was the last lecture, on philosophical consequences of relativity, at the Athenaeum (Fig. 12). The endocrinologist Gregorio Marañón (1887–1960) presided, and Einstein was introduced by the marine biologist Odón de Buen (1863–1945). In his introduction, de Buen made the rather startling proposal that Einstein head a joint Spanish-Mexican scientific commission to study the forthcoming solar eclipse in Mexico in September 1923. Einstein would remain for a year

EL PROFESOR EINSTEIN EN EL ATENEO

12. Einstein at the Madrid Athenaeum, March 8, 1923.

[37] *ABC* and *El Noticiero*, March 9, 1923.

at the head of the research group and, as a result, Spain would acquire scientific prestige. Spanish men of science, Buen indicated, were now on the verge of joining the scientific mainstream: "A generous and hopeful generation of researchers, who have the spirit to undertake the greatest scientific tasks, is now stirring in Spain." Such an enterprise, moreover, would be a lesson in teamwork and international cooperation which Spanish science would do well to adopt as a modus operandi. For, he warned, "There is not the best harmony among men of science here, and interests created around official scientific institutions are generally a hindrance and, what is worse, a source of discredit abroad." (Later, the Mexican government in fact issued Einstein a formal invitation, which he declined.)[38]

Einstein's Athenaeum speech on the philosophical consequences of relativity began by defining motion, indicating that all motions were relative and "there can be infinite systems of reference, without any having reason to be privileged." All natural phenomena are directed by identical laws when the respective motions refer to different systems. But since no one system of reference is privileged, things must be seen differently from the way Galileo and Newton saw them, "to formulate concepts which respond in a broad way to a general conception." He then moved on to discuss aspects of the Spe-

[38] Odón de Buen, "Un idea, antes de que marche Einstein," *La Voz*, March 9, 1923. On the Mexican government's invitation, see *The New York Times*, June 9, 1923. I have documented the entire episode in "Huellas de Einstein y Freud en México," *Tezcatlipoca: Anuario de Historia de la Ciencia y la Tecnología*, 2 (in press). The subject of a Spanish expedition to Mexico had already been broached in February by Rodrigo Gil, "El eclipse de Sol de septiembre próximo," *Ibérica*, 19 (1923), 125: "If the government wants to help him quickly and effectively, it might even obtain the material necessary to tackle, with good possibilities of success, the problem of the Einstein effect, in spite of the difficulties which the occasion will present in developing a broad program of research, which will in no way be unworthy with respect to [the research] of the advanced nations." Gil was a geographical engineer with the Magnetism Service. When the results of the 1922 Australian eclipse verified the "Einstein effect," interest in the 1923 eclipse was greatly diminished.

cial Theory (simultaneity, for example) and the General Theory: Euclidean geometry was invalid the moment a gravitational field influences solid bodies. The philosophical interest in this is the discovery that there is no absolute geometry.[39]

Most of Friday, March 9, was devoted to a tour of El Escorial and Manzanares el Real. "Trip to the mountains and Escorial. A marvelous day," Einstein noted. At six in the evening he accompanied Ortega to a public tribute to Einstein at the Residencia de Estudiantes, a unique residential college of the University of Madrid. Ortega began the session by comparing Einstein to Newton and Galileo (see discussion in chapter 5). Relativity, Ortega told the audience, constituted a new way of thought, the seed of a new culture, "the symbol of an entire age." Einstein (speaking in German with Ortega translating) replied by saying that he was more a traditionalist than an innovator (a theme he struck repeatedly throughout the Spanish tour). It was Maxwell who built theories with the smallest number of hypotheses and abstractions, his own criterion for theory construction. Relativity had not, he said, changed anything. It had reconciled facts that were irreconcilable by the habitual methods.[40]

The political overtones of Einstein's visit to the capital were subtle but unmistakable. Scientists, as Odón de Buen's remarks demonstrated, beheld in Einstein the redemption of Spanish science, even if he declined to lead it personally. Cabrera's characterization of Spanish science as embryonic immediately became a point of conflict in the liberal intelligentsia's battle with the Ministry of Public Instruction, which they had long regarded as an office for political neophytes and educational incompetents. At the Academy investiture, the current minister, Joaquim Salvatella (1881–1932), countered that

[39] *El Heraldo de Madrid* and *ABC*, March 9, 1923.

[40] I have followed the account in *El Sol*, March 10, 1923; another version, prepared in part on the *El Sol* text and in part on the original typescript, appears under the title "Mesura a Einstein" in *El tema del nuestro tiempo*, 18th ed. (Madrid: Revista de Occidente, 1976), pp. 189–193. See also Alberto Jiménez Fraud, *La Residencia de Estudiantes* (Barcelona: Ariel, 1972), pp. 36–37.

Cabrera "seemed too modest in his reference to Spanish science." When Salvatella went on to state, officiously, that he could make no predictions concerning the future acceptance of Einstein's theories, he was attacked editorially in *El Sol*: Ortega (the newspaper's editor!) had done well to elevate Einstein to the level of Galileo and Newton, "because if the world reads nothing more than the needless declaration of the Minister of Public Instruction, it will be said that we have not understood the theory."[41] In the context of the "Polemic of Spanish Science" (wherein conservatives held—liberal denials notwithstanding—that Spain had enjoyed a continuous scientific tradition, even if not at the European level), scientists translated Salvatella's remarks as meaning "We have enough science in Spain and we don't have to understand relativity (or be Darwinians, etc.) in order to be good scientists." In this way the "incomprehensibility" of relativity became a political symbol that the scientific community, in its recently won position of leadership among Spanish intellectuals, was obliged to combat.

The Madrid visit was now winding down. What remained was a weekend of unscheduled time, which left the Einsteins free to visit with family and to return to the Prado.

ZARAGOZA

On Monday, March 12, Einstein began a short visit to Zaragoza and was met at the station by the physicist Jerónimo Vecino, the University rector Ricardo Royo-Villanova, the German consul Freudenthal and the chemist Antonio de Gregorio Rocasolano. Vecino clearly was the instigator of the visit; he had given a course of ten lectures on relativity in 1921 entitled "Lectures on Matter and Energy."[42] For reasons that shall be-

---

[41] Joaquín Salvatella, in *Discursos*, p. 29; "La significación de Einstein" (editorial), *El Sol*, March 10, 1923.

[42] *Heraldo de Aragón*, March 2, 1923. See ibid., March 8, 1923. Vecino received a telegram that read: "Llegaré lunes rápido. Alberto Einstein."

come clear, however, the scientific focus of the visit was Rocasolano's laboratory.

Einstein gave only two lectures in Zaragoza, on special and general relativity, respectively, held Monday and Tuesday at 6 P.M. in the Sala de Actas of the Faculty of Medicine. On the first evening, the room "was completely filled with personalities of all classes and social status, and also some pretty young ladies and distinguished women adorned the severe hall with their presence."[43] After the lecture Rocasolano (1873–1941) praised Einstein, "pointing out the great work, not only didactic but also in research, carried out by the Faculty of Sciences of Zaragoza and based on the theories of Professor Einstein." He was referring to his own research on Brownian motion in colloids.[44]

On Tuesday morning, Einstein toured the city, visiting the cathedral of Pilar and its reliquary, the Lonja (medieval commercial exchange), and the Aljafería, the palace of the Muslim kings of Zaragoza. Einstein found the architecture impressive. He had enjoyed the art he had seen in Barcelona and Madrid, but the monuments of Zaragoza (according to a journalist's paraphrase) represented "a more robust and eloquent expression of [Spain's] regional physiognomy."[45] Lunch was at the Mercantile Casino, where guests, mainly university professors, had gathered at the invitation of the Academy of Sciences. After the banquet the philosopher Domingo Miral gave a short speech in German praising Einstein and alluding to "the affectionate care with which Zaragoza avails itself of German science, and, recalling Fichte's words to the effect that

[43] As stated on the course announcement, a copy of which was given to me by José Andreu Tormo. The course was a well-integrated introduction to both relativity and quantum theory, ending with a lecture on the physics of discontinuity and the concept of the universe in modern physics.

[44] Rocasolano reported on his research on Brownian movement in colloids in *Estudios químico-físicos sobre la materia viva*. Anales de la Universidad de Zaragoza, I (Zaragoza, 1917). See the detailed description, with photographs and floor plan, of Rocasolano's laboratory by José Albiñana, "Laboratorio de Investigaciones Bioquímicas," *Ibérica*, 16 (1921), 248–251.

[45] *El Noticiero*, March 14, 1923.

Germany had made patent its rights before God and history, he declared his confidence in the vitality of the German people." Einstein, who doubtless recoiled silently at these words and had by now finely honed his ability to deliver mundane comments on any subject, replied by noting that "to the present moment, he had perceived the palpitations of the Spanish soul only in Zaragoza."[46]

At the first lecture, *El Heraldo de Aragón* noted that "only a scant minority understood the bases and deductions of the theory of relativity, which was difficult to carry into the realm of popularization."[47] Not surprisingly, therefore, there were fewer in attendance at the second lecture, when Einstein spoke on the General Theory. (The third lecture was omitted in Zaragoza; Vecino must have anticipated that virtually no one there would be able to grasp the contents.) The rector, Royo-Villanova, who apparently had heard of Sagarra's evocation of the blackboard in Barcelona, announced afterward that "I have asked the Professor that he not erase the drawings made on the blackboards during his lectures and that he confirm their value with his signature, so that something lasting remains of Einstein's passage through the University. These will in due time be fixed and preserved, towards the end of our being able to show them to future generations as relics of this day."[48] The rector's high regard for Einstein provides an interesting insight into the phenomenon of civil discourse in Spanish science in the 1920s. A decade later, as the social and political tensions preceding the Civil War increased, the staunchly conservative Royo was to change his mind concerning the value of Einstein's contribution (see chapter 9).

[46] *El Heraldo de Aragon* and *El Noticiero*, March 14, 1923.

[47] *El Heraldo de Aragon*, March 13, 1923. José María Iñíguez Almech, who attended the lectures, recalled: "There was a large audience. But, inasmuch as the lecture was delivered in German and only twelve of the public, at most, understood anything of relativistic mechanics, it was merely an act of protocol" (personal communication, June 23, 1980). I assume that Einstein lectured in French, as in other cities.

[48] *El Heraldo de Aragón*, March 14, 1923. The blackboards did not survive, as far as I could determine.

After the meeting, Einstein attended a dinner at the German consul's house, where he played the violin, probably accompanied by the great pianist Emil von Sauer, who also happened to be in Zaragoza at the time.[49] Later, a local humorist remarked that "Einstein gave two concerts in Zaragoza. One at the Faculty of Sciences, interpreting 'Nocturns of Relativity,' and another, on the violin, in the house of the German consul." Still later Einstein, the consul, and Jerónimo Vecino repaired to the Teatro Principal to view a musical production entitled *La Viejecita*. The following morning Einstein visited Rocasolano's laboratory and then lunched with Sauer at the hotel, where, over dessert, Einstein had his picture taken with a child who had just danced the Aragonese *jota* for him. He then departed by train for the French border via Barcelona, where, according to the German consul's report, he spent another day, "although his presence there was not widely known and no special events took place."[50]

Sharply different assessments of the visit were recorded. In the opinion of Marcial del Coso (whose writing was typical of a genre of cynical humor characteristic of Spanish newspaper columnists of the 1920s), Einstein received all the honors of Aragonese hospitality, but nothing more: "Neither surprise, nor admiration, nor wonder, not even curiosity. Because Einstein had the bad luck to come to a land where his mysterious theories are better known than the cultivation of sugar beets."[51] The notion, already expressed in Catalonia, that certain cultures are innately attuned to the subtle theory, was a hallmark of the "popular" reception of relativity. But to the scientific community Einstein's visit was meaningful indeed. Not only did his presence honor Aragonese science, but link-

[49] Emil Sauer (1862–1942) was a student of Liszt who played frequently in Spain; see *Gran Enciclopedia Catalana*, XIII, 369.

[50] *El Heraldo de Aragón*, March 15, 1923. The German consular report of Einstein's stay in Zaragoza is reproduced in Appendix 2.

[51] Marcia del Coso, "Varios ejemplos clarísimos. Todos los aspectos de la vida son lecciones de relatividad," *El Imparcial*, March 17, 1923 (written at Zaragoza).

ing Zaragoza's name to Einstein's conferred prestige upon it—
a point made, in retrospect, by the great hydraulic engineer
Manuel Lorenzo Pardo in his annual report, as secretary, to the
Zaragoza Academy of Sciences: ". . . that the name of our city
is repeated in reference to his few attempts to popularize his
new ideas of the new system of interpeting the universe . . . is
one of the Academy's greatest satisfactions and also greatest
merits, perhaps the greatest it could adduce to gain general
esteem."[52]

[52] "Memoria . . . leída por el Secretario . . . 16 de Diciembre de 1923," *Revista de la Academia de Ciencias de Zaragoza*, 1923, 192.

FIVE

The Debate over Relativity
in the 1920s

DURING and after Einstein's visit, debate over the scientific and philosophical meanings of relativity involved virtually every sector of the Spanish intelligentsia. In this chapter we will consider scientific polemics and the philosophical and religious discussions.

A NEO-NEWTONIAN ASTRONOMER:
JOSEP COMAS SOLÀ

Comas was the leading anti-relativist in the higher echelons of Spanish science and the only one to propose an alternate theory. He was in the line of those who, like Walter Ritz, attempted to solve the contradiction between the classical theory of the relativity of motion and the electrodynamics of Maxwell and Lorentz in a way diametrically opposed to Einstein's approach. Instead of using the Lorentz transformations to bring kinematics into line with Maxwell's field equations, Ritz believed that electrodynamics and optics were in need of refinement. He was willing to discard the aether, but believed that the classical principle of relativity was satisfied by stating that the motion of light is also relative.[1]

Comas, who claimed to have formulated his own theory in 1914, favored reviving Newton's emission doctrine, which, he claimed, had been abandoned prematurely:

The introduction of an intermediary element of propagation of radiation (the aether) caused the loss of the fun-

---

[1] Paul Forman, "Walter Ritz," *Dictionary of Scientific Biography*, XI, 479–480.

damental principle of symmetry which Newton wanted to maintain with his theory of emission, a theory which was abandoned with excessive haste in the middle of the past century when [Jean] Foucault demonstrated that the speed of light is lower in a refractive medium than in a vacuum. The emissive theory, relativist in all the extension of the word, could then have been saved, but it was stifled, in spite of the insuperable obstacles that the acceptance of aether offered, by the wave-like nature confirmed in numerable experiments in optics and, later, in electromagnetism.

The classical experiment of Michelson and Morley demonstrating the independence of the speed of light in a terrestrial focus with respect to the movement of translation of the earth or solar system, was a mortal blow to the theory of aether. To explain that surprising result there was no other recourse, so neglected and disqualified was the emissive theory, than to substitute the postulate of the constancy of the velocity of light and of radiations in general, whatever might be the movement of source with respect to the observer. This is what Einstein did, giving rise to the so-called special theory of relativity. On the basis of said principle . . . all material explanation of phenomena is done away with, with the theory entering a purely mathematical terrain, at the cost of distorting the intuitive concept of relative velocities and of time.[2]

Instead of the aether theory, Comas proposed his "emissive-undulatory" hypothesis,

always admitting that the luminous elements follow the movement of the emitting body, the physical effects, as refer not only to radial motion but also to lateral, will be the same as in the aether hypothesis; the speed of light in all directions will be always the same with respect to the

[2] José Comas Solá, "Consideraciones sobre la relatividad," *Revista de la Sociedad Astronómica Hispano-Americana*, 15 (1925), 87.

emitting body; but, for whatever point in space, *the speed of light will vary with the movement of the emitting body.*[3]

Astronomically, Comas continues, it is impossible to prove the existence of an aether, and experiments on the earth's surface have demonstrated that it does not exist. The only solutions are the complete suppression of aether, as in special relativity, or the admission that physical energy is emissive, without ceasing to be undulatory. The particles emitted (which he called "electrons") are imponderable and without sensible mass, but they obey the law of inertia just as matter does.[4] Light, therefore, though it may have no mass, is still subject to Newton's laws of attraction, a universal law that is an intrinsic property of matter.

Comas believed that experimental proof of his contention would be forthcoming when the speed of light originating in stars and nebulae of great radial velocity could be measured directly. In the meantime, he repeatedly defended his theory in public forums, the Academy of Sciences of Barcelona in particular. In the session of February 16, 1922, Comas described an astronomical procedure for determining the absolute movement of the solar system, "in case this absolute motion exists." The method suggested was the use of the first satellite of Jupiter as a celestial clock, to test for absolute motion, "on the basis of variations of speed relative to light." He was answered by Ramon Jardí, who noted "conceptual contradictions [in Comas's emissive-undulatory theory] and noting that the immediate consequences of such hypotheses conflict with the principles of Mechanics." The session terminated with an "elegy" of Einstein's theories delivered by Terradas.[5]

Although the content of Jardí's and Terradas's rebuttal of Comas is not recorded, it must have been similar to the cri-

---

[3] "Consideraciones sobre la aberración de la luz," *Boletín del Observatorio Fabra,* 1 (1919–27), 31. The paper is dated April 18, 1919.

[4] Ibid., p. 32.

[5] *Boletín de la Real Academia de Ciencias e Artes de Barcelona,* 4 (1916–23), 515–516.

tique that Terradas, in his 1921 lectures, offered to Walter Ritz. Like Comas, Ritz held that the velocity of light was dependent upon that of its source, a hypothesis that, Terradas noted, "is sufficiently in contradiction with all the results of theoretical and practical research." De Sitter, specifically, had proven the contrary in 1913. Too many observed optical and electromagnetical phenomena (such as the Doppler effect and the aberration of light) were left unexplained by Ritz's hypothesis. The failure of Ritz's theories is important, Terradas continues, "because it suggests an idea which is also very important for the proper comprehension of the theory of relativity, that is, the fact that all *observed* phenomena are always linked to matter. The *field in the aether* is a fiction invented to describe as simply as possible the spatial and temporal relationships of phenomena in bodies."[6] On March 13, Comas replied to Jardí's attack, asserting that his objections to the logic of relativity were based on experiment.[7]

When Einstein lectured in Barcelona, Comas informed him of his "entire conviction" that the constancy of the speed of light was an erroneous interpretation of the Michelson–Morley experiment, that if light seems always to have the same velocity, it is because the observer "proceeds with the light," and that light, like all radiant energy, was ponderable and subject to gravitation and inertia.[8]

Throughout the 1920s Comas attacked relativity bitterly and claimed that his own theory was unfairly denied a hearing. He also insisted that he had anteceded Einstein in the conceptualization of light particles, and when Louis de Broglie won the Nobel Prize for the wave theory of electrons in 1929, Comas complained that his priority in this discovery had been ignored, both in Spain and abroad, and that de Broglie, who had published an article on his theory in *Scientia* in September

[6] "De relatividad," pp. 376–377.

[7] *Boletín de la Real Academia de Ciencias e Artes de Barcelona*, 4 (1916–23), 517.

[8] "Las conferencias del profesor Einstein," *Revista de la Sociedad Astronómica*, 13 (1923), 21; this article was originally published in *La Vanguardia*, March 14, 1923.

1926, had not bothered to cite Comas's explication of his own theory in the same journal some months before.[9] Comas, whose theory was based both on his own perceptions as an astronomer and on physical concepts current at the end of the nineteenth century, and whose knowledge of mathematics was limited, was never able to understand why his theory was ignored by the scientific elite.

COMAS, PLANS, AND THE AETHER-DRIFT EXPERIMENTS OF THE 1920S

Blas Cabrera noted that anti-relativists used two different kinds of argument: either they concentrated on possible errors in interpreting the experimental results of the General Theory, or else they proposed "semi-Einsteinian" alternative theories that led to the same results, without incurring a break with classical physics. The experimentalist perspective gave Cabrera answers to both arguments. The reasoning of those who pointed to possible errors in calculations of the perihelion of Mercury, the deflection of light, or in red shift would have force "had these not been three absolutely independent phenomena and, besides, the only predictable ones." (Pro-Einsteinians, he continues, are ever in wait of the next eclipse to provide a decisive proof; "but the most probable [outcome] is that scientific opinion will continue its evolution in favor of the new ideas.") As an example of a semi-Einsteinian argument, Cabrera cites Paul Painlevé, who wished to preserve absolute time, and hence "one can do without the modifications that special relativity introduced into Mechanics; but it is the case that the theory already has to its credit the support of experimental results of such importance that no physicist could doubt their exactitude."[10] Spanish engineers, such as Pérez del

[9] Comas, "Teoría corpuscular-ondulatoria de la radiación," *Boletín del Observatoraio Fabra*, 2 (1931), 25, and "Nueva teoría emisiva de la luz y de la energia radiante en general," *Scientia*, 36 (1924), 375–382.

[10] Cabrera, *Principio de relatividad* (Madrid: Residencia de Estudiantes, 1922), pp. 13–14.

Pulgar and Herrera (see chapter 6), were very partial to semi-Einsteinian arguments that sought to preserve this or that aspect of Newtonian physics or "intuitive" science (in the case of these two figures, a formulation that would give Einstein's results without having to admit the speed of light as the limit of velocities).

Comas, in addition to his countertheory, also popularized the results of aether-drift experiments, which sought to refine, improve upon, or explain the Michelson-Morley experiment in order to confirm the existence of aether and overthrow special relativity. This was a strategy typical of many anti-Einsteinians who pursued observations rather than attempting to attack the theory directly. The Australian eclipse results of 1922 convinced most of the nonideological doubters. Subsequently, aether-drag experiments became the principal recourse of the die-hard anti-relativists. Comas, as we have seen, was not interested in saving the aether; he believed, rather, that the drift experiments would have the double value of proving Einstein wrong and his emissive-undulatory theory right. Michelson's recent experiments proved his contention, he averred, as did those carried out in 1913 by Georges Sagnac. The latter, following a suggestion of Michelson, had measured on a rotating interferometer the change in position of the interference fringes of a split beam of light traveling a polygonal course in opposite directions, then recombined and focused on a photographic plate. Sagnac believed that the observed shift in the fringes proved the existence of the aether, but Paul Langevin showed that Sagnac's results were also predicted by relativity theory.[11] The other experiment mentioned by Comas was that conducted by Michelson and H. H. Gale to test the effect of the earth's rotation on the velocity of light, designed

---

[11] José Comas Solá, "Consideraciones sobre la relatividad," *Revista de la Sociedad Astronómica Hispano-Americana*, 15 (1925), 87–88. On Sagnac's experiments, see Loyd Swenson, Jr., *The Ethereal Aether: A History of the Michelson-Morley-Miller Aether-Drift Experiments, 1880–1930* (Austin: University of Texas Press, 1972), pp. 181–182.

to test Einstein's principle of the equivalence of inertial and gravitational masses.[12]

From the relativist camp Rafael and Plans provided Spanish readers with hard-hitting critiques of these experiments. At the Oporto meeting of the Spanish Association in 1921, Rafael presented an analysis of Augusto Righi's interpretation of the Michelson-Morley experiment. Rafael noted that Righi's comments had attracted scant commentary, "whether because they dealt with a subject already perfectly well understood, whose consequences were not thought to be modifiable by the considerations of a dilettante in theoretical physics . . . or because the experimental bases which gave rise to the admirable relativistic theories had already been confirmed by such diverse routes." According to Rafael, Righi's problems arose from his unwillingness to consider effects of relative rotation supplementary to those of the lineal movement of the two rays.[13] It is improbable, Rafael noted the following year, that relativity will be overturned by experiments of this type, because "the hypothesis of the fixed aether (which, as Einstein put it so well, is the equivalent of attributing physical properties to absolute space) has lost all authority."[14]

Reviewing Dayton Miller's interferometer experiments in 1927, Plans supposed that the Michelson-Morley experiment was "only too well known to the readers" of *Ibérica*, inasmuch as it constituted "one of the cornerstones of Einstein's work." He reviewed Miller and Morley's experiments of 1905 in Cleveland and those carried out by Miller alone, between 1921

[12] Comas, "Una experiencia notable," *Revista de la Sociedad Astronómica Hispano-Americana*, 15 (1925), 56–58. The Michelson-Gale experiment was designed to be a test of Einstein's "principle of equivalence" (of inertial and gravitational masses); Swenson, *Ethereal Aether*, pp. 207–208.

[13] Enrique de Rafael, "La teoría del experimento de Michelson," Asociación Española para el Progreso de las Ciencias, Congreso de Oporto, Tomo V: *Ciencias Fisicoquímicas* (Madrid, 1922), pp. 87, 105 ("Either Righi did not read, did not understand, or did not want to understand this observation").

[14] Enrique de Rafael, "Nociones de Mecánica clásica y relativista," *Anales de la Asociación de Ingenieros del Instituto Católico de Artes e Industrias* (hereafter cited as *Anales ICAI*), 1 (1922), 185–186, n. 1.

and 1925, at Mount Wilson Observatory in California, the latter executed at 1,850 meters of altitude to avoid registering a drag effect supposedly caused by the earth's surface. Both experiments produced results that supported a partial drag of aether by the earth, whose effect diminished with altitude. Although "the contradictors of the theory [of relativity] are clapping their hands" over Miller's results, there was no reason for it, according to Plans. Even if Miller were right, his results would not affect the General Theory, which has "so firm and logical a structure that it can probably resist such reverses." With regard to the Special Theory, he did not see Miller's results as definitive, particularly because the similarly designed experiments of F. T. Trouton and H. R. Noble, carried out with electrical charges rather than light rays, had also produced a null result and were "therefore favorable to the theory of relativity."[15] In a later article Plans stressed that "A dispassionate examination of these experiments points to the probable influence of temperature as the perturbing cause," a point later confirmed by Robert Shankland and collaborators.[16]

PHILOSOPHERS AND RELATIVITY

In Spain of the 1920s there was surprisingly little philosophical comment on the theory of relativity outside of a purely religious or theological context. Of the three principal figures discussed here, Manuel García Morente, José Ortega y Gasset, and Miguel de Unamuno, only Ortega's philosophy can be characterized as truly secular, but his arguments strike me as implicitly directed against misrepresentations of relativity commonly made by those approaching it from a religious perspective. Therefore there is a continuum of discussion between explicitly religious and secular commentators.

The first philosopher to exhibit interest was García Morente

[15] José M. Plans, "El experimento de Miller y la teoría de la relatividad," *Ibérica*, 27 (1927), 169–171.

[16] "Nuevas repeticiones del experimento de Michelson," *Ibérica*, 28 (1927), 94–95; see Swenson, *Ethereal Aether*, p. 244.

(1888–1942). In July 1921 *El Sol* published an extremely interesting interview with Einstein in London, just after the physicist and Erwin Freundlich, who was also present, had returned from a session with Whitehead in early June. Einstein
was in a philosophical mood, although he was careful to point
out that relativity did not, in itself, constitute a philosophy. He
explained that his theory was preferable to that of aether because it made fewer suppositions and went on to note that his
own philosophical evolution had gone from Hume to Mach to
the pragmatism of William James. He then inquired about
Spanish interest in relativity, and the journalist, Ricardo
Baeza, recalled two lectures on the subject by García Morente
at the Residencia de Estudiantes.[17] These lectures were apparently never published but must have been a byproduct of Morente's translation of Moritz Schlick's popularization of relativity, which was published in the spring of 1921.[18] This book
received a derogatory review in *Ibérica* by Enric de Rafael,
who deemed it far inferior to Freundlich's volume, available to
the Spanish public in Plans's translation. By comparison,
Schlick's book

> represents a step backwards . . . because a certain *enlight-
> ened* public prefers vagueness and five-minute explana-
> tions (lest I miss the train) rather than to have to make a
> serious study of tensorial calculus and the differential
> quadratic forms which enter into the exposition of Ein-
> stein's theory. And inasmuch as said *enlightened* public,
> pseudo-philosophical and pseudo-scientific, is the one
> which buys out editions, there is no help but to cook up
> relativity to their palates's taste.[19]

The volume's only saving grace, in de Rafael's view, were
eleven appendixes added by García Morente in order to bolster

[17] Ricardo Baeza, "Delante del Profesor Einstein," *El Sol*, July 3, 1921. The
interview took place in June.

[18] Moritz Schlick, *Teoría de la relatividad. Espacio y tiempo en la física actual*
(Madrid: Calpe, 1921).

[19] *Ibérica*, 16 (1921), 96.

the work's scientific value. The appendixes are particularly interesting because they demonstrate that Morente did indeed
have some mathematical background.[20] Another fragment of
Morente's commentary on relativity survives, but it is undoubtedly a late one, from the period after his ordination as
priest in 1940. It is a brief comment on a forgotten volume by
Felix Eberty (1812–84), *The Stars and the Earth* (1846), which
had been reprinted in 1923 with an introduction by Einstein.[21]
Morente notes that, because of the high velocity of light, the
order of time could be inverted, expanded, contracted, or
stopped in an eternal present for an individual invested with
the power to travel wherever in the universe he wished, at
whatever velocity, and endowed with the ability to perceive
details across immense distances. By such means, one could
observe the recent or distant past, and so forth. These tantaliz-
ing paradoxes serve as basis for speculation on the nature of
eternity as experienced by God, who would have all moments
of the past forever at his disposition.[22]

Ortega y Gasset (1883–1955) made two important contributions to the discussion of Einstein's thought. The first, his
famous essay on the historical significance of relativity, written in 1922, had its kernel in a carefully worded riposte to the
typical allegation of Catholic anti-relativists that relativity was
subjectivist and that it denied the existence of absolutes. The
second, his lecture introducing Einstein at the Residencia de
Estudiantes on March 9, 1923, is significant because it cast
Einstein's achievement in the context of the history of science,
in terms that a lay audience could comprehend. It should be
noted with regard to Ortega's approach to relativity that he

[20] Of particular interest are Morente's appendixes 1 (on Michelson-Morley), 6 (a discussion of Gaussian coordinates and the determination of distances
in Euclidean and non-Euclidean continua, and 10 (experimental proofs of general relativity). All the appendixes, plus related writings, have been reissued
recently as *Sobre la teoría de la Relatividad* (Madrid: Encuentro, 1984).

[21] Felix Eberty, *Die Gestirne und die Weltgeschichte: Gedanken über Raum, Zeit
und Ewigkeit* (Berlin: Gregor Rogoff, 1923).

[22] García Morente, "Einstein," *Arbor* (1979), 35–39.

did not communicate much with mathematicians nor have strong ties with them. On the other hand, Blas Cabrera was a member of his *tertulia*. Hence, Ortega's ideas on relativity emanated mainly from the discussion of the Special Theory.

In his essay on the historical sense of relativity, Ortega, implicitly addressing anti-relativists, criticized a main current of European thought which he called "utopianism." This value, characterized in "the enormous zeal for dominating the real" was, moreover, "specific to Europeans" and was a byproduct of European rationalism that insisted upon absolutes, in politics as well as in physics. (Ortega reiterated this view in later essays, without further allusion to relativity. In retrospect, it appears that Ortega was praising Einstein for having humanized physics, because the main defect of utopian rationalism was that it stressed the critical faculty at the expense of environmental and cultural limits to which people must ultimately adjust.)[23] The need for absolutes is at the root of scientific reductionism (Ortega does not identify it as such), which he ridicules by alluding to Jacques Loeb's exaggerated use of tropisms to explain animal and human behavior.[24]

Against the background of utopic-absolutist thought, Einstein was "a breath of fresh air." Anticipating a theme of the Residencia lecture, Ortega elaborates on the Kantian distinction between pure reason and sense data. He associates a physics that is purely geometric with Kantian pure reason and, by implication, with absolutism and utopism. An example is the Michelson-Morley experiment. Lorentz's solution was utopic, that of the old rationalism. But Einstein "inverts the inveterate relationship which existed between reason and observation," by insisting that geometry yield to observation.[25]

---

[23] See the succinct summary by John Butt, "José Ortega y Gasset," in *Makers of Modern Culture*, Justin Wintle, ed. (New York: Facts on File, 1981), p. 392.

[24] Ortega, "El sentido histórico de la teoría de Einstein," in *El tema del nuestro tiempo*, 18th ed. (Madrid: Revista de Occidente, 1976), pp. 149–151. Ortega was by no means alone in his criticism of Loeb; see *Dictionary of Scientific Biography*, VIII, 446.

[25] Ibid., pp. 152–154.

In the same essay, Ortega tackled the basic conservative philosophical objection to relativity—that it was subjectivist and denied the existence of absolutes. Einsteinian physics is indeed absolute, he explains, because it holds that physical laws are true, whatever the system of reference used. But it was not absolute in the aprioristic sense of the old rationalism. In classical physics our knowledge was relative because we never could achieve knowledge of the absolute. In Einstein's physics our knowledge is absolute; reality is relative.[26]

The Residencia lecture begins with a reflection on the role of science in Western culture. "This culture," he asserted, " has a characteristic discipline, which is its summit, its crown: physical science."[27] If there were some competition for a prize among all the cultures of the world, the West would have to present its physical science, "which approaches and almost realizes the ideal of knowledge, because it is at one and the same time exact and real and, for this reason, the purest theory, the most efficient technique, the purest contemplation and, at the same time, practical dominion over things." This synthesis of thought and action constituted, according to Ortega, the primary idiosyncratic characteristic of Western man. In the same line as the great figures of this tradition—Copernicus, Galileo, Kepler, and Newton—came Einstein. Newton was a systematizer who created the grammar of science: "Galileo created the most elegant anatomy of motion, so that one can say that Newton wrote his lengthy statement with letters discovered by Galileo." Einstein, on the other hand, had combined in his own work these two kinds of science: "Einstein invented the letters and, at the same time, is the sublime author of the paragraph." Just as Galileo and Newton created the molds in

[26] Ibid., pp. 142–143.

[27] I use the text of this important lecture, not included in Ortega's *Obras Completas*, as given in Ortega's newspaper *El Sol*, March 10, 1923. Certain of the themes of this lecture recur in the first chapter of *En torno a Galileo* [1933] (Madrid: Revista de Occidente, 1956). See an interpretation of Ortega's remarks by Alfonso Reyes, who was in the audience; "Einstein en Madrid," in *Simpatías y diferencias* (Mexico: Porrua, 1945), pp. 92–93.

which modern life had been formed, so will Einstein's work represent the beginnings of a new culture. In order to describe this revolution, Ortega did not think it necessary to dissect Einstein's science itself. (He believed it difficult to grasp, "although not as difficult to comprehend as is said.") There is an easier route, "which is to fix the significance of Einstein's intellectual act, to inquire into the intellectual faculties which most characterize him." In this light, one finds that Einstein's achievement lay in "a new kind of experience."

For Kant, Ortega explained, knowledge was a product of two factors: sense data and a priori thought. The history of modern physics can be described in terms of an oscillation between those two poles. In Descartes' system, the idea predominated over observation and experiment, inasmuch as the physical image of the world was to be constructed on a purely geometric basis. Kant, while recognizing the importance of both factors, "really conceded to experience the role of observing how the laws of geometry are fulfilled, so that decisive evidence continued to rest on reason." Einstein turns this state of affairs completely around:

> Einstein represents the opposite point of view. For him, what is strictly rational cannot arbitrate physical things, because mathematics is a formal science, not a science of things. Reason builds a repertory of ordinal concepts; but it is the experiment which selects the applicable order. Euclidean space is an order; experience, not theory, must decide whether or not it is applicable to the world. Einstein, then, symbolizes a shift towards empiricism.

To say, as many did, that Einstein simply confirmed Kantian doctrine on the subjectivity of time and space was simply wrong. Ortega's view of relativistic physics as empirical is almost unique among Spanish commentators, especially in view of the widespread misunderstanding of (not to say hostility toward) the role of mathematics in its formulation. Although Ortega does not here allude to the experimental proofs of Einstein's theories, his argument was perhaps the most clear articulation in Spain of the 1920s of the widespread feeling that such experimental confirmation legitimated a new system of physical thought, a sentiment that, I believe, underlay much of the popular enthusiasm for relativity.

Ortega was clearly drawn to relativity because it seemed to provide scientific substantiation for his own notion of "perspectivism," formulated in 1916. One of the properties of reality, he had said, is that things have perspective: "Perspective is the order and form that reality takes for the person contemplating it. If the place occupied by the contemplator varies, the perspective also varies."[28] The perspectivist argument, with its analysis of observation at least in a broad sense analogous to relativistic concepts, was really no more than a plea for cultural relativism, which led Ortega into platitudes on the equivalent values of different cultures, anticipating popular anthropological writings of a slightly later period. Ortega's essay has long since entered into the canon of relativistic philosophy. But at the time, it was harshly criticized by scientists and philosophers alike. Miquel Masriera, a physical chemist who had studied with Hermann Weyl, asserted that Ortega had, in fact, missed the main point about Einstein's critique of observation:

> What Ortega y Gasset has said about how relativity is essentially objectivist, anti-Kantian or vitalist is, as they say today, pure dilettantism. Relativity is chiefly characterized by not stating how objects are, but how we see them. Clearly, it establishes an absolute value, in the sense of the independence of each system of observation, and this is the so-called "interval" between phenomena. But this value can never have philosophical objectivity, for it has no tangible existence, but rather the character of a necessary abstraction, or, more concretely, of a mathematical indeterminant.[29]

---

[28] Ortega, "El sentido histórico de la teoría de Einstein," p. 147.

[29] Miguel Masriera, "El valor del Relativismo," *La Vanguardia*, February 4, 1925. Note that Masriera, in alluding to intervals, was attempting to move the philosophical discussion of relativity away from the Special Theory, where

For the conservative Catholic philosopher José Pemartín, Ortega's 1923 essay, in advancing the argument for perspectivism, had disfigured "not only the sense of Einstein's renovation, but even the nature of classical Mechanics, which he erroneously presents as something utopian, a pure entity of reason, arbitrarily imposed on reality by rationalist obfuscation." Ortega distorted classical physics by asserting that Galileo and Newton made the universe Euclidean simply because reason so dictated. But the truth is, says Pemartín, that they saw in Euclidean geometry a close fit with experimental data. Leverrier's discovery of Neptune with a pencil, later confirmed by observation, is a kind of Euclidean anticipation of Einstein's non-Euclidean feats. Also false was Ortega's interpretation of Einstein's achievement: that he had inverted the relationship between reason and observation and stood Kant on his head. Pemartín quotes Einstein himself to the end that time and space in a Gaussian four-dimensional system lose all physical reality. (There is a conceptual standoff here, replete with semantic "slides": since Pemartín may not have grasped the core of Einstein's critique of physical observation, there was no point in arguing what was real or not real in Einstein's system.) In a footnote, Ortega had alluded to the latent anti-causality of relativity theory. Pemartín thought it nonsensical to stress Einstein's anti-rationalism and, without alluding to Ortega's contraposition of Einstein and Descartes, he notes the profound Cartesian roots of the former.[30]

Another philosopher to comment on relativity was Ramiro Ledesma Ramos (1905–36), possibly the most incisive "popular" Spanish commentator on science in the late 1920s. This was the same Ledesma who, scant years later, was to found a

---

Spanish anti-relativists had been traditionally stuck, and into the General Theory. He was one of the few Spaniards to comment on the philosophical discussion of simultaneity, in which Henri Bergson was the principal figure, which he thought both philosophically and scientifically irrelevant.

[30] Ortega, "El sentido histórico de la teoría de Einstein," p. 156, n. 1. José Pemartín, "La física y el espíritu," *Acción Española*, 4 (1933), 144–146.

Fascist political movement. His wholehearted support of Einstein is eloquent testimony to the intellectual climate of the 1920s, so propitious for civil discourse in science.

Ledesma staunchly defended Einstein against the biologist Hans Dreisch's critique in a column published in October 1930. According to Ledesma, Einstein's theory was "the most fecund stimulus of all time" purely for its value in creating intellectual agitation, leaving aside the question of whether it was true or not. Because even if false, it demonstrated the weakness of classical concepts. Ledesma's critique of Dreisch's book, *Relativitätstheorie und Philosophie* (1924), was on philosophical rather than physical grounds. It appeared that Dreisch, ingenuous in Ledesma's view, wanted to deny to relativity its right to be a theory. Ledesma points out the circularity of the arguments of those who sought to attack the "logic" of the theory on metaphysical, not physical, grounds. To say (as many Spanish detractors of relativity did) that the Special Theory violates the principle of contradiction betrays a strange ingenuousness. The logic underlying c may be difficult for us at first, but once we have understood it, we recognize it as a concept of a higher order than the one it replaced.

But Dreisch was even more opposed to general relativity because it was built on "metageometrical spaces." He avoids dealing with the physical notion of gravitation just as, in his discussion of special relativity, he had avoided coming to grips with the new concept of mass. What Dreisch really reveals is an absurd predisposition against non-Euclidean geometry. To this line of argument, Ledesma rejoins that one geometry is no more or less valid than any other, only more or less accurate or appropriate. Therefore it is absurd to reject a given geometry as counterintuitive. In any case, Dreisch fell very wide of the mark because Einstein himself explains that space, whether Euclidean or not, is dependent upon the state of gravitation. Indeed, Ledesma continues, Einstein's theory of gravitation is his essential contribution to the new physics. In conclusion, Ledesma detects nothing new in Dreisch's argument, just the

165

old problem of failure to accept the mathematical method in physics.[31]

Ledesma's trenchant critique of Dreisch, particularly on the supposedly "metageometrical" concept of space attributed to Einstein, applies with equal validity to Eugeni d'Ors and other commentators who misconstrued the nature of Einstein's "geometry." To Ors (1881–1954), whose doctoral dissertation in philosophy had been on the topic "The Arguments of Zeno of Elea and the Modern Notion of Space-Time," the importance of Einstein was the rationalization of motion, a problem that had been disturbing thinkers since Zeno's time. Inasmuch as Einstein's universe was a finite one, the role of the infinite (as in Zeno's paradoxical continua) was severely reduced. Ors notes sourly that his thesis had lain unpublished for a decade.[32] But Einstein's geometry bothered him. During Einstein's appearance at the Residencia de Estudiantes, Ors remarked to the Mexican diplomat-writer Alfonso Reyes: "These are too many dimensions. This is to return Geometry to the barbarous state in which it was before Euclid derived the three symbolic commensurations—the only ones we need." Ors made the same mistake of conceiving relativity as pure geometry as others who tried to tag Einstein with a Kantian label.[33]

Allusions to Einstein in the writings of other philosophers of the 1920s are, on the whole, passing and undeveloped. In 1922 the French Hispanicist Marcel Bataillon suggested to Miguel de Unamuno (1864–1936) "the idea of writing something about spiritual time and Einstein's theory applied to history and how a subject which comes later in chronological

[31] Ramiro Ledesma Ramos, "Hans Dreisch y las teorías de Einstein," *Gaceta Literaria*, October 1, 1930; reprinted in *La filosofía, disciplina imperial* (Madrid: Tecnos, 1983), pp. 91–96. See also Ledesma's comment on Einstein's formula for the volume of the universe, "Nota de Matemática," in Tomás Borras, *Ramiro Ledesma Ramos* (Madrid: Editora Nacional, 1971), pp. 51–52. Dreisch had been a contributor to the anti-relativity volume, *Hundert Autoren Gegen Einstein* (1924).

[32] Eugenio d'Ors, *Nuevo Glosario*, 3 vols (Madrid: Aguilar, 1947–49), I, 796.

[33] Reyes, "Einstein en Madrid," p. 93.

time can be the precursor of something else in spiritual time—or better, in eternal time—even to the point where a grandchild could be the father of his own grandfather."[34] There are two other fleeting allusions to Einstein in articles written by Unamuno in 1922 and 1923. In the first, he notes that "When I hear people talking about Einstein's theories and the difficulty many have in understanding them, I think it is mainly because for the most part human intelligence lacks a sense of continuity, that is, a sense of the infinite and its function."[35] Most of us, he explains, live in the "Christian Democratic" concept of the universe, "based on atoms, vacuums, and the scholastic notion of causality." The Heraclitan concept of flux and continuity is foreign to such persons. In an article written soon after Einstein's departure from Spain, Unamuno discusses the replacement of the traditional measurement system by the metric system and notes that in certain Spanish agrarian regimes the unit of superficies is one of value rather than an abstract measure. A parcel that yields twice the value of another is accounted as its double in size. The rigid application of the metric system abolishes all this. "But then Einstein came along and where we used to refer to size, he has introduced value or valence. And goodbye meters and goodbye straight lines! And goodbye to everything metric and rectilinear, as in sociology."[36] In the first article he at least broaches the important question of the relation between classical and relativistic notions of causality. In the second reference he appears to convey the impression that Einstein's mission had been to humanize physics; however, the notion that the meter had been abolished

[34] Unamuno to Bataillon, August 1, 1922; quoted by Manuel García Blanco, En torno a Unamuno (Madrid: Taurus, 1985), p. 606. When Unamuno was exiled to the Canary Islands for his opposition to the dictatorship of Primo de Rivera, Einstein wrote him a letter of sympathy, according to José Rubia Barcía, "Unamuno the Man," in Unamuno, Creator and Creation, J. R. Barcía and M. A. Zeitlin, eds. (Berkeley: University of California Press, 1967), p. 16.

[35] Miguel de Unamuno, "Mi visita a palacio" [published May 12, 1922], Obras Completas, 9 vols. (Madrid: Escelicer, 1966–71), VIII, 467.

[36] "Aforismos y definiciones, V" [June 10, 1923], Obras Completas, VII, 1527.

(by the Lorentz contraction) was one of the commonest of "popular" interpretations of relativity and a favorite topic of cartoonists.

To close this discussion of the limited impact of Einstein on Spanish philosophers of the 1920s, it is interesting to note that Henri Bergson's critique of the relativistic concepts of duration and simultaneity stirred virtually no interest, although we have noted that Bergson's confrontation with Einstein at the Société Française de Philosophie in April 1922 was reported in the Spanish press. Bergson's critique centered on the train example given by Einstein in chapter 9 of his popularization, *Relativity, The Special and General Theory*, where there are two systems of reference represented by the embankment and the train, designated M and M'. Einstein denies, under the Special Theory, that two events (strokes of lightning at points A and B) perceived as simultaneous in one system are also simultaneous in the other. In Bergson's view one must suppose, however, that observations are actually made by one observer only ("Physicist," he calls him) who can only be in one system at a given time. If he is in system M, he cannot also be in M', and therefore nothing has really been observed in M'. Einstein's reply to Bergson was a model of concision: one cannot confuse physical time with philosophical time (which is both psychological and physical at the same time).[37]

One Spaniard who commented on this exchange was Miquel Masriera, who wrote a series of articles on special relativity in *La Vanguardia* in 1925. The first criticized both Bergson and Jacques Maritain as "psychological antirelativists," and the second outlined Bergson's critique of the train experiment and explained how relativity responds to the contradictions inherent in the classical concept of absolute motion. (A third article was his critique of Ortega's perspectivism, mentioned above.)[38] Masriera translated the three articles into German

[37] Bergson, *Durée et simultanéité* (Paris: Félix Alcan, 1922), pp. 135–139. *Bulletin de la Société Française de Philosophie*, 17 (1922), 107.

[38] Miguel Masriera, "El antirelativismo psicológico," "La verdad sobre Ein-

and sent them to his mentor, Hermann Weyl, who in turn forwarded them to Einstein. On October 7, 1925, Einstein replied to Masriera from Berlin, simply to repeat the argument given in chapter 9 of his *Relativity*, together with a diagram, concluding: "For years I have not responded to published works objecting to the theory of relativity. Life is too short to waste time on such matters."[39]

In his book, *Einstein and Saint Thomas*, Luis Urbano also gave an account of the Bergson-Einstein exchange in Paris, reproducing Einstein's reply almost verbatim. But Urbano rounded out his discussion of simultaneity by interjecting an example given by H. Thirring of the perception on earth of events taking place in stellar space, in this case the appearance of a new star in the constellation of Perseus in 1901. He concluded that simultaneity only made sense when admitting the constancy of the velocity of light and only if simultaneity be regarded as an observational rather than as a "real" phenomenon.[40] The stellar illustration, it seems to me, is inherently easier for those with little background in physics to grasp than is the train example because, as in Thirring's discussion of Perseus, events can be acknowledged to have occurred on different days (according to the earth's calendar), depending on whether they were observed in an Earth-Perseus system of reference or in a Perseus-Earth system.

Another clerical commentator, Ataulfo Huertas, followed Bergson's psychological-philosophical distinction between absolute and relative movement, the former being a metaphysical and the latter a physical concept. Huertas thought

---

stein," and "El valor del relativismo," *La Vanguardia*, January 7, January 15, February 4, 1925.

[39] Masriera, "De Einstein para mis lectores," *La Vanguardia*, October 9, 1925. A facsimile of Einstein's letter appears in Masriera, "La polémica con Bergson," *La Vanguardia*, March 14, 1979. In his letter, Einstein also comments on the Lorentz transformation, in the same terms as in chapter 11 of his popular volume.

[40] H. Thirring, *L'Idée de la Relativité* (Paris, 1923), chap. 6; Urbano, *Einstein y Santo Tomás: Estudio crítico de las teorías relativistas* (Madrid-Valencia: Ciencia Tomista, n.d.), pp. 78–82.

Bergson's assertion, in *Durée et Simultanéité*, that one could defend absolute motion within Einstein's system was a premature conclusion. It is obvious that there was no one philosophical interpretation of the implications of relativity, and Huertas's observation that philosophers of various schools saw in relativity the a priori confirmation of their own theories could well have been applied to Ortega.[41]

RELATIVITY AND RELIGION

Spanish theological discussion of relativity was extremely uneven in content—much of it opinionated, uninformed, and vulgar popularization. For reasons of consistency, however, I prefer to discuss it as a single body of literature, one colored by the scientific background of the various participants. Of the ten clerical writers most prominent in the discussion of relativity, seven had had varying degrees of scientific education. Luis Urbano (1882–1936), one of three supporters of relativity among the ten, received a doctorate in physics at Madrid, under Cabrera.[42] The Jesuit mathematician Enric de Rafael was a scientific and ideological ally of Plans and Terradas. The other relativist, Benjamín Navarro, science writer for the *Revista Calasancia*, had a doctorate in chemistry. Of the anti-relativists, Angel Rodríguez (1854–1935), director of the Vatican Observatory between 1898 and 1905, also held a doctorate in physics from Madrid. Teodoro Rodríguez (1864–1954) held a Madrid licenciate in physics and chemistry and was the author of a number of secondary-school science texts.[43] Ataulfo Huertas (1872–1936) appears to have studied science in Lou-

[41] Ataulfo Huertas, "La relatividad de Einstein," *Revista Calasancia*, 11 (1923), 374–375, 379–380 (commenting on Bergson, *Durée et Simultanéité*, p. vi).

[42] *Diccionario de Historia Ecclesiástica de España*, 4 vols. (Madrid: Consejo Superior de Investigaciones Científicas, 1975), IV, 2674–2676. In *Einstein y Santo Tomás*, p. ix, Urbano refers to Cabrera as "my illustrious professor" and to Plans as "my dear friend."

[43] *Diccionario de Historia Ecclesiástica de España*, III, 2103, 2105.

vain and was a member of the Mathematical Society in 1926. The Jesuit astronomer Josep Ubach wrote an anti-relativist tract. The three remaining figures were religious philosophers or theologians. Felipe Robles Degano (1853–1939) taught mathematics and philosophy at the Seminary of Avila. Bruno Ibeas (1879–1959) taught history of philosophy at the University of Madrid for a time, and Pedro de Medio taught moral philosophy in Avila.[44]

We might first inquire how relativity was presented to the readers of ecclesiastical publications, specifically the journals of the religious orders and, secondarily, in newspapers whose editorial policy was explicitly Catholic. The most influential religiously oriented journal was *Razón y Fe*, published by the Jesuits. During the 1920s it discussed relativity with considerably less frequency than other polemical scientific issues, such as Darwinism, which it opposed implacably. However, this comparative silence was probably due to the fact that *Ibérica*, the publication of the Observatorio del Ebro, under the direction of Enric de Rafael (until 1922) and with his frequent collaboration thereafter, was the Spanish journal most active in popularizing relativity and was also published under Jesuit auspices (although with no specific religious bias except in an occasional book review). Indeed, the only article on relativity to appear in *Razón y Fe* during this period was written by de Rafael himself.[45] In this essay he notes that scholastics had combatted Newtonian concepts of time and space (particularly as regarded the qualities of immensity, eternity, infinity, indestructibility, and so forth) as incompatible with their conception of God. Anti-Catholics of a skeptical bent had also opposed Newtonian absolutism. But these were metaphysical criticisms and mattered little in the face of the tremendous fit between the Newtonian system and observed facts.[46]

[44] Ibid., II, 1114 (Ibeas); II, 1107 (Huertas); III, 2097 (Robles); Urbano, *Einstein y Santo Tomás*, p. 7.

[45] Enrique de Rafael, "La teoría de la relatividad," *Razón y Fe*, 64 (1922), 344–359.

[46] Ibid., p. 345.

With this preamble, clearly designed to suggest an analogy between Catholic anti-Newtonianism and anti-relativism, de Rafael then describes the Michelson-Morley experiment, the Lorentz contraction, and the Special Theory (the latter mainly in Einstein's own words taken verbatim from Lorente de Nó's translation). It is noteworthy that de Rafael makes no attempt to hide his Einsteinian orthodoxy, agreeing with Einstein that since it is impossible to determine through psychology, physics, or metaphysics whether any privileged (absolute) system of reference exists in reality, then, following the philosophical rule of *Non sunt multiplicanda entia sine necessitate*, neither aether nor absolute time or space exist (or at least are not appreciable and hence have no use in physical explanation).[47] Finally, he notes four objections to relativity commonly voiced in religious circles. The first was that it destroyed classical science; "Nothing more exaggerated," was de Rafael's retort. In fact, in the context of their own times, Columbus, Copernicus, Galileo, and Newton were all more revolutionary than Einstein and such, indeed, "is the history of living science." The second objection, that relativity turns causality on its head and holds, for example, that an effect could precede a cause in time, is simply confused thinking. Time can never retrocess for any one individual. A distant observer, however, could perceive a cause and an effect as simultaneous occurrences. The third objection is that it is intolerable to do away with absolute space and time. Yet, de Rafael points out, such notions were never accepted until Newton and had never formed part of scholastic cosmology. In fact, scholastics "define an *imaginary* space with the real basis of the existence of extensive bod-

---

[47] Ibid., p. 357. The question of whether or not scholastic theology admitted privileged frames of reference was important to Catholic apologists of relativity such as Luis Urbano, who claimed that St. Thomas's position was congenial to that of Einstein. But Urbano was challenged on this point by Paolo Rossi, according to whom Thomistic cosmology has both a privileged movement (the uniform and regular motion of the first sphere) and a privileged position (the center of the universe), neither of which can be harmonized with Einsteinian relativity; *Rivista de Filosofia Neo-Scolastica*, 20 (1928), 131.

ies and which differs much less from Einsteinian space-time than from the absolute space and time of Newton." The fourth objection is diametrically opposed to the last, that relativity says nothing new but simply clothes old concepts in a new language. In denying the validity of this assertion, de Rafael seems to echo Blas Cabrera's assessment of the nature of popular fascination with Einstein's ideas. The truths of relativity "reside in everyone's subconscious, but we need a genius to analyze, specify and define them, and then they seem simple but continue to have extraordinary merit."[48]

Enric de Rafael's straightforward presentation, with no qualifications, was perhaps the most powerful single factor legitimating relativity among those sectors of Catholic opinion who expressed anxiety over the overturn of traditional physics. He was widely quoted in the popular press. For example, in a review of relativistic thought published in a newspaper in Zaragoza, de Rafael was quoted as affirming that there was no contradiction between scholastic philosophy and relativity. The anti-relativism of the day, according to de Rafael, confused *absolute* space and time with *real* space and time. Moreover, relativity was irrelevant to theology, and canons did not have to accept the theory in order to lead souls to heaven.[49] In exposing what might be called ingenuous realism, de Rafael identified the most salient feature of Spanish anti-relativism.

The *Revista Calasancia*, the journal of the Piarist Fathers, published four articles and a book review on relativity in 1922–23 whose stance ranged from favorable to hostile. First to appear was a summary of relativity by the journal's usual science commentator, Benjamín Navarro.[50] He notes, first, that during the past three years relativity had occupied the center of cosmological discourse and that the most select members of the physical and mathematical communities had attended Blas

[48] Ibid., pp. 358–359.
[49] Graciano Silván, "La actualidad científica," *El Noticiero*, March 14, 1923.
[50] Benjamín Navarro, "De relatividad," *Revista Calasancia*, 10 (1922), 38–47.

Cabrera's lecture course. In his own exposition, Navarro says he will follow E. M. Lémeray's *Le Principe de Relativité* (1916). Navarro here is wholly enthusiastic about relativity and strikingly critical of classical mechanics. For him the classical law of inertia was absurd, and he expressed wonderment that Newtonian-Galilean science had lasted as long as it had, given the instability of the edifice raised on such foundations. Aether was unnecessary, invented by physicists in order to hide their ignorance of the mechanisms of action at a distance. In discussing experimental proofs of the General Theory, he does not even enter the usual demurrer on the red shift, stating that Plans had given assurances that this phenomenon was on the way to confirmation. The following year, as if to compensate for Navarro's enthusiasm, the journal published chemist J. M. Goicoechea's critique (see chapter 6), but its readers also sampled an intermediate view in an extended philosophical commentary by Ataulfo Huertas.[51]

In the Augustinian journal, *Ciudad de Dios*, relativity was covered by Teodoro Rodríguez. The journal reprinted an article written during Einstein's visit, published in the Madrid Catholic daily *El Universo* and also serialized his book, *Relatividad, modernismo y matematicismo*.[52] Rodríguez was a throwback to the Darwinian wars of the 1870s and '80s. In his view, relativism, positivism, and evolutionism were all similar, all erroneous, all deniers of absolute truth, and all subsumed under the rubric of Modernism, a movement that saw its salvation in Einstein. For him, there was but one basic question: "Does objective, immutable truth exist, or is truth purely subjective and evolutionary?"[53] Underlying his answer is a profound distrust of modern science. Observation and experimentation are not the only criteria of truth, "as both conscious

---

[51] "La relatividad de Einstein," *Revista Calasancia*, 11 (1923), 241–254ff.

[52] Teodoro Rodríguez, "Relatividad y modernismo," *Ciudad de Dios*, 133 (1923), 293–302. "Relatividad, modernismo y matematicismo," ibid., 135 (1923), 42–67ff.; published as a book under the same title (Barcelona: Unión Librera de Editores, [1924]).

[53] "Relatividad y modernismo," p. 294.

and unconscious positivists assert"; indeed, they are the least certain ones. As proof of this contention, he lists a number of naturalists who had produced absurdities: Huxley's work on Bathybius; Haeckel's on embryos; Darwin's studies on the transformation of species; and a host of paleontologists who endeavored to prove the existence of Tertiary man.[54] Darwinism had, of course, become the model for any revolutionary scientific idea, but Rodríguez does not elaborate on how evolutionism contributes or complements the subversion of the axioms of classical physical science; he takes it as a given. Then he lists a number of mathematical and physical textbook axioms (for example, the sum of the angles of a triangle is equal to two right angles), which, in his view, represent timeless truths, and are not susceptible to alteration: "This firmness of physico-mathematical truths has been the dispair of evolutionists, Bergsonians and modernists."[55] Einstein's axioms can only seem plausible to one who believes that what is true today will be false tomorrow. There is no relative truth; there is only (absolute) truth, or error. Truth has many criteria, not only observation and experimentation, but revelation and others unmentioned.[56] Rodríguez's definition of truth as existing when there is conformity between reality and our concept of what that reality should be seems not only aprioristic but subjectivist as well. The "subjectivism" decried by Rodríguez would seem to be simply any departure from those truths held by some authorized body of dogma (or dogmatists) to be real. Rodríguez not only identifies Einstein as a "Kantian positivist" but also as a Hebrew or Israelite mathematician or physicist.[57] This reference gives him the dubious (and significant) distinction of being virtually the only Spanish commentator on Ein-

[54] Ibid , pp. 296, 299.
[55] Ibid., p. 295.
[56] Ibid., p. 301.
[57] Ibid., p. 302. See also Huertas's review of Rodríguez, in *Revista Calasancia*, 12 (1924), 874. Huertas underscores the ethnic referents and alludes to Rodríguez's waggishness in making such an identification ("su manera un poco zumbona").

175

stein in the 1920s who identified him as Jewish, rather than as German.

Rodríguez's views on "mathematicism" were entirely consistent with what other Spanish anti-relativists claimed—that mathematical formulas constituted a "kind of cuneiform writing, which few mortals know how to read."[58]

His book contains more of the same, and it is clear that he did not really understand the ideas in question. (He had already admitted to not having enough knowledge of "mathematical technique" to discuss Einstein's formulas.)[59] All that was left was the resort to ad hominem attacks and confused discussions. (Father Urbano pointed out that Rodríguez rejected an infinite universe as fantastic and also criticized Einstein's concept of a finite but unlimited universe as opposed to the principle of contradiction.)[60] Rodríguez was considered outlandish by other clerical anti-relativists (no doubt because, like many clerical anti-Darwinians of the previous century, he did not deal with the scientific issues under discussion squarely, but only with invective and innuendo). One shocked reviewer was Eustaquio Ugarte de Ercilla, a Jesuit psychologist and frequent contributor on scientific topics to *Razón y Fe*. Ugarte (b. 1865) was a leading clerical anti-Darwinian and was also opposed to Freud, in part because of the latter's evolutionary bias.[61] The linkage between these three ideas was taken axiomatically by most Catholic apologists.

Ugarte reviewed Rodríguez's volume together with the Spanish translation of a popularization of relativity by the German Jesuit Theodor Wulf.[62] Wulf intended only an exposition and made no judgment of the theory. But Ugarte seemed

[58] Ibid., p. 293.

[59] Ibid., p. 294.

[60] Urbano, *Einstein y Santo Tomás*, p. 11.

[61] See E. Ugarte de Ercilla, "El centenario de Darwin y el cinquecentenario del Darwinismo," *Razón y Fe*, 24 (1909), 167–182, and "La escuela freudiana y la metapsíquica," ibid., 73 (1925), 204–223, especially p. 220.

[62] *La teoría de la relatividad de Einstein*, Joaquín María de Barnola, S.J., trans. (Barcelona: Editorial Científico-Médica, 1925).

pleased to note that, according to Wulf, relativity was only a hypothesis. In contrast, Ugarte notes Rodríguez's unfairness in criticizing a relativist philosophy in Einstein's name when the scientist was clearly dealing only with physical and mathematical principles. Rodríguez's tone of condemnation seemed harsh and exaggerated to Ugarte, who declined to agree either that relativity was deterministic or that it led to "universal doubt."[63]

The *Revista Calasancia* published a far harsher critique by Ataulfo Huertas, who decried the intransigence of Rodríguez. For Huertas, subjectivism was no problem. The relativists held that only the interval was absolute, and on this basis they formulated laws of invariance, that is, laws independent of any particular system of reference. Hence relativity is, indeed, a science of the absolute. Our former absolutes must now be sacrificed on the altar of the "new Moloch," the tensor of gravitation. In an extended commentary, Huertas finds fault, in turn, with Rodríguez's physics, his metaphysics, and his theology. Rodríguez first reveals himself as a Newtonian, believing that the world moves within an unlimited reality, in a three-dimensional continuum formed by absolute space and time. But Rodríguez, unable to harmonize physical concepts of time and space with metaphysical concepts of matter and spirit, concludes that time and space are neither of those, but intermediate between them. Huertas asserts that he cannot follow Rodríguez in this peculiar manner of conceptualizing time and space. Neither did Rodríguez's critique of mathematicism make sense to Huertas. He favored an intangible Euclidean geometry; but, Huertas notes, physicists require a "physical geometry of measurement" and, for this end, Cartesian coordinates were insufficient and thus relativists had recurred to non-Euclidean geometry.[64] When Rodríguez claims that the relativistic pretension to "pass from mathematics to physics as if

[63] E. Ugarte de Ercilla, "Exposición y refutación de la relatividad," *Razón y Fe*, 73 (1925), 426–428.

[64] Ataulfo Huertas, review of Teodoro Rodríguez, *Relatividad, modernismo y matematicismo*, in *Revista Calasancia*, 12 (1924), p. 876.

the latter was subordinate to the former, and all mathematically deduced results are to be considered, for that reason only, to be objective realities, is manifestly false reasoning," Huertas retorts that relativists would agree that such a translation is impossible and that is why one cannot pass directly from Euclidean geometry to physical geometry.[65] Huertas notes that relativists tend to exaggerate their critics's lack of comprehension, but that the ivory tower in which they enclose themselves is not altogether impermeable. He could not himself *calculate* the tensor of universal gravitation, but was able nonetheless to follow others's explanations of the logical consequences of its guiding principles.[66] Rodríguez, not the relativists, appears to have succumbed to the modernist vice of mathematical determinism when he asserts that mathematical truths are more valid than, and even contradictory to, experimental results. Finally, Huertas believed relativistic cosmology to be more philosophically satisfying than the one whose departure was lamented by Rodríguez. Space was no longer amorphous, but well defined, "a universe everywhere endowed with a particular structure," characterized by a tensor of gravitation which "influences each point." Einsteinian space is more concrete than Euclidean space, and more complex.[67]

The critiques of Ugarte and Huertas are significant because they show that by the mid-1920s there was considerable distance between the position of Catholic apologists of the old polemical tradition who railed against the vice of modernism with little more than some threadbare rhetorical clichés and more sophisticated critics who felt obliged to air their doubts within the context of scientific or philosophical discourse.

[65] Ibid., p. 883. Interestingly, although Huertas thought the application of non-Euclidean geometry to problems of physics to be wholly logical, he believed that Einstein's own mathematics in the general theory had been "depasada ya matematicamente" by Elie Cartan and others (ibid., p. 880).

[66] Ibid., p. 877.

[67] Ibid., pp. 879, 881, 883–884. In his appreciation of Einsteinian space, Huertas says he follows the views of F. Renoirte, one of the few defenders of Einstein in Catholic journals.

Huertas was explicit on this point: innumerable Spaniards from every walk of life had given "loyal support to Einstein, both as a person and as a scientist. Fortunately, he was not the subject of political prejudices, which poison everything and which have often contaminated pure scientific research, to which Professor Einstein has dedicated the days and nights of his powerful genius."[68] On the other hand, immoderate Catholic critics (mainly clerical) of Einstein reserved their strongest invective for clerics favorable to the physicist. Thus Pedro de Medio attacked both Luis Urbano and Theodor Wulf as subjectivists and relativity as an imaginary mathematical theory, a reflection not of the real world of our intuition but of another forged in minds excessively enamored of mathematics.[69] In the Catholic daily, *El Debate*, Bruno Ibeas criticized a favorable report by Enrique de Benito, in the same paper, as precipitous.[70] Ibeas was another ingenuous realist who rejected a fourth dimension because the senses reveal to us but three. In the extremely clerical, traditionalist newspaper *El Siglo Futuro*, all columnists, whether clerical or not, opposed Einstein. Two pseudonymous columnists, Clarover and Betibat, made the similar point that Christ is superior to all philosophers, Einstein included, while Father Robles Degano commented that Einstein, who knew no metaphysics, had invented a "swindle" that will not alter scholastic metaphysics, no matter how many mathematical formulas he may use.[71] Such commentary in the Catholic press helps to locate the boundaries of civil discourse

[68] Huertas, "La relatividad de Einstein," *Revista Calasancia*, 11 (1923), 241.

[69] Pedro N. de Medio, "Un nuevo paladín del relativismo," *España y America*, 23 (1925), 97–112; "Resumen de los principales inonvenientes del relativismo," ibid., 24 (1926), 15–27.

[70] Bruno Ibeas, "El einsteinianismo y la venida de Einstein," *El Debate*, March 7, 1923, criticizing Enrique de Benito; "Las conferencias de Einstein. Notas de un oyente profano," in the same newspaper, March 6, 1923. Ibeas was the author of an anti-relativity tract entitled *Las teorías de la relatividad de A. Einstein* (Madrid: N. del Amo, 1922).

[71] Clarover, "La visita de Einstein," *El Siglo Futuro*, March 2, 1923; Betibat, "Saludando a Einstein: Chispazos racionalistas," March 13, 1923; F. Robles Degano, "La relatividad," March 21, 1923.

over science in Spain of the 1920s. Clearly the outer limit had pushed way to the right, where only retrograde extremists refused to participate. It is interesting to note that, when civil discourse began to break down in the 1930s, the nineteenth-century invective style became respectable again among more moderate Catholics, to be genuinely reborn as a hallmark of Francoist hostility toward modern science in the 1940s.

THE COMPARATIVE RECEPTION OF RELATIVITY

In a recent study of the reception of relativity in two contrasting political cultures (England and the Soviet Union), Loren Graham notes that interpretations of relativity may be in accord with extra-scientific attitudes of sectors of the society in question. Thus Graham concludes that Sir Arthur Eddington, a Quaker, sought to reconcile science with religion (for which there was ample tradition in British thought) and to reassure the public that Einstein's theory did not entail the abandonment of absolute values or standards. In the Soviet Union, the physicist V. A. Fock sought to make of relativity a defense of materialism.[72] In each case the ambient culture provided the scientist with a context for a distinctive reading of Einstein's theory.

Looking beyond the reception of relativity by individual scientists, one can also detect patterns of reception determined in large part by the specific profiles of national physics communities. In a survey of the comparative reception of special relativity until around 1911, Stanley Goldberg traced distinctive patterns in France, Germany, the United States, and England. In France, there was virtually no reaction to Einstein before 1910, in large measure due to the influence of Henri Poincaré, who believed that Einstein's work constituted a small and not very significant part of a larger theory developed by Lorentz and himself. As a result, there was scant discussion of the Special Theory until after Poincaré's death in 1911. (Even so,

[72] Graham, "The Reception of Einstein's Ideas," pp. 107–136.

Michel Biezunski has observed, the majority of French physicists avoided coming to grips with the physical meaning of relativity until after the *Second* World War.)

In Germany, by contrast, there was an intense discussion from the start. Although many German physicists opposed Einstein, only in Germany did his opponents understand the theory. The fact that anti-relativists took Einstein seriously assured the close examination, criticism, and, ultimately, the acceptance of relativity. In the United States, not only was relativity not taken seriously, there were few who grasped its meaning. There was a tendency to ridicule it as impractical and absurd. The appeal to practicality meant that the incomprehensibility of the theory was stressed, and W. F. Magie claimed in an important address to American physicists that a theory unintelligible to the common man could not be true. The first serious discussion, by G. N. Lewis and R. C. Tolman in 1909, attempted to show that the theory was indeed practical and based on experimental evidence. The British reaction was not so much a discussion of the merits of relativity as a reaction against its attack on the aether, which British physicists (following Lord Kelvin and Oliver Lodge) regarded as a mechanical object absolutely necessary to the correct explanation of electromagnetic phenomena.[73]

In considering the Spanish reception of relativity in comparative focus, we must first ask what there was in the common experience of so many men of the Catholic right (Terradas, Plans, Rafael, Luis Urbano) which made relativity so attractive to them. First, there was a social factor: to accept relativity was a way to embrace modern science without appearing to oppose traditional Catholic values (as had not been possible in the case of Darwinism). Second, there appears to have been a

[73] Stanley Goldberg, "In Defense of Ether: The British Response to Einstein's Special Theory of Relativity, 1905–1911," *Historical Studies in the Physical Sciences*, 2 (1970), 89–125, and "The Assimilation of Scientific Revolutions: The Case of Special Relativity in America," typescript (cited by permission); Michel Biezunski, "La diffusion de la théorie de la relativité en France," doctoral dissertation, University of Paris, 1981, p. 292.

contextual factor relating to the Catholic education of such persons. The Neo-Scholastic tradition had opposed Kantian notions of time and space in that Kant—decried by Rafael as the patriarch of modernist philosophy—had identified space and time as a priori categories of the human intellect. Ingenuous Catholic critics of relativity, confusing *absolute* with *real* space and time, assumed Einstein's position supported Kant's. "Nothing is more false," Rafael declared,

> there is nothing in common between the ideas of the patriarch of modernist philosophy and those of Einstein, but rather the impossibility of *directly* perceiving space and time as absolute in themselves, an impossibility which is not contrary to scholastic philosophy which sustained, against Newton, that space and time, as we conceive them, are not necessary, eternal and immediate entities, independent of God, however great his immensity (*sensorium*) and eternity (as Leibnitz reproved) might be, but simply entities of reason or the ideal, with their real basis in the existence of extensive, permanent and successive beings. This not only is not in contradiction with modern theories, insofar as these are the clear fruit of experience and represent a true advance, but they are sufficiently more in conformity with them than what has generally been admitted up to now.[74]

Einstein, in other words, had demonstrated the correct relationship between mind and reality.

Although Rafael was a Jesuit and Esteve Terradas a layman, the latter was just as avid a believer, according to the former's necrological sketch.[75] Terradas had a rather carefully worked-out theological justification for general relativity, stressing

[74] Enrique de Rafael, "De relatividad," *Ibérica*, 15 (1921), 91, n. 1. Emphasis is that of de Rafael. In his course notes, however, he indicates that both Einstein and Weyl had tied their theories to subjective a priori Kantian forms; *Anales ICAI*, 1 (1922), 187.

[75] Enrique de Rafael, "Juventud de Terradas," p. 10, where he describes saying the rosary with Terradas.

that its philosophical ramifications proceed directly from absolute differential calculus, whose "philosophical repercussions were totally unforeseen." If space and time are not real entities, but only fictitious ones "whose real basis is in the existence of extensive and changeable bodies" as many Thomist philosophers (for example, Suárez) have held, then there is neither a metaphysical nor a physical objection to admitting that the real properties of both depend on those bodies. If one further admits that there are as many true times as there are real movements, then "all the difficulties of a philosophical order which might be raised against the application of absolute differential calculus to the study of movement disappear as if by magic." These ideas, he states, are characteristic of the objectivist philosophers, especially the scholastics who follow Saint Thomas and Suárez:

> To admit the application of absolute differential calculus the subjectivists must overcome two difficulties, either because, like Descartes, they locate the essence of bodies in their extensions, or because they deny the absolute reality of extensive beings, as did Berkeley, Hume and Kant: 1) the explanation of the cause of variety in space-time and 2) the non-arbitrariness of the predicted results.[76]

Turning to cosmology, Terradas compares Einstein's notion of space as finite and closed—"but time is infinite, and the universe, cylindrical"—with de Sitter's conception, wherein not only space but also time was finite. He concludes, "Evidently Einstein's solution (possibly without his realizing it) is in greater conformity with the idea of eternity in Catholic philosophy."[77]

In embracing relativity, therefore, conservative Catholic scientists could do away with Newtonian absolutes by associating them with Kantian a priori categories while also opposing modernist philosophy at the same time. This was not a

[76] Terradas, "Relatividad," *Enciclopedia Ilustrada Universal*, L, 458–459.
[77] Ibid., p. 459.

point that was explicitly stressed, but it helps to locate the Catholic relativists in the intellectual spectrum between anti-Kantian traditionalists and secular modernists. Conservative relativists, Rafael in particular, took constant pains to reassure the Catholic public that there was no conflict between its values and Einstein's theories. But he did not belabor the point, probably because of the inability of anti-relativists to understand the material.

There was a significant fit between the conservative politics of the Catholic relativists and the official ideology of the dictatorship of Miguel Primo de Rivera, who seized power in September 1923, scant months after Einstein's departure. Over the next several years conservative scientists were in the ascendancy. Terradas, for example, gained the chair of Differential Equations in Madrid in 1928 and received other honors from the regime (as a result of which he was removed from the chair by the vote of liberal professors in 1931–32). In a series of articles written in 1926–27, José Pemartín, one of the regime's ideologues, joined in the common right-wing chorus of blanket condemnation of nineteenth-century philosophies. The relativism of the twentieth century, he argued, invoking Einstein's name, was preferable both in science and in politics to the absolute and categorical philosphies of the preceding century. In the revolt against the nineteenth century Einstein served the right's purposes.[78]

Finally, there are two generalizations that apply to all Spanish relativists, not just the conservative Catholic ones. First, there was no Maxwellian tradition in Spain to have impeded

[78] On Terradas, see Antoni Roca and J. M. Sánchez Ron, "La vuelta de Esteban Terradas a España," *Llull*, 6 (1983), 105–110. On Pemartín, see Shlomo Ben-Ami, *Fascism from Above: The Dictatorship of Primo de Rivera in Spain, 1923–1930* (Oxford: Clarendon Press, 1913), p. 183. The comparison with fascist Italy in the same period is interesting; cf. Barbara Reeves's account of Mussolini's favorable disposition toward relativity, "Einstein Politicized: The Early Acceptance of Relativity in Italy," in T. F. Glick, ed., *The Comparative Reception of Relativity* (Dordrecht: D Reidel, 1987), pp. 189–229.

the reception of relativity on dogmatic physical grounds. Indeed, as Antonio Lafuente has observed, "In Spain, the concerns of pre-relativistic physics were entirely unknown."[79] We can extend the argument by noting that conservative scientists of the late nineteenth century had opposed, as a matter of conviction, any kind of mechanistic world view, opposition to mechanicism having been a keynote in the Catholic rejection of natural selection. Two subissues, both identified by Lafuente, are related to this: first, there was no real debate over relativity among Spanish scientists; and second, dogmatic Newtonians did not on the whole come forward but remained silent instead.[80] A critical benchmark of scientific maturity is the capacity to contribute to research-front science, if only, at the minimum level, by having acquired the ability to offer constructive criticisms. One of the constraints that smallness imposes upon a scientific community is that the lack of large numbers of qualified workers makes it difficult for informed consensus over theoretical issues to arise, as it would in a scientific community where workers in any given field number in the dozens or hundreds instead of units. Gregorio Marañón noted this problem in 1922 in analyzing the way in which theoretical consensus was reached (in this case, imposed) in the emerging subdiscipline of endocrinology:

> In societies which are ahead of us scientifically, these strident polemics, in one way or another, take place on the fringes of a nucleus of positive density, constituted by the abundance of scrupulously collected facts, a true "center of gravity" of theory, which tends to even out the oscillations which proponents and adversaries imprint upon it.

[79] Lafuente, "La relatividad y Einstein in España," p. 588.

[80] Ibid., p. 589. Félix Apráiz, an electrical engineer who wrote an anti-relativity tract in 1921 (see Bibliography), was one of the few outspoken proponents of the mechanical theory of aether in Spain at this time. See his article, "La interpretación mecánica de los fenómenos eléctricos y magnéticos," Asociación Española para el Progreso de las Ciencias, Congreso de Oporto, tomo V, Ciencias fisicoquímicas (Madrid, 1922), pp. 73–108.

But we lack that nucleus of our own experimentation; that center of gravity is either lacking or very weak.[81]

Since subdisciplinary or disciplinary groups were so small, there was little or no competition for posts or prestige among their members, and leaders of research groups were able to impose their own theoretical perspectives without having to risk criticism. In the particular case of the reception of relativity, the control of key institutions and research groups in physics and mathematics by persons favorable to Einstein predetermined the range, depth, and rate of reception, not only in those fields but in the scientific community at large. In this process, networks of communication were all-important. The pattern of reception would have been strikingly different had Spanish mathematicians not had such close ties with their Italian counterparts, who had made a direct contribution to relativity theory, or had they elaborated ties with some other national group of mathematicians, less interested, or disinterested, in the problem, instead.

In the 1920s Spanish scientists in a number of disciplines began to have the feeling that they were active participants in scientific advance. Just as isolation and time lag had been leitmotifs of scientific stagnation, active communication with the research front was the clearest hallmark of the new state of Spanish science. Effective communication was not only a fact; it quickly became a symbol, a stamp of success, a corroboration of the leap to scientific maturity and excellence. For J. M. Plans the closing of the communication gap was the best proof of Spain's scientific maturity, and, in turn, the reception of relativity was proof that the gap had indeed been closed. Relativity had broken upon the world scientific scene at the precise moment when Spanish science was reaching maturity, and its reception became a powerful legitimation of that maturity. Plans made the point repeatedly. In his 1924 speech of acceptance of membership in the Academy of Sciences, he noted that

[81] Gregorio Marañón, *Problemas actuales de la doctrina de las secreciones internas* (Madrid: Ruiz, 1922), p. 9.

"Precisely in the theory of relativity and studies related to it do we count on a goodly number of enlightened Spanish cultivators; and we can affirm that the successive advances of the work of Einstein and his followers have reached us without delay."[82]

During Einstein's visit, Francisco Vera hailed Plans as a living refutation of those who thought that "in Spain scientific news is received with a lamentable lag."[83] Plans made the same point in his 1926 sketch of the recent history of Spanish mathematics. He attributed the fact that "things no longer arrive with delay" in particular to the efforts of Esteve Terradas, who played the leading role in organizing courses by Weyl, Levi-Civita, Einstein, and others. "Thanks to Terradas we are in contact with great foreign scholars and it can be said that we keep abreast of world science, in the disciplines just mentioned."[84] The scientific reception of relativity cannot be separated from the information network that informed it and which assured what in the Spanish context was a precocious assimilation of a new and difficult set of ideas.

[82] J. M. Plans, *Algunas consideraciones sobre los espacios de Weyl y Eddington y los últimos trabajos de Einstein* (Madrid: Real Academia de Ciencias Exactas, 1924), p. 42.

[83] Vera, "El doctorado 'honoris causa' y otras grandes menudencias," *El Liberal*, March 16, 1923.

[84] Plans, "Las matemáticas en España en los últimos cincuenta años," p. 173. From Plans's letters to Levi-Civita it is clear that, because of his ill health, he depended on Terradas to attend foreign congresses and to keep up contacts there.

SIX

Relativity and Spanish Engineers:
The Scientific Middle Class

SOCIOLOGY OF SPANISH SCIENCE

If the Spanish relativists pursued little or no research in relativistic physics or mathematics but limited their participation to "high-brow" syntheses, then we must ask at what audience these works were directed. The syntheses produced by Cabrera, Plans, and Terradas were aimed at integrating the theory of relativity with the corpus of mathematical physics available to Spanish readers. But at what level? These works, even Terradas's article for the popular Espasa-Calpe encyclopedia, were not for "popular" consumption. They were directed at persons with scientific education. But research scientists or those who held chairs of mathematics or physics hardly constituted an audience large enough to merit such an investment of energy. Rather, these syntheses were directed toward a specific audience, one both with mathematical education and one which had displayed consistent interest in relativity: the engineering community.

Such persons, who were instrumental in the reception of relativity in Spain, formed a distinctive and well-defined group of "consumers" of science in Spain, a group I call the "scientific middle class." This group is composed of persons with scientific education who may be interested in scientific research but who, in general, do not pursue it themselves. The leading sectors of the class in Spain of the late nineteenth and early twentieth centuries were physicians, engineers, pharmacists, and secondary-school science teachers. Less numerous components were scientifically educated priests, science writers, and technicians of one kind or another. It was the scientific middle class that provided the rank-and-file membership of all Spanish scientific societies, constituted the bulk of the readership of research journals, and was the main political constituency for science in Spain. Put another way, given the tiny size of the research-science establishment, science as an enterprise would not have worked at all were it not for the participation of this group.

Membership lists of Spanish scientific societies in the teens and twenties reveal a standard division between scientists and scientific middle-class members on an order of 40 to 60 percent, respectively. This was true of small societies such as the Spanish Society of Physics and Chemistry (346 members in 1920), as well as large ones such as the Spanish Association for the Progress of Science (770 members in 1912). Table 2 breaks down the membership of the former society in 1920 by category. Group I includes those identified as chairholders in faculties of science, pharmacy, or medicine; astronomers with titular positions in observatories; and those listing themselves as doctors of sciences, pharmacy, or medicine. Group II includes professors at technical or secondary schools, practical technicians, engineers, telegraph officials, factory owners, clerics, and those listed as holders of a licenciate in sciences. In addition, fifty-six individuals were listed with no title and are not included in the percentage breakdown.

There are obvious difficulties in categorizing society members in this fashion. There were chairholders who did no research and secondary-school science teachers and clerics who did. But, on the whole, such anomalies cancel each other out. I have included persons listing themselves as holders of the doctorate in category I because a number of these persons are identifiable as future chairholders or researchers. If this group were transferred to category II (Table 3, adjusted figures), the "middle class" would outnumber the "scientists" by 75 percent to 25 percent, and this figure may well provide a more accurate indication of the state of science in Spain at this time. The membership of the Spanish Association for the Progress of

189

TABLE 2
SPANISH SOCIETY OF PHYSICS AND CHEMISTRY
(MEMBERSHIP IN 1920, BY CATEGORY)

SCIENTISTS

Primary Disciplines		*Doctorates*	
Observatory astronomer	5	Doctor of Science	25
Professor, Faculty of Sciences	51	Doctor of Pharmacy	17
Professor, Faculty of Pharmacy	17	Doctor of Medicine	3
Professor, Faculty of Medicine	2		45
	75		
Total			120

SCIENTIFIC MIDDLE CLASS

Professor or director of an institute	27	Telegraph official	1
Professor, Industrial School	11	Factory officer	2
Professor, Commercial School	1	Cleric	11
Professor, Normal School	3	Licenciate in sciences	38
		Licenciate in pharmacy	3
Pharmacist, chemist, etc.	34		
Engineer			
Mining	14		
Civil	8		
Industrial	5		
Other	12		
Total			170

NO PROFESSION INDICATED			56

SOURCE: *Anales de la Sociedad Española de Física y Química*, 18 (1920), 277-291.

TABLE 3
SCIENTIFIC MIDDLE CLASS (SELECTED SPANISH
SCIENTIFIC SOCIETIES)

	Spanish Society of Physics and Chemistry 1920		Spanish Society of Physics and Chemistry (adjusted) 1920		Spanish Association for Progress of Science 1912	
	No.	%	No.	%	No.	%
Scientists	120	41.4	75	25.9	239	42.7
Scientific Middle Class	170	58.6	215	74.1	321	57.3
Total	290	100	290	100	321	100

SOURCE: Table 2 and "Lista de los miembros que componen la Asociación Española para el Progreso de las Ciencias en 31 de julio de 1912."

Science in 1912 breaks down in a similar fashion (Table 3, column 3). Less information was provided in its membership list, making it more difficult to score; but application of the same criteria produced comparable figures.

In summary, only about one-quarter of the members of Spanish scientific societies were likely to hold major positions in academic science and can be reasonably presumed to have kept abreast of current developments in their fields and to have maintained a respectable level of research activity. The other three-quarters were involved in practical pursuits generally unrelated to pure research.

The scientific middle class played a prominent role in the reception of new scientific ideas in the late nineteenth and early twentieth centuries. Perhaps 90 percent of the Spanish literature on Darwinism in the period 1868–1900 was produced by members of this class. Indeed, they led researchers in this instance (stimulating demand for evolution-oriented research, insisting that the issue be discussed publicly) because in many cases those who held chairs of natural history were afraid to risk their positions by being openly favorable to Darwin. The Darwinian wars were fought by secondary-school teachers, physicians, and social scientists, not by biologists.[1]

The same was largely true of the reception of relativity. The rank and file of the "consumers" of Einstein's theories in Spain of the 1920s were engineers. They were the readers of the works of Plans, Cabrera, and Terradas and they predominated at their lectures as well as at those of Einstein; it was they who were responsible for a large portion of the popular scientific literature.

For the purposes of the present discussion, I have extended my basic definition of "scientific middle class" to include research scientists outside of the three receiving disciplines. The level of discussion of relativity by Spanish chemists, for example, had more in common with the engineering discussion than with the mathematical or physical.

MATHEMATICS AND ENGINEERING EDUCATION

The involvement of engineering schools in higher mathematics should not be surprising. In the first place, there was a long tradition, well antedating the nineteenth century, that united the careers of mathematician and military engineer. The leading figure of nineteenth-century Spanish mathematics, José de Echegaray, was a highway engineer and also, from 1855, professor of mathematics at the Special School of Highway Engineers. Later he became professor of mathematical physics at the University of Madrid. Mathematicians of the early twentieth century followed the same pattern. There were very few

---

[1] Thomas F. Glick, "Spain," in T. F. Glick, *The Comparative Reception of Darwinism* (Austin: University of Texas Press, 1974), pp. 307–345.

jobs for higher mathematicians in Spain. Since most of the available chairs were in engineering schools, those who wished to compete had to become engineers as well, inasmuch as one could not, by statute, teach in an engineering school without being a member of the corps in question. This regulation worked in two ways. Either it forced good mathematicians, like Rodríguez Bachiller, to obtain an engineering degree in order to enhance career prospects as a mathematician, or it meant that courses would end up being taught not by a professor, but by a simple instructor (*encargado de curso*) with unstable tenure.[2]

Right before World War I an ample debate took place among European engineers on the role of mathematics in engineering education. The cry "Enough Mathematics" was raised by prominent engineering pedagogues in Germany and England, and the issue was fully debated at the London Congress on Engineering Education in June 1911. One Spaniard who commented on this polemic before the war was Enrique Jiménez, Jesuit and professor of calculus at the Catholic Institute (I.C.A.I.).[3] Jiménez was in general agreement with those who criticized teaching overly abstract mathematics to engineering students: What need was there for an engineer to be able to expound the intricacies of Riemannian space? (This was an interesting allusion, because José A. Pérez del Pulgar, the Institute's director, had studied non-Euclidean geometry in Germany and had written extensively on the subject.) On the other hand, Jiménez noted, Charles P. Steinmetz had devoted an entire chapter of his volume on alternating currents to a study of imaginary quantities, "observing that numerical calculation of electrical machines is impossible without the aid of this theory." Jiménez concluded that one wanted pure mathematicians

---

[2] See Anon., "La enseñanza de la ingeniería," *Madrid Científico*, 33 (1926), 95–97. Rodríguez Bachiller took an engineering degree at the Escuela de Caminos.

[3] Enrique Jiménez, "Las matemáticas y la ingeniería," *Electricidad y Mecánica* (Valencia), 10, no. 4 (April 1914), 4–10 (reprinted from *Ibérica*).

to teach the basic courses, including analytical geometry and infinitesimal calculus. The I.C.A.I., with its own textbook, was teaching calculus at the proper level, he believed, which was equivalent to that in Germany or France. He also noted that Spanish secondary education in mathematics lagged well behind the English standard: "Only with difficulty could even one in a hundred of our youths satisfactorily resolve the problems posed to mathematics students in [English] secondary schools." It is interesting to note that Spain was the only country in Europe (and still remained so in the 1930s) where candidates for entrance to engineering schools were required to study mathematics privately in order to meet admission standards, so poor was the quality of secondary education.[4]

During the 1920s Julio Rey Pastor was the leading spokesman among Spanish mathematicians for the inclusion of higher mathematics in engineering curricula. In the middle years of the decade he polemicized with Pedro González Quijano (also a relativist), who wanted more practice and less theory in the education of engineers. Rey Pastor argued that the "Enough Mathematics" program had proven a failure and that modern mathematics, abstract yet appropriate to the ends of engineering education, should be taught in the schools "by people who know it well, whether they are engineers or not."[5] He endeavored to make a distinction between "theoretical" and "professional" engineers, claiming that the former required more specialized mathematical training. But those in the engineering schools who favored extensive mathematical training favored it for all students.[6] In fact, Rey Pastor had the jobs of his students to protect; the pure mathematicians won the day and continued to win appointments to chairs in the engineering schools (a similar outcome held for physicists as well).

[4] Paulino Castells, "La preparación matemática en la carerra de ingeniero," *Memorias de la Academia de Ciencias y Artes de Barcelona*, 22 (1932), 475.

[5] Julio Rey Pastor, "Sobre enseñanza técnica y espíritu de cuerpo," *Madrid Científico*, 32 (1925), 337–340.

[6] Castells, "Preparación matemática," p. 469.

ENGINEERING COURSES AND TEXTBOOKS

The presence of relativity in engineering-school curricula both explains the great interest that engineers graduating between 1918 and 1925 had in the topic and also their ability to discuss it. Best documented is the course entitled "Notions of Classical and Relativistic Mechanics" that Enric de Rafael gave at the Catholic Institute of Arts and Industries in the academic year 1921–22. Rafael noted that all engineering schools offered a standard set of courses in the first year, but that upper-level courses were elective. In deciding whether to offer an elective course on some classical problem or on some current topic, he chose the latter because "living science is more pleasing and, in a certain sense, more instructive than dead science." In any case, in order to teach modern science one had to review the bases of the classical as well. For Rafael, a mathematician, the decision to teach relativity was even more severe because of the deficient physics education of his students: ". . . inasmuch as my students did not know classical mechanics, I had to begin by explaining its essentials to them." Rafael, of course, betrays his personal predilection by identifying nonmodern science as "dead." Nevertheless, in view of his students' lack of preparation, the first third of the course was devoted to a discussion of Newton's three laws and their analytical exposition, together with a "light philosophical discussion of them." The second trimester would be devoted to the Special Theory and the third to the General.[7]

[7] Enrique de Rafael, "Nociones de Mecánica clásica y relativista," *Anales de la Asociación de Ingenieros del Instituto Católico de Artes e Industrias,* hereinafter cited as *Anales ICAI,* 1 (1922), 20. The recommended reading for the course included Born's synthesis *Die Relativitätstheorie Einsteins*; Chwolson, *Traité de physique*; the anthological volume *Das Relativitatsprinzip,* by Lorentz, Einstein, and Minkowski; and Plans, *Nociones fundamentales de le Mecánica relativista.* For the General Theory, Rafael recommended Eddington's *Report on the Relativity Theory of Gravitation,* Weyl's *Raum, Zeit, Materie,* and recent lectures by Terradas, Cabrera, and Levi-Civita (see chapter 1) as soon as published. De Rafael notes that he had attended most of the these lectures. Cf. his postcard to Levi-Civita, May 12, 1921, in which he states: "Quanto mi sia stato gradito l'avere

Rafael believed that some background in the philosophy of science was also necessary in view of the philosophical implications of relativity, especially for Catholics. Here, he detected a problem of interpretation:

> The philosophy of science is extremely difficult, because professional philosophers are, in general, not only ignorant, but also incapable of correcting their ignorance of scientific matters owing to their lack of education and their biases of school; on the other hand, scientists are easily led astray in philosophical matters, of which they are ignorant not only of the terminology, but, very often, of the formal significance and analytical linkages (even though twentieth-century science, not only mathematics, but physics, has evolved in this direction with great force and quite some success), which those not habituated to philosophical disputation are unaware of.[8]

For general philosophical background, he recommended the works of the nineteenth-century Spanish philosopher Balmes and, for philosophy of science, volumes by Pierre Duhem and Henri Poincaré.

Rafael's elective course, devoted wholly to relativity, was untypical. More common were short courses, such as that given at the Escuela de Caminos in the spring of 1923 by Pedro Lucía Ordóñez of the class of 1918. The eight lessons covered the principle of relativity in classical physics, the Special Theory, tensorial algebra, relativistic mechanics, the General Theory, tensorial analysis, and finally Einstein's theory of gravi-

---

presso parte alle sue sceltissime lezzione" (Tullio Levi-Civita Papers, Duplicate Archive, Robert A. Millikan Memorial Library, California Institute of Technology, Pasadena). Four popularizing books were also recommended to the students: Lorente de Nó's translation of Einstein; Erwin Freundlich, *Los fundamentos de la Teoría de la gravitación de Einstein*, J. M. Plans, trans. (Madrid-Barcelona: Calpe, 1920), with a prologue by Albert Einstein; and García Morente's translations of volumes by Schlick and Born.

[8] De Rafael, "Nociones de Mecánica clásica y relativista," *Anales ICAI*, 1 (1922), 21.

tation and its consequences. The lectures were attended by "an abundant group of civil engineers and of students and professors of the School, who departed well satisfied by them."[9]

Relativity was diffused in engineering schools less in monographic courses and more by its inclusion in the normal curriculum, particularly in courses on rational mechanics. At the School of Industrial Engineers of Barcelona, Ferran Tallada taught rational mechanics from a relativistic perspective and so did his counterpart in Madrid, Carlos Mataix Aracil. Mataix wrote a textbook, *Mecánica racional*, whose first edition contained no discussion of relativity. In 1923 he published a twenty-three-page supplement on the new topic, introducing students to relativity in classical mechanics, the Special Theory, the problem of simultaneity, the universe of Minkowski, the Lorentz transformations, local time, longitudinal contraction, mass and energy, and general relativity (in that order). The supplement ends with five questions or problems dealing with special relativity (in particular, those of its results, such as the contraction of length and dilation of time, which do not appear in classical mechanics), the answers to which require, at the most, an elementary knowledge of the calculus.[10]

Similarly, the professor of applied optics at the School of Industrial Engineers (Barcelona), J. Mañas y Bonví, included an ample discussion of relativity in a supplement to his textbook, *Optica aplicada*, a volume that developed out of Mañas's course on "Applications of Light," first given in 1910. In preparing his course notes for publication he was in close communication with Terradas, who read the manuscript and provided Mañas with "whatever notes he deemed of interest for this type of book." It is interesting that when the first edition was

[9] *Revista de Obras Públicas,* 71 (1923), 32; *Anuario de la Escuela Especial de Ingenieros de Caminos, Curso de 1922–1923* (Madrid, 1924), p. 60. In the same vein was a lecture on relativity by the engineer Cantos Abad at the Industrial Engineering School of Madrid on April 27, 1923; see *Madrid Científico,* 30 (1923), 151.

[10] Carlos Mataix Aracil, *Primeras nociónes de mecánica relativista* (Madrid: Koehler, 1923).

ready for publication in November 1913, Terradas did not feel that relativity need be discussed. The second edition, of 1915, contained a number of supplements, including one on relativity, and apparently reflected Terradas's conviction that special relativity had by then become germane to much of contemporary physics. Mañas's supplement is a completely adequate discussion of the Special Theory, set in the context of the history of aether theory and stating that Einstein had denied the existence of the aether "as a fixed medium for the transmission of electromagnetic actions."[11]

In Madrid, relativistic mathematicians trained in Rey Pastor's Mathematical Seminar dispersed to technical schools: Puig Adam to the Catholic Institute, Fernando Peña to the Special School of Forestry Engineers, and, at the end of the decade, Rodríguez Bachiller to Emilio Herrera's Advanced School of Aerotechnics. All retained their early interest in relativity, and it can be presumed that they communicated it to their students throughout the decade. In 1932 Peña returned to the Mathematical Laboratory to give a short course on relativity. Peña, moreover, was the author of a chapter surveying special and general relativity included in the Spanish translation of William Watson's *Curso de Física*, a popular textbook during the 1920s.[12] At the Naval School, relativity appears to have cropped up in mathematics courses by 1920, although in 1922 critics of the school's mathematics program complained that current instruction was inadequate to "the new and already popularized doctrines of Einstein, through inability to form a clear idea of the four dimensions of space-time."[13]

[11] J. Mañas y Bonví, *Optica aplicada* (Barcelona, 1915), "Suplemento a la página 427," p. viii.

[12] *Revista Matemática Hispanoamericana*, 2nd series, 7 (1932), 150–151; Peña, "Bosquejo de la teoría de la Relatividad," in William Watson, *Curso de Física* (Barcelona: Labor, 1925), pp. 867–886.

[13] A. Azarola, "El estudio de las Matemáticas en la Escuela Naval Militar," *Revista General de Marina*, 86 (1920), 441–453; L. de Saralegui, "Nueva orientación de los estudios en las carreras de la Armada," *ibid.*, 90 (1922), 357–362. See comment by J. J. Tato Puigcerver, "Una nota sobre la *Revista General de*

Relativity was, therefore, not only a lively topic of discussion among professors of engineering and their students in the early 1920s, but it insinuated itself into the texture of engineering education in a variety of ways, relating, I think, to the reigning ambivalence regarding the proper level of mathematical training. For one thing, knowledge of relativity conferred prestige. Tomás Rodríguez Bachiller recalled that he was once unable to complete on time an assignment for the course on bridges at the Escuela de Caminos, whose professor was Santos María de la Puente. La Puente excused him while, at the same time, he declined to repeat the favor for another student in a similar predicament. When the latter demanded an explanation for the professor's unfair decision, La Puente replied: "Because Sr. Bachiller knows relativity and you don't!"[14] In intercourse between the "Two Cultures," engineers, with considerable justice, believed themselves to be the proper intermediaries in the explanation of pure science to the reading public. One author complained of glib writers in the popular press who spoke of absolute differential calculus and relativity with greater surety than Levi-Civita or Einstein. Rey Pastor, it seems, had accused Spanish intellectuals of cultivating form rather than substance and had thus raised the polemical issue of the dividing line between educated persons and the scientific community (*inteligentes y técnicos*).[15]

The encounter of Spanish engineers with relativity seems roughly comparable to that of their compatriots in England. At the University of Bristol, according to a recent memoir by P.A.M. Dirac, engineering students had little direct exposure to (general) relativity except for a series of lectures by a professor more interested in the philosophical than in the physical side of relativity. Direct access to the General Theory came only in 1923, with the publication of Eddington's volume *The*

Marina y la recepción de la relatividad en España," *Llull: Boletín de la Sociedad Española de Historia de las Ciencias*, 3.1 (October 1980), 137–138.

[14] Tomás Rodríguez Bachiller, interview, Madrid, April 10, 1980.

[15] Federico de la Fuente, "La ciencia aplicada y la ciencia pura," *Madrid Científico*, 32 (1925), 321–322.

Mathematical Theory of Relativity, on the basis of which, according to Dirac, "it was possible for people who had a knowledge of the calculus, people such as engineering students, to check the work and study it in detail. The going was pretty tough. It was a harder kind of mathematics than we had been used to in our engineering training, but still it was possible to master the theory. That was how I got to know about relativity in an accurate way."[16] In Spain, Plans's 1924 manual on absolute differential calculus assumed a role analogous to that played by Eddington's book in England.

POPULARIZATION BY AND FOR ENGINEERS

In the next chapter, I document the widespread perception that any attempts to popularize relativity had failed. But that conclusion was a specific reflection of Einstein's visit, when the net was cast too broadly, and a negative or defensive judgment was reached by intellectuals with no scientific background. In fact, popularization by engineers or engineer/mathematicians was highly successful. Manuel Velasco de Pando's lecture on relativity were, according to all reports, electrifying. Velasco de Pando (b. 1888) was a Sevillian industrialist who had received his engineering degree from the School of Industrial Engineers in Bilbao in 1910. Throughout the 1920s he lectured on relativity in Seville and Bilbao, and the impact of his presentation on one listener, Antonio Fernández Barreto, the military governor of Seville, is an explicit description of the process by which demand for popularization occurs, an eloquent testimony to Velasco's prowess as a popularizer, and also a reflection of Spanish awareness of Einstein's activities prior to his trip to Spain:

> It happened when Einstein went to Paris to discuss his theory of relativity with the French scholars at the Col-

---

[16] P.A.M. Dirac, "The Early Years of Relativity," in Gerald Holton and Yehuda Elkana, eds., *Albert Einstein: Historical and Cultural Perspectives* (Princeton, N.J.: Princeton University Press, 1982), p. 82.

lège de France. I believe Painlevé was his principal antag-
onist. Einstein triumphed over all his opponents. Well
then, I, who read all about this in the popular press, conceived the intent to find out about Einstein's theory, and I
read Normand and Abbé Moreux, but I confess that,
from that time on, I doubted I knew enough mathematics, because I was unable to understand in what that doctrine consisted. For that reason, it was a revelation to me
to hear Sr. Velasco deliver, first in the Academy of Letters
and later in the Athenaeum of Seville, in the Farmers' Circle, a number of lectures on Einstein's theory of relativity,
which were attended by such a throng of people that, especially in the Athenaeum, the lecture hall and adjoining
rooms were filled and there were even listeners on the
stairway. . . . From that time on I admired Sr. Velasco de
Pando, first as a mathematician of stature, for having assimilated such difficult theories, and then as a popularizer
for having known how to expound them in a form accessible to all.[17]

Velasco inaugurated the lecture series of the Association of Industrial Engineers of Bilbao in 1924 with two lectures on relativity, subsequently published.[18] Since Velasco had been
elected to the Academy of Exact Sciences as a mathematician,
his lectures presumably bore the imprint of his mathematical
sophistication.

Other engineers popularized on more or less the same level
as those without scientific education. An example is Salvador
Corbella Alvarez's *La teoría de Einstein al alcance de todos* (Barcelona, 1921). Corbella, a civil engineer, stated that his approach, based on simple examples that any reader "with com-

---

[17] *Palabras pronunciadas por el Excmo. Sr. Gobernador Militar de Sevilla, D. Antonio Fernández Barreto* (Seville, Camara de Comercio, Industria y Navegación, 1929), p. 5; cited by Antonio Lafuente García, *Introducción de la relatividad
especial en España*, Memoria de Licenciatura, Universidad de Barcelona, 1978.
On Velasco de Pando, see *Enciclopedia Ilustrada Universal*, LXVII, 613.

[18] Manuel Velasco de Pando, *Relatividad general y restringida* (Bilbao, 1924;
2nd ed., 1926). I have been unable to consult this text.

mon knowledge" could grasp, could also be appreciated by persons with background in higher mathematics but who wished a quick introduction to Einstein based on simple and familiar illustrations of the Special Theory, a short discussion on the similarities of the effects of gravitation and acceleration, and an exposition of the significance of the 1919 eclipse results. On a similar level was a lecture by José Ochoa y Benjumea, another civil engineer, who sought to explicate relativity "without the aid of mathematics" at the Fomento de Trabajo of Villanueva y Geltrú on March 30, 1924.[19] Ochoa's lecture is particularly successful in distinguishing between geometric relativity and relativistic mechanics. From a geometric standpoint, he argued, the Ptolemaic earth-centered view of the universe is as valid as the Copernican. A relativistic theory is one that admits no absolute motion. In so orienting his lecture, he was able to avoid the troubling question of whether time and space were *real*, at least until the very end of his discussion when he responds affirmatively, but only if their relativity be taken into account. The reality so keenly desired was not, he points out, a Euclidean one, as was the hall in which he was speaking. He closes on a Machian note: "What I can measure exists."

I have mentioned that the syntheses of Cabrera, Plans, and Terradas were directed primarily at engineers with substantial mathematical background, rather than at the general public. Engineers also had access to shorter pieces of popularization published in their professional journals. An example is an article by the mathematician Ferran Tallada in *Técnica*, the official organ of the Association of Industrial Engineers of Barce-

---

[19] José Ochoa y Benjumea, *El espacio y el tiempo desde Newton a Einstein* (Barcelona: Bazar Ritz, 1924). Most of Ochoa's concrete examples were drawn from Paul Kirchenberger's *¿Qué puede comprenderse sin matemáticas de la teoría de la relatividad?*, J[osé] de la Puente, trans. (Barcelona: Juan Ruiz Romero, 1923). This was one of the most widely circulated "low-brow" popularizations in its German and French editions. De la Puente, a physicist, was a disciple of Blas Cabrera.

lona.[20] Like many mathematicians—Rey Pastor is another example—Tallada popularized on two levels: in newspaper articles for those with nonscientific background (see chapter 7), using familiar geometrical examples and avoiding mathematical reasoning and language, and, for engineers who were able to digest mathematical formulas. As Tallada noted:

> The journal of our Association has a vacuum to fill, the more so owing as much to the fragmentary and incomplete exposition of works appearing in other publications, as to their excessively vulgar and therefore scientifically undemonstrative form, in some cases, and in others because they recur to abstractions to which technicians in general are unaccustomed. Hence there reigns in many of them a certain disorientation of ideas which, nevertheless, could still be mastered by applying no other mathematical equipment than that which is imparted in the courses of our School.[21]

The heart of Tallada's article is a presentation of Lorentz's transformation formulas, followed by equations governing longitudinal contraction and the slowing down of clocks. Indeed, the exposition is wholly devoted to the Special Theory and concludes on the reassuring note that classical mechanics still has "a practical rationale" inasmuch as the velocities it deals with are small compared to that of light.

Engineers were fascinated by relativistic oddities and paradoxes, which gave them an opportunity to approach the new concept through concrete problems that were amenable to analysis with the conceptual tools they already possessed. Two polemics from the 1920s are typical: first, because of the characteristic level of discussion; second, because Emilio Herrera (1879–1967), who delighted in scientific controversy, was the protagonist of both; and third, because both controversies fol-

---

[20] "Fundamentos del principio de relatividad," *Técnica*, 45 (1922), 237–244.
[21] Ibid., p. 237.

lowed the same pattern: an impasse between Herrera and his antagonists, resolved by an appeal to Sir Arthur Eddington.

The first polemic, waged on the weekly engineering page of *El Sol* during the first three months of 1920, concerned the weight of light. The point of departure was an article by R. Izaguirre describing in some detail the eclipse observations of May 1919 and consequent proof of Einstein's theory of gravitation. (For this reason alone the article is an important one in the popularization of relativity in Spain, although it contains the curious error that Michelson and Morley had deduced the Lorentz contraction from their own experiment. The description of the eclipse results, however, is both detailed and accurate.) The article concludes by noting that Newton had considered it possible that light has weight and that Henry Cavendish had calculated such a weight in 1795. The weight of solar light falling on the earth's surface in one day, Izaguirre tells us, is 160 tons. Therefore, "If we wanted to buy light by weight from an electric company, its price would not be lower than 5,000 millions of pesetas per kilogram."[22]

Herrera then entered a lengthy comment that is also an interesting piece of popularization because he begins by comparing Newton and Maxwell's theories of light with those of Einstein and continues with an easily intelligible capsule description of the General Theory. Ignoring Izaguirre completely, Herrera then notes that Eddington had published an article in the *Illustrated London News* stating that 160 tons of light fall upon the earth daily.[23] He suspects that Eddington "had committed the vulgar error of confusing a pressure (force times surface unit) with weight by unit of time of the supposed

[22] "La pesantez de la luz," *El Sol*, January 9, 1920.

[23] November 20, 1919. The same information appears in Eddington, *Space, Time and Gravitation* (Cambridge, Eng.: Cambridge University Press, 1920), p. 111: "It is legitimate to speak of a pound of light as we speak of a pound of any other substance. The mass of ordinary quantities of light is however extremely small, and I have calculated that at the low charge of 3d. a unit, an Electric Light Company would have to sell light at the rate of £140,000,000 a pound. All the sunlight falling on the earth amounts to 160 tons daily."

ponderable substance capable of producing this pressure by its impact." Herrera then repeats the calculation, positing that 160 tons of light fall upon the earth each ten-thousandth of a second (the pressure being the same for that time period as for a twenty-four-hour period), concluding that "the value of light turns out to be 5.80 pesetas per kilo, lower than the current price of most staple goods."[24]

In the next round Izaguirre, joined now by his colleague M. Correa, asserted that he had not read Eddington's article but had arrived at a similar result based on known data. They reproduced their calculations, arriving at 163 tons, more or less Eddington's figure, using Einstein's famous formula, $E = mc^2$.[25] But, on the same page, Herrera then rejoined that "not all that weight is light," and that, in fact, one had to distinguish between light per se and radiant energy in the form of heat. While the polemic continued, Herrera wrote to Eddington to inquire whether the astronomer meant that the 160 tons were composed entirely of light, or of "the entire calorific radiation of the sun." If the latter were true, then how much weight could be assigned to light alone? On February 29 Eddington replied that his reference was to the totality of solar radiation, and he estimated the light component to be approximately half of the total.[26]

The newspaper editorialized that Eddington's reply had been adverse to the arguments of his young defenders who had proven to be "more Eddingtonian than Eddington himself." To assign a value of 50 percent to the portion of light included in the total of solar radiation seemed excessive, continues the editor of the engineering page, because S. P. Thompson had estimated that the energy of the solar spectrum dissipated in heat was approximately five times as great as that released as

[24] "¿Compraremos la luz por kilos?" El Sol, January 16, 1920.
[25] "El peso de la luz," El Sol, February 13, 1920.
[26] Herrera "Estrambote luminoso. Una carta del profesor Eddington," El Sol, March 12, 1920. Herrera's letter and Eddington's reply appear in Spanish translation. I assume that both of the original letters were written in English.

light.[27] There had been a semantic problem underlying the polemic since, by convention, English physicists used the term "light" to designate the sum of solar radiation. Nevertheless, the polemic bore one consolation—"the immense progress of education in Physics in Spain. Thirty years ago we would not have found in this land of the chick-pea persons capable of discussing these difficult questions with the causal knowledge displayed by our collaborators, nor any newspaper capable of interesting its readers with such disquisitions."[28]

Herrera and Eddington were the central figures of another polemic, this one pursued in the pages of the prominent engineering journal, *Madrid Científico*. The December 23, 1922, issue of the British journal *Nature* contained "A Relativity Paradox" submitted by an anonymous reader. The paradox had to do with the possibility of exceeding the velocity of light, a problem of particular interest to those who, while admitting the practical physical impossibility of velocities greater than that of light, were still reluctant to deny their theoretical possibility. In this problem (Fig. 13) an observer A has two immensely long, rigid triangles.

> A signals to B by sliding the two triangles together, in the direction of the arrows; the point X, where the two sides intersect, moves towards the observer B, who receives the signal when he observes the point of intersection pass over him. If the angle at X is 10″ and the triangles are moved together at a speed of ten miles a second [about half the speed of light] (an absurdly small speed for a relativist), the signal will be transmitted to B with more than twice the speed of light.

Eddington replied in *Nature* that the questioner assumed that when A tugged at the bases of the triangles, the apices would begin to move instantaneously, when in fact the impulse

[27] The editor cites the French translation of Thompson's *Light Visible and Invisible*, i.e., *Radiations visibles et invisibles*, trans. L. Dunoyer (Paris, 1914), p. 330.

[28] "Para terminar," *El Sol*, March 19, 1920.

A Relativity Paradox.

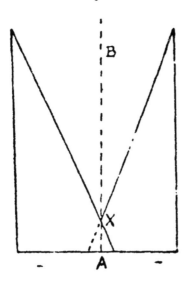

13. "A Relativity Paradox."

would travel from base to apex at the speed of elastic waves in the material, far less than the speed of light. "After the lapse of sufficient time," Eddington continued, "the two triangles would move uniformly and as a whole; and the mechanism provides a good illustration of a recognisable point moving much faster than light." But relativists do not object to such a proposition, because X does not constitute a signal. "The time of signalling from A to B must be reckoned from the moment that A gives the impulse to the mechanism."[29]

Upon reading this exchange, Herrera thought Eddington's defense weak and wrote (via *Nature*) to Eddington, opining that the velocity of point X must be less than that of light, because the angle of intersection changes during the propagation of the impulse from the base to the vertex of the triangle. Ed-

[29] *Nature*, 110 (1922), 844.

dington replied that Herrera had failed to understand what he had meant by "after the lapse of sufficient time," stating that the triangles would suffer a deformation but with sufficient time would return to their original form. Herrera dissented and wrote to Eddington again, while the Spanish journal awaited his reply, having already lost hope that Spanish relativists such as Terradas, Plans, González Quijano, and Vicente Burgaleta "might resolve our doubts."[30]

Manuel Lucini soon entered a short comment. It was erroneous, in his view, to introduce the physical and mechanical properties of ordinary matter into purely theoretical geometrical discussions. Indeed, no physical phenomenon was under discussion. Observers see the triangles, but deformed; this in no way contradicts Einstein but is in perfect accord with the Special Theory. Moreover, if X is a material point of mass M, an infinite force would be needed to move the triangle. There is no paradox here, because the movement of X at a velocity greater than light is kinematically possible in four-dimensional geometry.[31] He was followed by Vicente Burgaleta, admittedly pleased to see himself numbered, in the previous issue, among the relativists, "in whose ranks I never thought to include myself, not through disdain for theories which I admire, but because of my lack of preparation to tackle them." Indeed, the repeated use of the train example to explain relativity must have induced the editors to include him in the list—he mused—because he was, by profession, a railroad engineer! The problem, according to Burgaleta, lay in Eddington's use of the term "signal," which would seem to imply the necessity of a velocity less than c. But if by signal we mean any phenomenon that serves as a point of reference for the measurement of time, then it is reasonable to consider X a signal. He concluded that "the existence of velocities greater than that of light is undeniable, without that constituting a serious objection to the theory of relativity, but rather to certain forms of expression

[30] *Madrid Científico*, 30 (1933), 33–35.
[31] "Sobre 'una paradoja relativista,' " *Madrid Científico*, 30 (1933), 52.

of which relativists are very fond, no doubt owing to their habit of astounding the world with their conclusions.[32]

SPANISH ENGINEERS, RELATIVITY, AND THE PHILOSOPHY OF SCIENCE

Herrera was involved in one other noteworthy polemic involving relativity in 1923, once again in the pages of *Madrid Científico*. The issue was the role of intuition in science and his statement here, using relativity as a case in point, was simply a reiteration of his presentation to the 1921 congress of the Spanish Association for the Progress of Science, at Oporto.[33] For Herrera science was a process that involved the alignment of observed with intuitive truths. To explain a phenomenon is to establish, by means of logic, the relationship between it and intuitive knowledge. This relationship provided Herrera with the basis for a model of scientific change:

The great scientific revolutions which have occurred in the intellectual history of the world have developed in the connection established by reason between observed facts and intuition. At times, the observation of a new fact turns out to be incompatible with a scientific theory whose logical branching we believed to have been solidly established in the trunk of intuition, and this branching had to be destroyed or substituted for by another which, originating in the same trunk, could encompass the new fact. But there has never been a case which led to the destruction of the very roots of the trunk. The usually-cited precedent of resistance that other scientific innovations have suffered until they obtained the abolition of firmly-established beliefs is not applicable in this case. The struggle has always been between differing explanations of ob-

[32] "Una paradoja relativista," *Madrid Científico*, 30 (1933), 67.
[33] "La intuición y la ciencia," *Madrid Científico*, 30 (1923), 17–19; Herrera, *Algunas consideraciones sobre la teoría de la relatividad de Einstein* (Madrid: Imprenta del Memorial de Ingenieros, 1922).

served facts or between different modes of connecting these facts with intuition. But never has anyone tried to explain a fact by destroying intuition, the only point of support of any explanation.[34]

Herrera, doubtless because of his own theories of hyperspace (see below), had no problem with general relativity; it was the Special Theory that offended intuition:

> Relativistic theory contains brilliant ideas which are compatible with classical or intuitive science, such as the equivalence between accelerations and gravitational fields. Other consequences, such as the deflection of light and the deformation of space in gravitational fields, the modification of Newton's law, the limitation of the volume of physical space, and so forth, in accord with experience, are also explicable in intuitive science by merely admitting the existence of geometrical dimensions outside of our three-dimensional space. This is not counterintuitive, because the number 3 of the dimensions of physical space which we perceive is an experimental or empirical fact, while intuition indicates to us the possible existence of an Absolute Space, exterior to ours, without limit of dimensions. The only consequence [of Relativity] that our most intimate intuitive convictions reject is the invariance of the speed of light as it is [currently] posed.[35]

Would it be possible, he asks rhetorically, to conceive of a new kind of educational process to modify our intuition so that we could accept what now seems absurd? To Herrera, an unfettered human intuition seemed to have been mankind's most powerful weapon in unlocking the secrets of nature. He felt constrained, restricted by the growing number of universal

[34] *Algunas consideraciones*, p. 12.

[35] Ibid., p. 19. Cf. Loyd Swenson, *The Ethereal Aether* (Austin. University of Texas Press, 1972), p. 185. The eclipse of 1919 made it possible for many to accept general relativity, vaulting over doubts that remained over the Special Theory.

constants, of which c was only the latest in a series that included Avogadro's number, Planck's constant, and others. This feeling of constraint was offensive to his personal cosmology, doubtless inspired in part by religious sentiment:

It is an unsustainable error to suppose ourselves in possession of numbers which limit and regulate creation. These figures can only signify qualities peculiar to our own physical space-time, a minute element of the total Universe, which in its entirety is infinite and inconceivable. . . . To call these numbers universal constants when they are inherent only in the hyperspheroid which constitutes our own physical space, is as ridiculous a pretension as would be that of the inhabitants of Madrid, who having never left the city, considering the 40°24' of the altitude of the pole over its horizon a universal constant.[36]

Accepting the explanatory value of relativity, he still wondered what the cost of abandoning intuition would be.

The origins of Herrera's ingenuous intuitionism are obscure, linked vaguely to Descartes, to Kant, and to the contemporary French discussion of intuition's role in science in the works of such philosophers as Henri Bergson and Edouard LeRoy, although not to any one of them in particular. The only authority cited by Herrera in his 1921 Oporto paper was Henri Poincaré, in specific his notion of the "direct intuition" of Euclidean geometry.[37] Poincaré indeed believed, as Herrera also seems to have, that spaces of greater than three dimensions could be perceived intuitively, so that intuition could even play a role in Hilbert's axiomatic geometry, which had been deliberately designed to do away with any intuitive basis.[38]

[36] Ibid., p. 24. On scientists' disquiet over the absence of "free parameters" in Einstein's theoretical formulations, see Eisenstaedt, "La relativité générale à l'étiage," sec. 5.

[37] Herrera, *Algunas consideraciones*, p. 12.

[38] Poincaré, "¿Porqué el espacio tiene tres dimensiones?" in *Ultimos pensamientos* (Buenos Aires: Espasa-Calpe, 1946), pp. 72–74 (article originally published in the *Revue de Métaphysique*, 1912). Herrera summarized his philosoph-

Herrera was answered by another engineer with interests in relativity, Pedro Lucía. For Lucía, the problem was simple: the intuitive view of science "almost no longer exists," a victim of the modern theory of knowledge. Scientific axioms are not intuitive and evident, as Herrera had it (Herrera had said: "All human knowledge, its diverse methodological orders which constitute the sciences, have as their foundation an intuitive basis which appears clearly to the spirit as a set of indemonstrable truths which, according to their grade of evidence, receive the name of axioms or postulates."),[39] but rather, "The axioms which serve as the basis of a purely analytical science are established by the scientist in some way arbitrarily, paying heed to but one basic principle: that between them there exist no logical contradiction, nor should any appear in a theory based upon them."[40] Indeed, mathematical theories are construed in a purely logical region, precisely to get away from intuition and its so-called evidence. (Here we can denote Lucía's dependence on German thought, particularly on the works of David Hilbert and Moritz Schlick, both cited, who were exponents of a self-consciously anti-intuitive science.)

Geometry, Lucía continues, is not today a science of space, but simply the understanding of relations existing between certain ideas (such as point, plane, and so forth), which have been determined by the axiomatic method. In mathematical physics, relationships are among symbols. Hypotheses and theories of these relationships are formed and then tested by experiments, and in all this the role played by intuition is nil. Intuition has been, at best, a *grosso modo* aid but is totally inadequate when dealing with modern science. Relativity is not the first anti-intuitive science, and science, in sum, should not fear to lose what it never had.[41]

Some weeks later, Lucía spelled out in greater detail his ideas

---

ical concepts in "Intuición, ciencia y conocimiento," *Madrid Científico*, 30 (1923), 102–103.

[39] Herrera, *Algunas consideraciones*, p. 8.

[40] Lucia, "La intuición y el conocimiento," *Madrid Científico*, 30 (1923), 86.

[41] Ibid., p, 87.

on the nature of scientific laws as symbolic languages. In his view, "The physical sciences are constituted only by establishing a rigorous *one-to-one* relationship between the distinct elements which make up natural phenomena and certain rigorously defined *symbols* or *scientific ideas*." Each natural element has a specific symbol, and parallel to the real world, therefore, is another composed of scientific symbols. Consider, he continues, the relationship between a geometrical figure and its analogy in physical space. One hundred years ago no one (for example, Kant) doubted the equivalence of the two. Noting the existence of many ideal geometries, Lucía then asks:

> Which geometry can best be utilized to establish a one-to-one relationship with our physical space? In this case, only experience can respond. This experience—recall the transcendental observation of stellar rays near the Sun in the eclipse of 1919—may tell us that our real space is one with metric properties which vary according to the gravitational potential at the point in question. Such is, in effect, the result found by Einstein in his theory of relativity, and which seems to have been verified by the observation just mentioned and by the other proven consequences of the theory.[42]

Geometry cannot be intuitive, Lucía implies, but must be related to specific experimental data. Scientific laws are *not* natural laws that derive from the nature of things. Nature only supplies the facts whose elements are measured and then symbolized by science: "Scientific-physical laws, so-called, are the symbolic garb with which we dress the facts of experience, in order to capture them, predict them with the least mental effort." The relations between such symbols is a totally human artifice and is purely conventional; Einstein sought the mathematical expression of reality in non-Euclidean geometry, but

[42] Lucía, "Valor y significación de las leyes científicas," *Revista de Obras Públicas*, 71 (1923), 162.

the choice is wholly determined by what, "by reason of simplicity, is most appropriate in each case."

It is interesting to note that Herrera's Oporto presentation (discussed in greater detail later in this chapter) should have passed unnoticed in the scientific press, but that his abridged version, published in *Madrid Científico* should have attracted a sophisticated, well-reasoned reply, countering Herrera's French-inspired views with contemporary statements of the anti-intuitive position drawn from German mathematics and philosophy of science. In common with many engineers, Lucía was a well-read scientific intellectual, and he was a member of Ortega's philosophical tertulia. In Spain of the 1920s it was not uncommon for a younger engineer to be both broader and more philosophically sophisticated in his scientific reading than many a professor of physics.

Departing from a philosophical stance similar to that of Pedro Lucía was a paper (doubtless first a lecture) by Joan Rosich, professor at the Industrial School of Tarrasa, who, like Lucía, set the problem of relativity in the context of the nature of scientific theories, clearly with an eye toward defusing the ambivalence, or fears, of his readers. Hypotheses and theories, he begins, rise in opposition to reigning conceptions. As these accumulate, tendentious explanations are devised to make the new hypotheses agree with the reigning concepts. "The acceptance of truths or concepts have been combatted and retarded in the name of common sense, which later have been defended in its name."[43] When he was a student, he recalled, the concierge at his school had believed the earth to be flat: what is common sense to some is absurd to others. Both relativity and quantum physics had been opposed because they appeared to be contrary to common sense. He thought he discerned some characteristics common in the reception of revolutionary scientific ideas. For one thing, persons directly involved in some area of research where the new idea arises

[43] Rosich, *De las hipótesis y teorías en las ciencias físicas. Algunos antecedentes de la teoría de la relatividad* (Tarrasa: Escuela Industrial, [1922?]), p. 3.

tend to be most hostile. People outside chemistry accepted Lavoisier earlier than did chemists of his generation. The resistance to new ideas within science itself is extraordinary: "This resistance is singularly extraordinary in scientists who are specialized in experimental research of a determined order." Such persons, "upon acquiring great intensity of penetration, lose in conceptual breadth."[44]

The result is that scientists, threatened by new ideas, throw up a wall of defense, generally subconsciously. The defenders of the new idea, for their part, make things as hard as possible for their cause. Instead of attempting to explain the connections between old and new theories, they exaggerate differences and "meticulously search for paradoxical effects. . . . This is the fate of the theory of relativity, since it lends itself so easily to paradox and affectation and which seems expressly designed to enliven the scientific curiosity section of some illustrated weekly."[45] Any scientific change tends to be "propitious to alarming auguries," a point made by other commentators.[46] The difficulty of assimilating any new theory, relativity included, was, for Rosich, "a normal process."[47] It is noteworthy that many Spanish commentators on relativity—Herrera, Rey Pastor, Rosich, Marañón—had clearly defined views on the nature of cognitive change in science, and the latter two authors saw this process as an interaction between in-

[44] Ibid., p. 8.

[45] Ibid., p. 9. See Eddington's explanation of the prevalence of paradox in relativistic explanation: "The relativist is sometimes suspected of an inordinate fondness for paradox; but that is rather a misunderstanding of his argument. The paradoxes exist when the new experimental discoveries are woven into the scheme of physics hitherto current, and the relativist is ready enough to point this out. But the conclusion he draws is that a revised scheme of physics is needed in which the new experimental results will find a natural place without paradox"; Space, Time and Gravitation (Cambridge, Eng.: Cambridge University Press, 1920), p. 27.

[46] Cf. F. Tallada, "La contracción de los cuerpos en movimiento relativo," La Vanguardia, March 24, 1923: "Scandal . . . is a phenomenon which occurs whenever a new theory is born."

[47] Rosich, Hipótesis y teorías en las ciencias físicas, p. 10.

tellectual and social phenomena.[48] The current vogue of
Kuhnian explanation has tended to dissuade investigation into
pre- or proto-Kuhnian accounts of scientific change. The
prevalence of such expositions in Spain of the 1920s demonstrates not only widespread awareness of the prevalent atmosphere of change in science but also concern for explaining it.

Emilio Herrera, the Fourth Dimension, and General Relativity

The new geometries of the nineteenth century were reflected
in a popular literature of hyperspace (a space of four or more
dimensions), which mixed (or confused) non-Euclidean geometry with that of *n* dimensions. The former fueled speculation about the possible curvature of space, while the latter
was applied to the question of the number of dimensions of
space. In *n*-dimensional geometry, the fourth dimension was
simultaneously perpendicular to the other three. In some writers, this fourth dimension had supernatural properties associated with the "astral plane" of the theosophists. In much of the
popular discussion by such authors of Charles Howard Hinton
and Esprit Pascal Jouffret, the fourth dimension was purely
spatial, not temporal. Yet when relativity began to be popularized, it was inevitable that four-dimensional space-time
should be confused with (or in some cases conflated with) already venerable popular notions of the fourth dimension. Einstein himself noted that the fourth dimension had a mysterious
appeal to nonmathematicians, something akin to the occult.[49]
In the 1920s Emilio Herrera combined an old fascination with
hyperspace with a newly awakened interest in relativity. Be-

[48] On Marañón's views on the sociology of the reception of relativity, see
chapter 5.

[49] Linda Dalrymple Henderson, *The Fourth Dimension and Non-Euclidean
Geometry in Modern Art* (Princeton, N.J.: Princeton University Press, 1983), p
299. Henderson (p. 25) notes the prevalence of "hyperspace philosophy,"
which posited a belief on the fourth dimension without requiring any empirical proof for it.

cause he wrote on both subjects in popular and scientific media, he played the role of linking up "hyperspace philosophy" with relativity, particularly on a popular level.[50]

In his *Memoirs*, Herrera recalls that he had been introduced to the geometry of *n* dimensions by an alcoholic lieutenant while on war duty in Morocco.[51] By 1913 his ongoing study began to deepen into a cosmological theory. Herrera begins his 1916 formulation of the *Relation of Hypergeometry to Celestial Mechanics* by noting that gravitational forces observed in space do not appear to obey Newtonian laws. Furthermore, the observed anomalies of celestial mechanics (he had in mind the bending of light rays in a gravitational field) suggest that the metric of space might well be non-Euclidean or curved, rather than Euclidean and straight. He then explains gravitation by asserting that the space of our universe (which he assumed was filled with aether) must rotate, engendering centrifugal force in all bodies that it contains. The fourth dimension, in his view, was not time but another direction, simultaneously perpendicular to all directions of three-dimensional space, that is, a "vortex" in which our universe rotates. (A fifth dimension would involve the placement of this four-dimensional universe within a five-dimensional one, and so forth.) From these considerations he deduced that our universe is a closed, spheroid hyperellipsoid with a limited volume, and that the centrifugal force produced by its rotation "would create the transversal deformation of space, giving rise to the apparent attraction among all bodies and the bending of the geodesic lines of space and of the direction of a ray of light, on passing through a gravitational field," among other phenomena.[52]

---

[50] See Herrera's popular articles: "La cuarta dimensión: El tiempo," *El Sol*, October 15, 1920 (a popularization of relativity), and "La cuarta dimensión: El hiperespacio," *El Sol*, October 22 (a recapitulation of traditional hyperspace philosophy, following Hinton and Jouffret, in which the fourth dimension is purely spatial).

[51] Emilio Herrera, *Flying: The Memoirs of a Spanish Aeronaut* (Albuquerque: University of New Mexico Press, 1984), p. 25.

[52] Relación de la hipergeometría con la mecánica celeste," *Memorial de Inge-*

Although not so identified, such notions were of Cartesian inspiration, in that Descartes explained gravitation by the action exerted on the aether (or "second element") by the vortices of celestial bodies, causing centrifugal force of unequal distribution. Celestial masses were pushed toward the center of the vortex.

It is probable that Herrera did not become aware of general relativity, which likewise pictured a curved, non-Euclidean space and accounted for the bending of light rays, until the results of the 1919 eclipse observations had been publicized. After at least a year of familiarity with the theory, he presented his reactions at the Oporto meeting of the Spanish Association in June 1921. It was obvious that Herrera found Einstein's ideas congenial with his own: "The principal advance of the relativistic theory and the origin of its successes is, besides the energetic concept of mass, the four-dimensional conception of the physical universe as a continuous entity, according to time, of spaces deformed by the presence of the masses contained within them."[53] He added, in a brief reference to his previous articles, that the modification of the Newtonian law of gravitation and the bending of light in gravitational fields could also be explained without recurring to Einstein's theory, by simply admitting the rotation and deformability of our three-dimensional physical space within an extension of a higher order. Still, on a cosmological level, he could not disagree with the General Theory.

He could not, however, completely accept the Special Theory, at least not some of its philosophical ramifications, as he understood them. Herrera and other Spanish engineers trained in the tradition of nineteenth-century electromagnetism were unable to disentangle the kinematics of light from its dynamics, which depended upon its interaction with the aether. Therefore this aspect of Einstein's theories, and this one alone,

nieros de Ejercito, 71 (1916), 371–383, and summary of Herrera's views, which practically did not change thereafter, in "El universo y la hiperdinamica," Anales de la Sociedad Española de Física y Química, 32 (1934), 121.

[53] Algunas consideraciones, p. 22.

struck Herrera as being counter to intuition (see discussion earlier in this chapter). All of the other consequences of relativity, except for the constancy of the speed of light, were comprehensible intuitively, so long as one admitted the existence of geometric dimensions exterior to our own space and time. The acceptance of a Cartesian universe, potentially infinite through the continual superposition of additional vortices, was a way to retain a universe traditionally conceptualized in Catholic theology as infinite, while at the same time admitting that Einstein had provided the most accurate description of the tetradimensional subset that was, in itself, closed and finite. Only in this way could Herrera bridge the gap between classical and modern cosmologies.

VARIETY OF APPROACHES TO RELATIVITY
BY SPANISH ENGINEERS

The range of comment on relativity from within the Spanish engineering community was staggeringly vast. Here we will review a number of examples. In spite of the efforts of Tallada and Terradas, the Industrial Engineering School of Barcelona can also be viewed as a focus of resistance to relativity. Already in 1922 Ramon Vilamitjana i Masdevall (probably one of the three academicians who voted against Einstein's election to the Barcelona Academy of Sciences) had, on the request of "distinguished alumni of the School" offered the judgment that the Special Theory was "a monstrous absurdity," on the basis of a critique of relativistic interpretations of the Lorentz transformations and the problem of simultaneity. Using the usual train example, he concludes that the time of one observer exists only in his imagination and is a totally subjective process. For him, relativists committed the sin of attributing objective reality to what were merely hypotheses.[54] Also opposed to relativity was Josep Tous i Biaggi, nearing the end of a forty-three-

---

[54] Ramon Vilamitjana, "¿Teoría de la relatividad?" *Técnica* (Barcelona), 45 (1922), 92–94; "La cinemática relativista," ibid., pp. 124–126.

year teaching career at the same school when he delivered a lecture on "The Principle of Contradiction in Non–Euclidean Geometry and the Principle of Relativity," at a session of the Academy of Sciences of Barcelona held on June 8, 1926.[55] The discourse begins with a study of cultural, conventional, and psychological notions of unities of time to demonstrate; first, the relative and subjective nature of duration, and next, to introduce his argument that the principle of contradiction (that a thing cannot both be and not be) is not axiomatic in non–Euclidean geometry nor in infinitesimal analysis. The principle becomes irrelevant when one introduces imaginary elements. He then introduces the Special Theory by way of showing that there is something fixed (the velocity of light) in a conceptual system that appears to deny fixity. In his discussion of the aether, although he noted that some relativistic systems deny its existence (because of the difficulty of explaining the Michelson-Morley experiment), he asserted that aether cannot be done away with, although he admitted that its distribution can vary through space. Still, Tous saw the positive value of the Special Theory in elucidating classical problems of relative motion.[56] Like many detractors of relativity, Tous continued to believe that the limit of the speed of light "will vary with the medium of propagation."

Tous's discussion of the General Theory is totally confused; he did not understand Einstein's theory of gravitation and held out the belief that someone might yet discover that the velocity of gravitation is superior to that of light, which would introduce a new relativistic dimension.[57] Finally, he notes that another result of the development of relativistic theories had been "to expand mathematical symbolism in physics," but this "has been abused by giving excessive importance to mathematical formalism as if, instead of formulas being a way of representing natural phenomena, insofar as these are quantifiable,

[55] *Memorias de la Real Academia de Ciencias y Artes de Barcelona,* 3rd epoch, 20 (1926), 17–42.

[56] Ibid., p. 17.

[57] Ibid., p. 24.

the contrary was true, that natural phenomena and even the Universe were nothing more than a reification of mathematical formulas." The prurience of wishing to explain all natural phenomena in mathematical symbolism had led to an "exaggerated subjectivism."[58] These were standard objections of the older generation, wedded to nineteenth-century physics and lacking the mathematical sophistication necessary to grasp the new language of physics.[59]

We have noted the significance of Enric de Rafael's course at the I.C.A.I. But relativity was also an interest of that institution's director, José A. Pérez del Pulgar and, as also noted, of Vicente Burgaleta, who also taught there. Pérez del Pulgar studied non-Euclidean geometry in Göttingen with Felix Klein and David Hilbert in the summers of 1907 and 1908. He may well have heard of the Special Theory at that time, although there is no direct evidence. In 1923 he and Burgaleta collaborated on a study of the mechanics of Einstein-Minkowski, endorsing its general value while withholding assent from a number of particulars.[60] All of the profound sciences, they begin, striking a familiar chord, were born when the public was poorly prepared to understand them. The devisers of new theories proceed more by intuition than by logic and arrive at the truth by a kind of jump. Warning the reader not to be bothered by the paradoxical nature of certain relativistic principles, they firmly assert: "For us there is no doubt. The Mechanics of Einstein-Minkowski is the true one. The only analytical mechanics which, far from being in disagreement with classical mechanics, includes it as one of so many possible systems. But it has greater analytical reach and, as a result, it manages logically to give the complete theory of phenomena

[58] Ibid., p. 25
[59] Tous was an influential figure, said to have taught in the course of his long career "the great majority of Spanish industrial engineers"; *Enciclopedia Universal Ilustrada*, apéndice 10, p. 575.
[60] Pérez del Pulgar and Vicente Burgaleta, "Observaciónes sobre la mecánica de Einstein-Minkowski," *Anales ICAI*, 2 (1923), 480–494; 3 (1924), 485–496.

which classical mechanics does not explain." They did not, however, agree with all relativistic points of view, particularly the electromagnetic theory of light currently adopted in relativistic thought. This theory was neither in conformity with observed facts, they said, nor was it basic to relativistic mechanics or to its principle consequences such as the deformation of the advance of the perihelion of Mercury, the deflection of light rays in a gravitational field, or the red shift. The basis of their objection was that, like many Spanish opponents of special relativity, they believed that light, under certain conditions, could be propagated with a speed greater than that "with which a bundle of parallel rays is propagated in a vacuum." They did not agree that the speed of light was independent of the source that emits it.[61]

I have noted in passing the interest of military engineers in relativity and the fact that their uniforms made their presence at Einstein's lectures conspicuous. It is interesting to note how many prominent military engineers of the older generation were ready to embrace Einstein's theories when their civilian colleagues were more reticent. Carlos Banús, general of engineers, noted in an article written in 1924, when he was past the age of seventy, that Einstein "in special relativity, does away with the aether; but in the [General Theory] has had to seek a substitute for it: space. By this means he substitutes a geometrical concept for an aether endowed with mechanical properties." The coexistence of two fields complicates the structure of the universe. Banús believed that the aether concept should be retained, while recognizing the conceptual change at the same time.[62] Given his age, Banús was astute in realizing that the mechanical model of the aether had to be dropped. His

---

[61] The question continued to bother Pérez del Pulgar. In a subsequent article, "Portée philosophique de la théorie de la relativité," *Archives de Philosophie*, 3 (1925), 106–140, he insisted that relativity was based on an erroneous theory of the propagation of light, that there was no maximum velocity in nature, and that Einstein's gravitation theory was the best part of his work.

[62] Carlos Banús, "La vieja y la nueva física," *Madrid Científico*, 31 (1924), 321–322.

strategy of finding properties of aether that would be consistent with relativity was common among European physicists of the older generations.[63]

Of the same generation was Nicolás de Ugarte, the engineer and mathematician who had served as the Academy of Science's referee for Plans's 1919 manuscript. Ugarte was wholly sympathetic to relativity and was most likely addressing colleagues of his own generation when he advised that partisans of the old physics should not be seized with panic and, moreover, "We ought not be so wedded to the old, nor deny the possibility of greater progress, lest we incur the ridicule in which powerful minds have fallen, such as Mr. Thiers, when he refused to support Stephenson's invention even after having seen the railroad at Liverpool." Ugarte, we recall, was critical of Plans's original draft because it was purely theoretical and omitted discussion of experimental proofs of the theory. Nevertheless, he acknowledged that "The real importance of relativistic Mechanics, born of the failure of attempts to verify the uniform movement of a body, such as the earth, through experiments carried out on it, is, up to now, more qualitative than quantitative; that is to say, it does not suppose numerical results, but rather concepts." It is not therefore necessary to introduce profound modifications in classical mechanics; the new concepts would make themselves felt through a period of slow gestation.[64]

Numerous other military engineers made comments, in general far-fetched, on relativity. Enrique de Paniagua commented on an experiment devised by Einstein to define the coordinates of time and space in a field that turns. Rafael Aparici proposed that the existence of a second, opaque sun could explain the nature of gravitation as well as Einstein's theory did. And military engineers read and commented on Federico Can-

[63] Cf. Swenson, *Ethereal Aether*, p. 182.

[64] "Informe . . . sobre la Memoria . . . ," *Revista de la Real Academia de Ciencias Exactas*, 2nd series, 19 (1920–21), 238–239. Most of these same points are reproduced exactly in Ugarte's article, "Las teorías relativistas," *Madrid Científico*, 31 (1924), 178–179.

tero Villamil's book, which ostensibly demonstrated the applicability of relativity to aeronautics.[65]

The broad discussion of relativity among Spanish engineers supports my conclusion that rank-and-file support for Einstein's theories in Spain was centered primarily in this group. This was partly because in Spanish engineering schools, mathematicians were the intellectual pace setters. We might compare the general enthusiasm of Spanish engineers to the hostility toward relativity of German engineers, such as those at Göttingen, whose horrified reaction to a lecture by Einstein was recorded by Hyman Levy:

> I remember watching the engineering professors who were present and who were, of course, horrified by his approach, because to them reality was the wheels in machinery—real solid entities. And here was this man talking in abstract terms about space-time and the geometry of space-time, not the geometry of a surface which you can think of as a physical surface, but the geometry of space-time, and the curvature of space-time. . . . This was all so abstract that it became unreal to them. I remember seeing one of the professors getting up and walking out in a rage, and as he went out I heard him say, "Das ist absolut Blödsinn" ("That is absolute nonsense").[66]

Einstein himself had said, in Madrid, that to understand his theory one needed, as a minimum, "the scientific education and technical preparation of an engineer."[67] He could not have

---

[65] Enrique de Paniagua, "Un comentario sobre relatividad general," *Madrid Científico*, 30 (1923), 83–85; Rafael Aparici, "La gravitación ondulatoria o sea una opinión más acerca de las teorías de Einstein," *Madrid Científico*, 32 (1925), 22; Federico Cantera Villamil, *Aviación y relatividad* (Madrid: Gráficas Reunidas, 1923), and "De relatividad," *Madrid Científico*, 30 (1923), 150, a short version of the appendix to his book, pp. 121–140. See Paniagua's critique, "La relatividad y la realidad," *Madrid Científico*, 30 (1923), 193–195.

[66] Ronald Clark, *Einstein: The Life and Times* (New York: World, 1971), pp. 153–154.

[67] Ataulfo Huertas, "La relatividad de Einstein," *Revista Calasancia*, 11 (1923), 245.

found a more willing audience than the Spanish engineering community. In countries where engineers were markedly hostile to relativity one must look for sources of hostility not in mathematical education, which was more or less standard in Europe, but in attitudes toward utilitarianism, of applied versus pure science, and in the practical values of "speculation." Certainly, a passion for the abstract, most clear in Emilio Herrera's popular writings on the fourth dimension, seems to have been a commonplace among Spanish engineers.

Relativity's capacity to generate prestige among engineers is wholly logical in the context of popular reception. The Einstein myth, which did not distinguish between the special and general theories, held relativity to be incomprehensible. In fact, special relativity was easy to understand, and, inasmuch as many undergraduate engineering students learned its rudiments through textbooks, they must have been fully aware of having mastered a concept widely believed to be incomprehensible. The sense of having overcome the incomprehensibility of relativity both conveyed prestige and built confidence, which was subsequently heightened when large numbers of engineers studied absolute differential calculus and gained thereby at least a minimal understanding of the more difficult General Theory.

Horacio Bentaból's Anti-Relativity Crusade

Although he was an engineer, Horacio Bentaból requires a separate discussion because he was an outcaste and, as a result, did not use the usual engineering forums (from which his access was barred) to promote his anti-relativity message. Bentaból was convinced that the Spanish public would only accept an innovation if promoted in the name of foreign science. In Spain, "there is no nonsense that people will not believe and accept if . . . it is presented as something discovered or applied abroad by real or supposed eminences with exotic surnames, unpronounceable if possible."[68] Indeed, as a critic of Einstein, it was not enough for Bentaból to list his own accomplishments; no one would take him seriously unless he was able to cite an "exotic name."[69] Accordingly, Bentaból invoked the name of Henri Bouasse, one of the most widely cited anti-relativists.[70]

When he attended Einstein's four lectures in Madrid, "armed with paper and pencil," Bentaból felt the pressure to conform to the general sentiment of deference to Einstein's genius, for fear of losing face:

> I am the first to applaud these manifestations of deference and admiration, and I applauded from my seat in the university halls. . . . I had to applaud, and we all applauded because if we in Spain had not extolled Mr. Einstein as much as in other countries, or more, what would not have been said here and abroad regarding our lack of culture?
>
> Those who heard nothing more from him than the pet word *alors!* repeated every eight or ten words, and a couple of times the adjective *privilégié*, referring to certain coordinate axes, after applauding the master heatedly, departed (or, better said, *we* departed, because I attended all four lectures), peeved and dejected. But the believers—those who understood nothing, not even that Mr. Einstein had said nothing comprehensible—fearful of losing credit in the public eye if they confessed their disillusion-

---

[68] Horacio de Bentaból, *Observaciónes contradictorias a la Teoría de la Relatividad del profesor Alberto Einstein* (Madrid: 1925), p. 57. Biezunski notes a similar attitude in the French commentary on Einstein. if a Frenchman had come forward with this theory, he would not have been believed; see "La diffusion de la théorie de la relativité en France," doctoral dissertation, University of Paris, 1981, pp. 118–119.

[69] Bentaból, *Observaciones contradictorias*, p. 23. Cf. the similar view expressed by Josep Escofet (see chapter 8) that seeking support of foreign experts was particularly characteristic of Spanish anti-relativists. There was good reason for this. The "antis" lacked the mathematical preparation to criticize Einstein on his own level.

[70] See H. Bouasse, "La question préalable contre la théorie d'Einstein," *Scientia*, January 1923.

ment, invariably replied to those who asked their opinion of the lectures:

—Magnificent!   Admirable![71]

He compared this situation to the joke about a priest who intimidated his parishioners by telling them that sinners would be unable to hear his sermon. He then mouthed the sermon inaudibly, but afterward the congregants told each other how admirable a sermon it had been.

Bentaból attacked Einstein's theory from a number of perspectives, all of them ingenuous, to say the least. He repeatedly attacked Einstein at the axiomatic or definitional level, stating that it was impossible to comprehend what exactly was meant in relativistic literature by basic terms such as "time," "space," "absolute," and so forth. Indeed, the questions that Bentaból would have posed personally to Einstein, had not his intent to confront the visitor been frustrated, were all at this level (for example, "What do you mean by the word *Space*?"), a convenient example of the circularity (and mutual incomprehensibility) of arguments between proponents of different paradigms. There was at least a failure of communication.[72]

Whether Bentaból's general criticism was just or not, his refutations of specific points fell wide of the mark. He opposed the Lorentz contraction, but clearly did not understand it because he believed that the contraction was supposed to occur "on approaching a source of light"; in that case, movement away from a light source would produce a corresponding enlargement. His understanding of the relationship between matter and energy was also eccentric. There could be no con-

[71] Bentaból, *Observaciones contradictorias*, pp. 9, 47.

[72] Ibid., p. 85. Cf. Biezunski, "La diffusion de la relativité," p. 130, citing French criticism of Einstein for not using words with their habitual meanings. Bentaból had read the Spanish translations of books by Freundlich and Eddington, but his reading of Einstein was limited to a little-remarked 1920 Leiden lecture, *Aethere und Relativitätstheorie* (Berlin: Julius Springer, 1920). Bentaból read the French translation (Paris: Gauthier-Villars, 1921) and claimed that Einstein had now resuscitated the aether concept. On the significance of this lecture, see Swenson, *Ethereal Aether*, p. 187.

stant relation between *e* and *m*, because the relation varies with the type of matter and its temperature. Like many other anti-relativists, he rejected the notion of the constancy of *c*, a concept he labeled "Einstein's eternal nightmare." Finally, and most bizarre of all, was his refutation of the experimental proofs of the General Theory. The red shift was unproven; the deformation of the perihelion of Mercury was less than what Einstein predicted; and finally, he had already explained the deflection of light in the gravitational field of the sun in his observations of the eclipse of 1905, when he "demonstrated" the existence of a lunar atmosphere and proved that this atmosphere influences the apparent location of stars near the sun through the effects of refraction.[73]

The published version of Bentaból's lecture reproduces audience reaction in parentheses and provides some interesting insights into the nature of his listeners. In the introduction to the printed version he noted that "to the contrary of what the admirers of Professor Einstein's doctrines are accustomed to say, no profound or special mathematical preparation is needed to understand them perfectly."[74] Midway through the lecture, when he asked if those in the audience with no mathematical education could follow him, there ensued "great and deafening applause."[75] Later, when he asserted that relativity was a tissue of errors "which the Creator luckily took no account of in those remote days of the Creation," there was "great and prolonged applause and voices of approbation among the public."[76] One can assume a preponderance of conservative upper-class persons in this audience, perhaps frustrated by their inability to follow public discussions of relativity, through lack of scientific education, and who also may have perceived a threat to established concepts of time and space, which, Bentaból assured them, were not in any danger. Indeed, Bentaból's summation of the place of relativity in

[73] *Observaciones contradictorias*, pp. 4, 50–51, 78, 92, 117–118.
[74] Ibid., p. 3.
[75] Ibid., p. 37.
[76] Ibid., p. 71.

modern culture could well have struck a concordant note with extreme traditionalists. Einstein's theories

> were a consequence and manifestation of the errors of scientific and philosophical disorientation building for more than a century which in the field of science produced the ill-named non-Euclidean geometries, relativity and several very notable, although not recommendable, books by certain scientific eminences. In politics, the many errors propagated with various nicknames: the crime of Sarajevo, the frightening European war, the Russian Revolution, and the great political and social events which we have witnessed both before and after 1905 in a formidable ebullition. In art, cubism, and in religion, the most complete disorientation.[77]

Among the usual list of modern horrors recited by traditionalists, only psychoanalysis and Kantian modernism are omitted by Bentaból.

Bentaból was a laughable figure, "an outlandish type" as Tomás Rodríguez Bachiller recalled him a half-century later, and journalistic reaction ranged from pique to mirth. Francisco Vera listened to his Athenaeum lecture "stupefied," concluding that Bentaból's presentation was "a case of literary measles."[78] Writing with tongue in cheek, Lotario asserted that Bentaból had merited the greatest elegies for his many academic titles and engineering achievements. But his major contribution was as a polemicist: "Interest in bullfighting declines daily and we citizens of Madrid have to satisfy our polemical spirits with something. The gladsome words spoken in *tertulias* by relativists and Bentabólists would be something to see!" Bentaból should follow Einstein in his travels, wherever he speaks, to destroy his theories.[79]

That would have been the end of Bentaból, had he not en-

---

[77] Ibid., p. 86.

[78] Francisco Vera, "Sarampion relativista," *El Liberal*, March 14, 1923.

[79] Lotario, "Relativismo y Bentabólismo," *El Heraldo de Madrid*, March 15, 1923.

tered into an odd alliance with Comas Solà—odd because it underscored Comas's isolation from the scientific community on this issue, and had a total lack of detachment or objectivity in his position on relativity, which assumed phobic dimensions. Probably the reason for their alliance was that both resented having been denied the opportunity to confront Einstein publicly. On March 1, 1923, Bentaból had written to the rector, Carracido, "begging him to invite Mr. Einstein to a public debate with me with respect to his theories." Carracido failed to respond.[80] Comas then published an apologia by Bentaból, asserting what Comas had always claimed—that Einstein would be defeated in a public forum because his ideas were "based on errors." Bentaból noted that he had always combatted wrong ideas, including relativity, just as he had sustained unpopular ones, such as the existence of a lunar atmosphere of great diameter.[81]

CHEMISTS, PHYSICIANS . . . AND ARTISTS

In the scientific middle class, it was not one's profession but one's education that conveyed title to speak with credibility on scientific subjects. As examples, we may take two scientifically trained artists who confronted relativity in the 1920s. The first was the Catalan composer Jaume Pahissa i Jo (1880–1969), who had studied exact sciences and architecture at the University of Barcelona. He wrote a volume on relativity in 1921, and his prowess as a popularizer was noted during Einstein's visit to Barcelona.[82] The second example is the aristocratic artist and writer Eusebi Güell i López, Viscount of Güell (1877–

---

[80] Bentaból, *Observaciones contradictorias*, pp. 10–11.

[81] "Relatividad—Un concurso patriótico," *Revista de la Sociedad Astronómica de España y América*, 13 (1923), 29–30.

[82] Jaime Pahissa y Jó, *Idea de la teoría de la relatividad de Einstein* (Barcelona: La Publicidad, 1921). For biographical details, see *Gran Enciclopedia Catalana*, XI, 69. Pahissa wrote operas, symphonic pieces, and chamber music and was also a music critic for several newspapers. On his activities during Einstein's visit, see chapter 3.

1955). Güell had studied engineering in London and had a life-long interest in geometry.[83] In the 1920s he was interested in parallelism and Euclid's fifth postulate. For Güell this problem could be approached either through non-Euclidean geome-tries, which contained no postulate of parallelism, or through Euclidean geometry, which replaced Euclid's postulate with others. In explaining his intuitive approach to Euclidean ge-ometry, he had to deal with relativity by, in effect, denying any direct relationship between geometrical and physical space. For Güell, "geometric space is an ideal space, purely rational and constructed by the mind," and therefore problems of ab-stract geometry do not affect physical space directly.[84] Güell presents a capsule history of relativity, from the Michelson-Morley experiment, through the Lorentz contraction and Ein-stein's solution to the problem. (It was current in populariza-tions of this period to assume that the Special Theory had been a deliberate response to the contradiction posed by Michelson-Morley.) Güell borrowed his description of the General Theory from Lorente de Nó's translation of the appropriate chapter of Einstein's own manual.[85] The mathematician and engineer Antoni Torroja criticized Güell's volume as an exces-sively intuitive construction of Euclidean geometry, and sug-gested that Güell's money would have been better spent on fi-nancing the research of others.[86] That an artist and scientific dilettante could have one's views on relativity taken seriously and reviewed in respectable journals was a normal aspect of the

[83] See his book, *New Basis for the Foundation of Geometry* (Manchester, Eng.: Palmer House, 1900).

[84] *Espacio, relación, posición (Ensayo sobre los fundamentos de la geometría)*, 2nd ed. (Madrid: Nuevas Gráficas, 1942; first edition, 1924) p. 102.

[85] Ibid., p. 69. Cf. Einstein, *La teoría de la relatividad al alcance de todos*, 3rd ed. (Madrid, 1925), pp. 90–93. In this chapter, Einstein identifies space as being Gaussian, not Riemannian. Bentaból, who had read Güell but not Ein-stein, picked up the concept of the "mollusk" from the former and ridiculed it (*Observaciónes contradictorias*, p. 74). For Bentaból, space had *no* properties at all (ibid., p. 33) and therefore could not be described in this way.

[86] *Ibérica*, 22 (1924), 399–400, review of Güell.

interaction among the different scientific "classes" in Spain in the 1920s.

Spanish chemists produced a highly idiosyncratic commentary on relativity. Chemistry in the 1920s was not fully institutionalized as an academic discipline; the most prestigious chairs were in the schools of pharmacy, and the vast majority of Spanish chemists were also pharmacists. Since the cognitive distance of Spanish chemists from the primary receiving disciplines seems greater than that of many engineers, it is logical to discuss their response here. It is significant that the doyen of Spanish chemists, Eugenio Piñerúa Alvarez (1854–1937), nearing the end of his career at the time, expressed himself favorably to relativity. Piñerúa was an analytical chemist, professor of general chemistry at the University of Madrid since 1899. Like most Spanish chemists, he had a pharmacy degree, and his first professional position had been as pharmacist in the Provincial Hospital of Oviedo. His essay on relativity was derived from two lectures organized by the Asociación Nacional de Químicos, given at the Faculty of Sciences at Madrid on January 14 and 19, 1923, clearly in anticipation of Einstein's visit. The lectures were subsequently published in *La Farmacia Epañola*, a scientific middle-class journal read by professional pharmacists as well as chemists—indeed the most widely circulated Spanish pharmacological journal of the period.[87] Piñerúa's presentation may be characterized as conservative, cautious, but on the whole accepting of the theory. He appears to have read only popularizations, citing not only Normand and Nordmann but also Bruno Ibeas's religiously oriented anti-relativity treatise. He notes the rapidity of the diffusion of relativistic ideas, asserting, rather floridly, that "there is no example in the history of science of any theory which has diffused in a similar time span, except for the theory of colloidal states."[88] Piñerúa's choice of illustrations is somewhat atypical

---

[87] Eugenio Piñerúa, "Nociones acerca de la Teoría de la Relatividad," *La Farmacia Española*, 55 (1923), 241–243ff.

[88] Ibid., p. 242 (emphasis Piñerúa's).

of Spanish relativity popularizations. Although he does use the train example, he lays much greater stress on the perceived experimental basis of the theory in the Michelson-Morley experiment. Rather than present a series of paradoxes and analogies, a number of popularizers, including Piñerúa and Güell, thought it less confusing to present a linear sequence from Michelson-Morley through its explanation in the Lorentz contraction and Einstein's resolution of the problem, leaving the reader with the strong impression of a direct causal link between Michelson-Morley and the Special Theory.[89] Einstein, in solving the problem posed, "conceiving it as an *apparent* disparity caused by poorly defined concepts of the measures used in our calculations," had forced the revision of all our traditional views of time and space. Like many commentators of the period, Piñerúa assumed that Einstein held the Lorentz contraction to be apparent, not real. (There is ample evidence that Einstein did in fact believe the contraction to be real). His chemist's sensibility rejected, however, what he perceived as the energetic monism of the Special Theory, which perceived all bodies as composed of energy, eliminating matter. This was pure illusion, he claimed, the equivalent of the energetic monism of Ostwald.[90]

In Spanish commentary on relativity, rejection of the equivalence of matter and energy was a predictable indication that the author was a chemist. This is how I identified José Maria

[89] Ibid., p. 293. On the prevalence of this use of Michelson-Morley, see Swenson, *Ethereal Aether*, pp. 159–160, 188. Piñerúa's presentation of Michelson-Morley is inaccurate. On pp. 292–293 he states that although aberration phenomena indicate that the earth does not drag aether, Michelson-Morley had confirmed the drag effect. However, Michelson and Morley's 1885 drag test confirmed Fresnel's stationary aether; there is a partial drag, but it is so fractional that, they concluded, matter in motion does not disturb the aether by passing through it (Swenson, *Ethereal Aether*, pp. 86–87).

[90] Piñerúa, "Nociones," p. 276. Piñerúa (p. 308) found the 1919 eclipse results a convincing proof of general relativity, but he was ambivalent about the broader meaning of the theory, rejecting Einstein's fusion of time and space as stripping this concept "of the characteristics which are proper and peculiar to them."

Goicoechea y Alzuarán, a pharmacist who wrote an anti-relativist essay in the religious journal *Revista Calasancia*, which nevertheless contained an account of Einstein's theories on a par with some of the better popularizations. Goicoechea's article was in two parts, representing the printed versions of two lectures given at the Madrid Athenaeum on June 8–9, 1923. The first was a straightforward exposition of relativity for those who lacked mathematical background.[91] (The Athenaeum, the site of Bentaból's lecture as well, was apparently a preferred locale for nonmathematical presentations of relativity.) Calling Einstein an "unrivaled mathematician," he notes that relativity had followers in all social classes, who could now pass for intellectuals simply by mentioning it. He notes that the two concepts of classical mechanics most modified by relativity were the addition of velocities and the constancy of mass. For Goicoechea, Einstein's concept of the universe was his greatest contribution. If, in accordance with Newtonian mechanics, masses attracted each other in inverse relation to the square of the distance separating them, their gravitational potential would be null and the universe would quickly die, making impossible the eternal universe of Catholic theology. Einstein's curved universe solved this problem.[92]

The second lecture, however, was sharply critical, although Goicoechea began by admitting the intellectual challenge presented by Einstein, whom he admired "for his power of demolition," whereby many classical physical concepts were placed in jeopardy.[93] Prior to his entry into the relativity debate, Goicoechea had received ample criticism for his odd "cosmo-material" chemical philosophy of "universal tri-dynamism," wherein all natural phenomena were explicable in terms of three natural forces: "cosmic force," matter, and solar energy.[94] Relativity in itself presented no problems; the Span-

---

[91] Goicoechea, "Las teorías de Einstein sin matemáticas," *Revista Calsancia,* 11 (1923), 468–489.

[92] Ibid., pp. 488–489.

[93] "Crítica de las teorías de Einstein," *Revista Calasancia,* 11 (1923), 563–585.

[94] See the critique of Goicoechea's book, *La nueva física-química. Exposición*

ish philosopher Jaime Balmes, in his denial of absolute space and time, was in fact more radically relativistic than Einstein. However, Balmes is preferred to Einstein because he preserves the concept of the aether—"the substantive continuity of the universe." He then follows with a number of assertions supportive of aether mechanics: light is simply the undulatory manifestation of weightless aether and itself has no weight (although it can exercise mechanical pressure); cathode and beta (radio) waves are also manifestations of the aethereal medium and have no material mass; and if light curves when it passes the sun, it is only due to the "mechanical collison of certain waves with others of equal aethereal nature." It is a pity that Goicoechea did not identify his sources, because he was one of the very few Spanish exponents of aether mechanics who participated in the Einstein debate. In a passage on the absurdity of equating mass with energy, Goicoechea gives wholly chemical examples in order to refute Einstein's famous formula. On balance, then, he accepted the principle of relativity and Einstein's conception of the universe, while rejecting his theory of gravitation.

Physicians were also members of the scientific middle class, but few had the requisite mathematical background to comment on relativity in an informed way. One who attempted to integrate relativistic perspectives into human physiology was Roberto Novoa Santos (1885–1933), an inventive and novel medical thinker who had been sensitized to relativity through Mach's physiology, particularly his notions regarding the sensation of time: "Mach observes that there is no symmetry at all in the domain of time and rhythm, for physical time, just like physiological time, always flows in a single direction." In physiological time, there is "a personal, internal clock," which functions automatically on the basis of anabolic and catabolic phases of the nutritive process. It is this inner clock "which

---

fundamentada de la teoría cosmo-material (Durango, 1918), in *Ibérica*, 9 (1918), 240. The reviewer takes him to task for the needless substitution of his three terms to describe properties such as gravity, weight, and affinity.

ought to be taken as a system of reference to measure the duration of an event." But since reaction time varies from individual to individual (the internal clock does not run at the same velocity for all), "We can affirm once again the concept of temporal relativity and this, with absolute independence of any external system of reference chosen to measure the duration of the event. By taking these systems as a foundation, the physicist Einstein formulated the concept of the relativity of the space-time complex."[95] In the case of different individuals stationed the same distance away from a light signal, it is clear that each will perceive the signal differently, depending on the reaction time of each to visual stimuli. In the nervous system of the individual, there must be some strict spatial-temporal correspondence among given "cronaxic points" rather than a simple progression of engrams, or else there would be no sense of time. Biological time, in Novoa's construction, must accordingly be understood in a spatial-temporal sense.

Mach's physiology was also the point of departure for the relativistic philosophy of science espoused by the physiologist August Pi Sunyer (1879–1965), according to whom the advance of science had aroused awareness of the relativization of time and space. "Relativization is one case more of the interdependent organization of concepts corresponding to phenomena or series of phenomena similarly interdependent." For Pi Sunyer, relativity's great achievement was the realization that the intellectual world corresponded more exactly to reality than the sensorial world.[96]

Geometrical Practitioners

At the very lowest level of the "scientific middle class" we can detect another distinctive type of assimilation of relativity. In the journal published by the Aids and Auxiliaries of the Corps

---

[95] Roberto Novoa Santos, *Physis y psyquis* (Santiago de Compostela, 1922), pp. 32–36.

[96] August Pi Sunyer, "Filosofia i ciència experimental [July 1928]," in *Conferències filosofiques* (Barcelona: Ateneu Barcelonès, 1930), p. 240.

of Civil Engineers of the State, we find a curious tribute to Einstein signed by Pablo Pulido, an aide in the land-survey office in Cuenca. Pulido's notes, reflecting his attendance at Einstein's lectures in Madrid, constitute a peculiar mixture of references: non-Euclidean geometry, Emilio Herrera's ideas on hyperspace, and the theosophist notions of Mario Roso de Luna.[97] Clearly, at the level of practical geometry, one could acquire some notions of hypergeometry, but without the ability to interpret them in the context of the new physics.

[97] Pablo Pulido, "Einstein en España: Algunos apuntes relacionados en la teoría de la relatividad," *El Auxiliar de la Ingeniería y Arquitectura*, 3 (1923), 84–85. For an example of a popular science treatment in which the fourth dimension *is* the astral plane of the theosophists, see Edmundo González-Blanco, *El universo invisible* (Madrid: Mundo Latino, 1929), p. 733. González-Blanco also cites Güell in his discussion of the fourth dimension.

SEVEN

The Slave at the Sermon: Einstein and the Spanish Intelligentsia

Contexts of Popular Reception

The distinction between the "scientific middle class" and the educated class in general was frequently a tenuous one. It is clear that some educated laymen may well have caught the broader implications of relativity in a much more sophisticated and profound manner than did geometrical practitioners. The reason for distinguishing between the two is that the intellectual referents of each group were quite different. A practical geometrician, no matter how limited his understanding of Einstein may have been, still understood that his message bore some relationship to non-Euclidean geometry or even the ideas of Emilio Herrera. He knew this because he read the typical organs of the scientific middle class: engineering and popular science journals like *Madrid Científico*, the science page of *El Sol*, and so forth. In this chapter we consider the reception of relativity by the educated, literate public, the manner in which information was made available to it and the nature of its commentary.

It is difficult, if not impossible, to gauge the amount of physical knowledge a given commentator may have actually possessed; it is also difficult to tell from written texts whether a lay commentator was actually making a physical point or whether he may have arrived at a reasonable-sounding statement through some kind of unwitting semantic congruity. To probe the popular literature for signs of physical knowledge can only produce a null result and ends up telling us little about the nature of the discussion. If one concentrates less on the

traces of physics, and more on the contexts of these statements, then it is possible to ascertain the social and intellectual
functions of such writing, whose importance transcended the
meager amount of physical information conveyed. We want to
know whether an attempt was made to articulate Einstein's
ideas to a lay audience within a broadly scientific context by
relating relativity, for example, to well-known events in the
history of physics, or whether information was presented
without such articulation, in the form of one-line clichés. The
latter might contain some physical information (for example,
that "light has weight") or none at all (see Table 4). Such responses, though vastly different in content and quality of information, represent attempts to assimilate the theory within
some definable intellectual context. Other comments, however, made no attempt at assimilation at all. Two types of such
comment appear consistently in the Spanish literature. The
first is what might be called "Two Cultures" journalism: in-
sistent carping about the incomprehensibility of the theory, associated with ambivalence or hostility to science in general and
mathematics in particular, characterized by expressions of in-
security and self-doubt when faced with the inability to understand Einstein. These themes occurred in other countries, of
course, but in Spain the incomprehensibility issue was inextricably linked to widely held doubts concerning the ability of
the society as a whole to produce or assimilate modern science.
Second, there was a more specific working out of the same
themes in a symbolic or ideological response to Einstein,
which played itself out, as more than one journalist noted, in a
kind of civil war among intellectuals: the scientists (and their
boosters) versus the nonscientific intelligentsia. For the former, Einstein was made to represent the hope of Spanish science; for the latter, he was simply an embarrassing reminder of
Spain's presumed cultural decadence. In general parlance,
expressions like "popular" reception or "popular" science refer to the reading public, which, needless to say, is a distortion
of the term given a situation of high illiteracy. This chapter is
a discussion of how intellectuals reacted to Einstein's persona

TABLE 4
POPULAR RESPONSES TO EINSTEIN

A. ASSIMILATION

1. Within the context of science
   a. articulated ("popular science" level; "Einstein without mathematics"; attempts to relate relativity to the history of physics, etc.).
   b. disarticulated (bits and pieces; clichés such as "light has weight"); impressionistic.
   c. understood incorrectly (for example, confusion of real with absolute space), or impressionistically (relativity includes *any* concept of space or time which is odd or contrary to common sense); semantic slides.
2. Within some other conceptual framework (for example, philosophical relativism).

B. NON-ASSIMILATION

1. "Two Cultures" journalism (relativity is important, but commentator is unable to understand it).
2. Symbolic or ideological response (relativity and/or Einstein are used as code words for good or bad).
3. A priori rejection (some theologically inspired responses).

C. PSEUDO-ASSIMILATION

Confusion of relativity with philosophical relativism or with some other current scientific idea—for example, "relativity glands."

and how they handled the incomprehensibility issue. In chapter 8, I consider some of the ways in which relativity was adapted to literary or artistic purposes and some of the forms the discussion took as it filtered from the intelligentsia into other, more "popular" domains.

THE PRESS AND POPULARIZATION

The primary burden of popularization fell upon the daily press. The papers did yeoman service in commissioning popularizing articles, but they had to face the popularization issue in a more direct way when obliged to cover Einstein's lectures. The press, clearly, recognized its responsibility to report what Einstein said. According to the director of *El Diario de Barcelona*, Joan Burgada (1870–1946), Einstein had undertaken his arduous traveling venture in order to explain his ideas personally:

> It is clear that this puts the press under an obligation; but this is limited to pointing out the presence of the scholar and outlining, modestly and in a semi-official way, the gist of his lectures. The obligation of the press, then, in situations like these, is to place itself in the vanguard of the masses, not in order to explain the cause, but to describe the effect—not to investigate the substance, but to demonstrate the phenomenon.

Scientific commentators could deal with relativity's substance.

> The rest of us can only note the "external movement" of those manifestations of a science that is so far removed from the common domain, and even from that of non-specialized intellectuals. We are not dealing with "news," with vulgarization and popularization, but with the unique contributions of a specialist, although for both things an exact knowledge of the material is necessary. We are faced furthermore with the problem of technical terminology, without which there is no way to express the ideas precisely. It isn't that newspapers do not make frequent errors in the topics most often discussed in their columns (politics, literature, sociology and sometimes philosophy); but, even when they are wrong, they manage to speak of these topics in a way that enlightens the public. . . . We do not discuss mathematics or natural science as much, because it cannot be done without a knowledge of the correct technical terminology, which is, of course, the province of its practitioners.[1]

[1] J. Burgada, "Einstein en España," *El Diario de Barcelona*, March 10, 1923.

Burgada's comments are valuable because they constitute the only published statement made by a newspaper director on editorial policy regarding coverage of Einstein's visit. There may well have been some self-justification in Burgada's statement because it was made well after Einstein's stay in Barcelona, and he had had ample time to legitimate his newspaper's course of action.

Other newspapers had less clearly formulated policies. *ABC*, for example, changed policy in the middle of its coverage. Before Einstein's first lecture, the paper announced a practical position: it would be content to explain part of the theory correctly rather than present a vulgar synthesis of the whole (in 1922 the newspaper had already published a popularizing series on the subject by "Wahr").[2] After the second lecture, however, the newspaper abandoned even this modest position: "[The second lecture] is absolutely inaccessible even to a public of extensive education, and we sincerely believe that a large-circulation daily should not be a substitute for scientific journals. A newspaper can not talk of axes of coordinates, of quadratic forms, of geodesics, or of transformation formulas."[3] A point-by-point explanation, *ABC* concluded, would teach experts nothing, nor would it enlighten the public.

In practical terms, newspapers had only two options: either send a regular reporter, with no special training in science, simply to report external details, or send a scientist to comment on the substance of the lectures. In Barcelona, *El Noticiero Universal* chose the first option: "What did Dr. Einstein say? If it was difficult for the professor to resolve the complicated problem of relativity, it is even harder for a 'cub reporter' to transfer to his notebooks the concepts expounded by the famous lecturer."[4] The result was a wholly external report, detailing Einstein's manner of speech (for example, "he spoke so slowly" that he seemed less like a professor and more like a

[2] *ABC*, March 4, 1923.
[3] *ABC*, March 6, 1923.
[4] *El Noticiero Universal*, February 28, 1923.

student undergoing an oral examination), his method of using the blackboard, or the fact that he drank no water during his presentation.

The opposite tack was taken by *El Debate* and *El Liberal* in Madrid. The former paper promised[5] and then delivered the best possible scientific abstracts, those which it commissioned Tomás Rodríguez Bachiller to prepare. The latter's coverage of Einstein was in the hands of Francisco Vera, mathematician and historian of science, one of the best popular science writers in Spain at the time.

A number of journalists specifically addressed the issue of popularization in newspapers and came to the conclusion that the press had failed to provide effective popularization of Einstein's theories. Manuel Graña noted how curious it was that Spain, a country that had been so little interested in scientific speculation, had been so moved by Einstein. Stranger yet, he continued, was the fact that once popular curiosity had been aroused, the press had not been successful in conveying to the public the tangible results of the theory.[6] The truth is, however, that such tangible evidence as the 1919 eclipse results had, in fact, been adequately presented to the reading public. Graña, however, had less obvious consequences in mind, such as the mystery of the Lorentz contraction and Einstein's "marvelous explanation" of it. It is not that a person grows large by the mere fact of proceeding in such a direction, Graña explained, but that there is no other way of measuring it: "Einstein teaches us to correct ourselves by thinking, not by measuring."

Others noted the insufficiency of the popularization effort. Luis Araquistáin observed that Einstein was himself a mediocre expositor of his own theories, possibly because he had to express himself in French. For Araquistáin, however, Einstein's failings in this area were beside the point. All great ideas needed to be interpreted to the public by persons trained to do

[5] *El Liberal*, March 4, 1923.
[6] *El Debate*, March 14, 1923.

so: legal ideas by politicians, religious ideas by apostles, art by critics, and so forth. Without such interpreters, the human meaning of relativity might well go unnoticed, and that would be a pity:

> How sad it is that an actor worthy of the role of the great physicist in the scientific drama of understanding the movement of matter did not know how to communicate to us the emotive or universally human part of reason transcending itself, recreating itself in order to comprehend better the great world of mechanics! Because in every science there is a dramatic background, not only with regard to its origins or objectives . . . but also its very process. The drama of the principle of relativity is that of reason resolving itself, finally, to understand the universe on the basis of fixed points that exist only in the unreal world of thought. Relativity is—if we can mix such contradictory terms—the greatest fulfillment, up to now, of rationalism.[7]

A similar conclusion was reached by Lucanor, columnist for *La Epoca*, in reply to criticisms of the type voiced by Araquistain. Lucanor noted that Einstein had departed, "leaving in all of us the profound shock of curiosity."[8] Yet no one, except for specialists in mathematical physics, has a clear idea today of what he said. The "cultural *middle class*" had not achieved "that intelligible vision." Some have said that Einstein was not a good lecturer and that he had trouble with French. This is true, but one expects high scientific speculation to be abstruse. Then Lucanor drew an analogy familiar to all of his readers. There are, he says, different levels of knowledge, some more accessible to others. For example, "When, years ago, our grand-

[7] Luis Araquistáin, "Einstein o la razón estremecida," *El arca de Noé* (Valencia: Sempere, 1926), pp. 90–91.

[8] Lucanor, "Después de oír a Einstein," *La Época*, March 16, 1923. Lucanor's use of the word *estremecimiento*, together with allusions, suggests that he was glossing Araquistáin's article.

fathers enjoyed participating in those scientific and philosophical polemics which Darwin's evolution stimulated
everywhere, the index of knowledge required to emit a judge-
ment was shorter or perhaps easier. . . . And any M. Homais
felt himself equipped with sufficient baggage to explicate the
relationship between man and monkey." Biology was one
matter; but "the opinion of the uninitiated is irrelevant when
geometrical forms are concerned." There has existed, therefore, a need to "socialize these conquests, [but that] was not
Einstein's task but that of popularization." Spaniards, Lucanor
duly notes, once had a great popularizer, José de Echegaray,
and the French still had one in the person of Henri Bergson.
Lacking such figures, "the public of Madrid listened to Einstein bereft of any teacher or guide." It is interesting that Monsieur Homais, the sententious pharmacist in Flaubert's novel,
Madame Bovary, was used as a symbol of a certain kind of popularization of science: the discussion by practitioners, frequently provincial, who were part of the scientific middle class
and who followed recent developments in science.

My review of a great deal of popularizing literature indicates
that the Spanish reading public had access to the same kind of
information, at the same order of difficulty, as was available to
other European publics in the same period, and that the perceived lack of intermediaries was highly colored by subjective
factors related to the issues of comprehensibility and the role
of science in Spanish culture. Before exploring these factors,
let us first review the literature of popularization.

POPULARIZING BOOKS AND ARTICLES

From the point of view of the scientific popularizer, relativity
presented a basic problem. As Jerónimo Vecino explained it to
the readers of *El Heraldo de Aragón*, he had been asked by the
newspaper to give a *clear idea* of the theory: "But I face a great
problem: the impossibility of using in a newspaper article the
mathematical language required for a clear and precise expo-

sition of relativistic doctrine. A mathematical formula is a condenser of ideas which common language cannot express."[9]

As a physicist himself, Vecino was bound to perceive the intranslatability of mathematical language as a key problem. Popularizers based in the scientific middle class did not have this problem and, in general, finessed the mathematical issue completely. Jaume Pahissa was clear on the subject. For him, relativity was more than just a great and new law; but like the Newtonian theory of gravitation, it represented the perfection of a method that permits the explanation of all physical phenomena, those of classical physics as well as new laws: the mathematical method. "Inasmuch as physical laws and phenomena are symbolized by mathematical equations and formulas," he elaborated, "working through these symbols in accordance with advances in higher mathematics, we can arrive at new equations and formulas which, when they are integrated and also considered as symbols, may lead to the discovery of unknown consequences which experience can later confirm."[10] Pahissa concluded his description of the theory by noting that this had been Einstein's procedure in explaining the deformation of the velocities of planets through the influence of the sun,[11] the observed deformity in the perihelion of Mercury, and the weight of light—inconceivable if one believes that light is a vibration in a hypothetical aether.

Only such concrete information conveyed meaning to the

[9] Jerónimo Vecino, "La teoría de la relatividad de Einstein," *El Heraldo de Aragón*, March 14, 1923.

[10] Miguel-Emilio Durán, "Einstein en Barcelona: La teoría de la relatividad y la música," *Las Provincias*, March 6, 1923.

[11] It is worth pointing out that Pahissa, as well as many others, showed in their writings that although they accepted the experimental consequences of general relativity (that is, the three classical tests), they nevertheless had not understood Einstein's theory properly; they continued to use concepts (such as the weight of light, the change in the velocity of light, and so on) that have meaning in Newton's theory of gravitation only. Araquistáin, "Einstein," p. 92, was apparently referring to the same question when he asserted that, according to Einstein's theory, "the velocity of light changes, not because of the mobility of the aether, but according to each gravitational field."

general readership of daily newspapers who, by and large, were willing to take on faith the veracity of the mathematical infrastructure that made such explanations possible. All other popularization followed a contrived pattern, which I doubt the average reader could follow (although those who were able to do so need not have had any scientific education). The basic strategy was to follow Einstein's examples and present simple, exemplary problems of trains, elevators, and other familiar phenomena. Some endeavored to make such examples more palatable by acculturating them. "Let us suppose," wrote Vecino for the benefit of the citizens of Zaragoza, "that the Plaza de la Constitución be measured between the Café Gambrinus and the Hotel Europea." The ground measurement will be different from the figure achieved by a person in an airplane who would derive a smaller distance—the smaller, the faster he flies.[12]

Spanish readers had access to a great number of popularizing articles between 1920 and 1923. These were written by Spanish scientists, such as Vecino, Blas Cabrera,[13] or Ferran Tallada,[14] or by foreign scientist/popularizers such as Charles Nordmann.[15] There was also popularization by anti-relativity scientists, both foreign and Spanish, such as Josep Comas Solà,[16] Camille Flammarion (who was not wholly opposed but tended to follow Sir Oliver Lodge in a skeptical attitude),[17] and other foreign anti-relativists paraded by Comas through the

[12] *El Heraldo de Aragón*, March 14, 1923. An analogous example, with Catalan referents, placed two clocks in Barcelona and Villanueva; José Ochoa, *El espacio y el tiempo desde Newton a Einstein* (Barcelona, 1924), p. 20.

[13] See the section of the introduction to his book, *Princìpio de relatividad*, reproduced in *El Imparcial*, March 3, 1923.

[14] See Tallada's three-article series in *La Vanguardia*, March 4, 13, and 24, 1923.

[15] Nordmann, "Una revolución en la ciencia. Teorías de Einstein," *Redención* (Alcoy), March 8, 15, and 22, 1923.

[16] Comas, "Las conferencias del profesor Einstein," *La Vanguardia*, March 14, 1923.

[17] Flammarion, "La doctrina de Newton y las teorías de Einstein," *Madrid Científico*, 27 (1920), 154–155.

pages of his journals, including Julius von Sittert and Charles Lallemand.[18] There was commentary by members of the scientific middle class, such as the engineer Manuel Moreno-Carraciolo or the chemist Emilio Hunolt.[19] There were also pseudonymous series of articles of science writers, such as the one in *ABC* by "Wahr," presumably not a Spaniard, or another by "Rigel," a Spaniard, in *El Heraldo de Madrid*.[20] Wahr's last article was devoted wholly to experimental evidence supporting the General Theory.

Two other "popular" sources require discussion. The first is the *Enciclopedia Universal Ilustrada*, the famous Espasa-Calpe encyclopedia that began to appear before World War I and continued issuing new volumes throughout the 1920s. One must assume that all the articles containing references to relativity were written by Terradas, who was assigned many articles dealing with physics, mathematics, and engineering topics.[21] The first mention of relativity occurs in the article "Electricidad," probably written in 1913, where there are passing references to the electrodynamics of moving bodies according to the theory of relativity (vol. 19, p. 555) and a cross-reference to "Relatividad" after a discussion of Lorentz's electromagnetism equations (p. 625). In the article on the aether ("Eter," vol. 22, pp. 1178–82), Terradas notes (p. 1180): "At the time when this article is written [1923], the modifications which the the-

[18] For example, von Sittert, "Teoría general de la relatividad y el espectro solar," José Sagrista, Pbo., trans., *Revista de la Sociedad Astronómica de España y América*, 12 (1922), 8–12; Lallemand, "La teoría de la relatividad," ibid., 16 (1926), 9–10.

[19] Moreno-Caracciolo, "La teoría de la relatividad," *El Sol*, October 8, 1920; Hunolt, "Sobre las teorías de Einstein," *El Sol*, July 6, 1922.

[20] Wahr, "La teoría de la relatividad," *ABC*, January 19, February 3, 10, and 18, 1922; Rigel, "Einstein y la relatividad," *Heraldo de Madrid*, March 2 and 3, 1923.

[21] Julio Rey Pastor, "Esteban Terradas, su vida y su obra," in Real Academia de Ciencias Exactas, *Discursos pronunciados en la sesión necrológica en honor del Excmo. Sr. D. Esteban Terradas e Ilia* (Madrid: 1951), p. 64, indicates the enormous number of articles written by Terradas for the Espasa-Calpe, seventeen on scientific topics between Ac and As alone.

ory of relativity has introduced into our notion of aether is of great interest." This major understatement would seem to have been an interpolation added in a re-edition to an original article written at least ten years before. In the article "Física" there is a capsule history of relativity (vol. 23, p. 1586), citing Einstein's papers of 1905 and 1911. Terradas notes that relativity and quantum theory are "of such transcendence that it would not be presumptuous to consider our epoch as the golden century of physics." Another early mention of Einstein's name comes in volume 24, in the article "Fotoeléctrico," where there is a similar cross-reference and also an allusion to Einstein's formula on the absorption of light quanta (p. 668). The biography of Minkowski in volume 35 (p. 683) refers to "the geometric endowment which the principle of relativity gave to the vectorial calculus of four dimensions." I have already alluded to Terradas's fifty-page summary of relativity theory (not a popularization, but rather a synthesis addressed to the scientific middle class, as well as to scientists and science students) in volume 50, which is followed by a biography of Einstein, both written in 1923.[22]

The second source is *Ibérica*, a popular science journal published by the Jesuits' Ebro Observatory. *Ibérica* offered its readers a running commentary on relativity beginning in 1916, with a report on a lecture by Blas Cabrera on "Energy, Mass and Gravity," which included a discussion of Lorentz-Einstein transformation equations.[23] The 1919 eclipse observations were discussed both before and after they took place.[24] Articles by Plans, Terradas, and Cabrera (or reports on their lectures) appeared throughout the decade. In 1921–23, Enric de Rafael provided reviews of the most significant books to appear on

---

[22] The third appendix volume of the *Enciclopedia Universal*, pp. 65–68, has a list of "the most important investigations and discoveries in the natural sciences" from 1100 B.C. to 1925. This list includes natural selection (Darwin and Wallace), but neither Einstein nor relativity.

[23] *Ibérica*, 5 (1916), 175.

[24] "La luz y la gravitación universal," *Ibérica*, 11 (1919), 118; "Desviación de la luz por el Sol," *ibid.*, 13 (1920), 323 [June 12].

relativity in Spanish: Plans's translation of Freundlich; Moritz Schlick's *Teoría de la relatividad*, translated by García Morente (Madrid, Calpe, 1921); Plans's *Nociones fundamentales de Mecánica relativista*; and Lorente de Nó's translation of Einstein's popular volume.[25] The journal kept its readers up to date on further astronomical observations relevant to the General Theory, including four separate articles on the solar eclipse of September 21, 1922.[26] Also of interest was a report on the discussion of relativity that took place at the International Congress of Philosophy, held in Naples in May 1924, with papers by Elie Cartan, Hadamard, Nordmann, and others.[27]

Besides articles, there were also popularizing books. Two of the most widely circulated were Salvador Corbella Alvarez's *La Teoría de la relatividad de Einstein al alcance de todos* (Barcelona, 1921) and Pelayo Vizuete's *Einstein y el misterio de los mundos* (2 vols., Madrid, Arte y Ciencia, 1923–24). Vizuete (b. 1872) was a popularizer who had written on a vast number of scientific themes. Vizuete's first volume was an introduction to the study of planetary systems and the principles of movement which did not in fact mention relativity. The second was a discussion of special relativity, and the third (which never appeared) promised an exposition of the fourth dimension.

The Comprehensibility Issue

Bearing in mind not only the easy availability of accounts of relativity putatively within the grasp of the educated layman, not to mention works of a more technical nature, already discussed, aimed primarily at the scientific middle class, we can now turn to the debate over the comprehensibility of Einstein's theories. I am interested here in how persons of differing educational backgrounds perceived the problem, and I will

[25] *Ibérica*, 15 (1921), 63 (Freundlich); 16 (1921), 96 (Schlick); 16 (1921), 351 (Plans); 16 (1921), 400 (Einstein). Earlier, De Rafael had reviewed the French translation, from the German tenth edition, of the same work: 15 (1921), 288.
[26] *Ibérica*, 18 (1922), 357–358; 19 (1923), 296; 20 (1923), 228; 21 (1924), 134.
[27] *Ibérica*, 22 (1924), 101–102.

not attempt to determine the clarity, or lack thereof, with which ideas were expressed.

The Spanish reading public had been put on notice by Einstein himself, who declared to Andrés Révéscz that popularizing books aimed at "le grand publique" may be interesting but did not contain his theory.[28] Similar statements by scientists were, of course, legion—for example, Einstein's own assertion that it was easier to learn the mathematics necessary to understand the theory than it was to attempt to understand it without mathematics. Another cliché, repeated ad nauseam in Spain, was that which held that only a handful of the initiated could understand relativity. Replying to a friend who asked him what he had "gotten cleanly" from Einstein's lectures, the newspaper editor Josep Escofet replied, "What I got cleanly is that there aren't a half dozen persons in Spain capable of following Einstein, with his sublime calculations, without tiring, without giving up before covering half the distance."[29]

Most of the discussion of the theory's comprehensibility originated with literati. One common note they struck was that of embarrassment, as in Josep Maria Sagarra's open confession: "I attended Einstein's lectures knowing that I would understand little of his explanations, half afraid of playing the ridiculous role of falling asleep. I went into the lectures without saying a word to anyone, as if I'd be ashamed."[30] Rafagas, a Zaragozan columnist, was even more blunt: "We went to hear him . . . in the certainty of not being able to understand him; we felt ourselves to be as wretched as the lowest of household bugs." Inability to understand "truths born yesterday" filled him with anguish. Yet, "if anything consoles us, it is the spectacle presented by many persons interested, intrigued by the scholar's theories. They waved their arms about like blind people. But doesn't this same gesticulation indicate that they

---

[28] *ABC*, March 4, 1923.
[29] José Escofet, "Einstein y los matemáticos," *Las Provincias*, March 18, 1923. Escofet (1884–1939) was codirector, with Gaziel, of *La Vanguardia*.
[30] Josep Maria Sagarra, "Einstein," *La Publicitat*, March 4, 1923.

seek light?"[31] The most evocative self-deprecatory image to appear in the comprehensibility discussion was that of the "slave at the sermon." Joan Colominas Maseras used this image to describe audience reaction to Einstein's lectures in Barcelona: "We must confess that many of the listeners got from the lecturer's explanation what the slave got from the sermon, as demonstrated by the looks of exhaustion which surprised us on a great number of familiar faces, and sighs of liberation exhaled from many breasts upon hearing the closing words."[32]

Other commentators were just as frank but more petulant. In Carles Soldevila's view, Einstein's popularity was explained by the public's blind faith in science. What did Einstein do? "The immense majority of the people of Barcelona are totally ignorant of it. The rest are divided among those who said that Einstein was a great genius, inventor of an enormously curious and transcendent theory, and the minuscule group of specialists who had read the theory and understood it." Indeed, it was difficult to find a similar case of celebrity in all the history of science. The press and popularizing books had brought his name before the public. And so, in sum, "Einstein is famous because a few hundred mathematicians have believed him worthy of so being." The rest of the public must take it on faith alone, because "The theory of relativity, in spite of all attempts to popularize it, is something that the good burghers who walk along the streets reading the paper can never understand."[33] For Soldevila, then, there was a connection between Einstein's popularity, "almost divine," the incomprehensibil-

[31] Rafagas, "Lecciones de humildad," *El Heraldo de Aragón*, March 14, 1923.

[32] Juan Colominas Maseras, "Einstein en Barcelona," *El Pueblo*, March 2, 1923. After hearing Einstein's lecture, he remarked: "Hemos de confesar que muchos de los concurrentes sacaron de las explicaciones del conferenciante lo que el negro del sermón. . . ." The metaphor was also used by C. Sánchez Peguero in *El Noticiero*, March 13, 1923.

[33] Carles Soldevila, "La popularitat d'Einstein," *La Publicitat*, February 25, 1923. Cf. Bentaból, *Observaciones contradictorias a la Teoría de la Relatividad del profesor Alberto Einstein* (Madrid, 1925), p. 27, who decried "the accreditation granted in advance to the scholar whom no one understands."

ity of his theory (which made its acceptance an act of faith), and a kind of conspiracy on the part of scientists to bring about this result. A similar interpretation was made by the humorist Julio Camba who observed that everyone admired Einstein, but few, including himself, knew why. He supposed that the inventors of absolute differential calculus had jobbed their invention to Einstein just to assure it would be used (and, it followed, so that no one else but they would be able to understand him).[34]

So Einstein engaged in a kind of magic show or confidence game wherever he happened to visit: "He arrives at a place . . . opens the package of three lectures and once again begins to speak to his presumed clients." He was, in the opinion of Tomás Gómez de Nicolás, in danger of converting himself into a music-hall act, and who is to prove him wrong? The public has no choice but to go along with what he says. The subtitle of Gómez's article, "Rejoice that we aren't scholars!" well symbolizes the anti-intellectualism of much journalistic commentary on Einstein.[35]

Some press commentary focused upon the incomprehensibility of mathematics:

> What a great pity that scientific language needs to use expressions that are cabalistic for nearly all mortals, especially in the case of Einstein, who is at the leading edge of mathematics; because if scientific education were not required to understand him, the teachings of this revolutionary of the ideas of space, time and motion, and of classical physics and geometry, would find their proper place in the open, where the multitude, educated and not, would crowd around to hear him and to learn that about nothing in this wretched human life has the last word been said.[36]

[34] Julio Camba, "Los admiradores de Einstein," *El Sol*, March 6, 1923.
[35] Tomás Gómez de Nicolás, "La relatividad de los valores, Alegrémonos de no ser sabios," *El Imparcial*, March 10, 1923.
[36] *Heraldo de Aragón*, March 8, 1923.

Jaime Mariscal de Gante was even more specific. Most people will never be able to understand relativity fully because, as Einstein himself had said, one must know absolute differential calculus to do so. Spaniards need not feel bad about this, he noted parenthetically: Cajal probably does not know it either!

> It is more: it is quite possible that there are many engineers that understand differential calculus; but those that have mastered it are probably few. Our astronomers are those who have studied it most, and just as the enjoyment of the universe, of following the swift path of the stars, are reserved for them, they are also the ones who will be able to ponder and measure the doctrines of Einstein along with the privileged few who devote themselves to the cultivation of these sciences.

For the rest of us mortals, Einstein is simply the discoverer of a doctrine that revolutionized the science of space in the same way that Galileo had. That is why we respect and venerate him, even though we cannot understand him.[37]

Some even gloated over the failure of comprehension, as did Miguel de Castro, who hailed "the defeat of the pedants." Many of our "celebrated writers" had attended the lectures. Like Goethe or Renan, they "strive . . . to merge science with literature . . . and they present themselves, as always, as know-it-alls, armed with muskets of perspicacity to hunt the swiftest and fleetest falcons crossing the limits of their minds. But, of course!" The hunters had been unable to bag anything. These "scholars of the moment," as it turned out, were defeated, unable to divulge the principles of the new theory in the press. The intellectuals had failed as intermediaries. And thus if the people "have understood nothing cleanly, it is not their fault."[38]

Others felt that something had been learned but were at a

---

[37] Jaime Mariscal de Gante, "La doctrina de la relatividad," *La Voz Valenciana*, March 6, 1923 [dated Madrid, March 5].

[38] Miguel de Castro, "Einstein y los madrileños o la derrota de los pedantes," *Las Provincias*, March 11, 1923.

loss to explain how. According to one writer, "They go to hear him, they are bored, but they continue to admire him." This common reaction amounts to a case of collective suggestion, whereby all agree to "to deify a man without knowing why."[39] For another, Einstein had simply created the illusion of accessibility, by allowing his listeners to participate in his discovery through the communication of arcane facts.[40] Another, commenting on the popular misunderstanding resulting in Japan from the similar spellings of the words for relativity and sexual intercourse, believed that the Spanish language would enhance the intelligibility of Einstein's ideas: "Here in Spain the richness and flexibility of our language does not permit the public's disenchantment after the divulgation of Einstein's theories; but they might well feel deceived had they heard his lectures, since they had nothing in common with what was heard on streetcars or in cafés."[41] Still, the "slave at the sermon," with this expression's connotation of passivity and the ignorance and simplicity of a child, and of the inability to understand because of a cultural screen, was still one of the metaphors most commonly used by journalists to describe the educated public's incomprehension of Einstein.

The nearly unanimous opinion of press commentators that the theory of relativity was unintelligible to all but a handful of mathematicians and physicists was, in the main, supported by anti-relativist scientists. Josep Comas, for example, alluded to the incomprehensibility issue in a newspaper article commenting on Einstein's lectures in Barcelona. To him, the public's boundless curiosity to inform itself on the theory of relativity

[39] C. Sánchez Peguero, "Un aspecto minísculo de la relatividad," El Noticiero, March 13, 1923.

[40] Regina Lamo, "Interpretaciónes sentimentales: Einstein el precursor," El Diluvio, March 2, 1923.

[41] Rigel, "Einstein y la relatividad," El Heraldo de Madrid, March 1, 1923. The confusion in Japanese between "Sotai-sei" (relativity) and "Aitai-se" (sex between lovers), both written with the same characters, is discussed by Tsutumo Kaneko, "Einstein's Impact on Japanese Intellectuals," in T. F. Glick, ed., The Comparative Reception of Relativity (Dordrecht: D. Reidel, 1987), p. 363.

was "a notable phenomenon of collective psychology which certainly has no equal in the history of mankind. . . . It was a mute agitation of anxious and disoriented spirits. On the one hand, it was said that 'everything was relative,' with many believing that Einstein's theory could be reduced to that simple and current phrase, to which no one could object." The avidity of the Spanish public, most of whom had never opened a mathematics book nor shown the least interest in physics before Einstein's visit, was extraordinary, all because they wanted to "acquaint themselves with such sensational revelations." But the worst of it, Comas continues, is that the public is now even more disoriented than it had been before, "because of 'not having understood anything.' " People felt defrauded because they were unable to find the revelations they had hoped for. To reduce their expectations, Comas would tell people that the theory had no practical relevance whatever to the daily course of human life. (Indeed, Einstein frequently made the same point.) But the public was not wholly at fault for its own disorientation (through its failure to comprehend): "A good part of the blame was owing to the system adopted to expound such theories in a small number of lectures, as if some very complicated mathematical theories could be compared to a ballad sung by a tenor in a world competition."[42] Comas had objected to the closed nature of the lectures, partly because some questioning might have cleared up some common misconceptions, but also because he had been denied the opportunity to refute Einstein publicly.

Others in the scientific and intellectual world (a minority voice in this case) took exception to the incomprehensibility argument. To Antoni Rius, a young Catalan chemist at the University of Zaragoza, the incomprehensibility of relativity was simply an invention: "At the conclusion of the festivities occasioned by Einstein's visit to Spain, journalists, and even many professors, shameful to say, have invented the excuse

[42] José Comas Solá, "Las conferencias del profesor Einstein," *La Vanguardia*, March 14, 1923.

that the theory of relativity is impossible to understand: an excuse both hypocritical and conducive to their forgetting the existence of that formidable scientific advance the same day that Einstein crossed the Pyrenees."[43]

More complex was the analysis of the philosopher Rafael Selfa Mora, for whom declarations by intellectuals that they had, or had not, understood Einstein tended to be, in both cases, largely self-serving:

> As for those who say they have understood Einstein, unless they are among the exceptionally privileged group in Spain who are learned, we doubt their comprehension. On the other hand, those who say that the number who can listen to him is limited to half a dozen also make us doubt their competence. Both groups disguise their thoughts or dress them with pedantry; the first, so self-possessed that they believe they can transmit to us the knowledge they have been able to capture from the complexity of higher understanding which the theory of relativity supposes; the second, because they either include themselves among the half dozen, excluding the rest, or they disguise their sin by saying that what is difficult to master in totality is absolutely incomprehensible, and thus they justify the passivity of their intellects or the incapacity of their understanding.
>
> The majority of intellectuals have not been able to understand wholly the scientific exposition of the professor; but most have been able to listen to him and understand him. A theory only accessible to eight or ten human minds, given the current progress of science, would have little or no value.[44]

[43] Antoni Rius, "Albert Einstein," *Revista del Centre de Lectura* (Reus), 5 (1923), 87.

[44] Rafael Selfa Mora, "La sed intelectual," *El Luchador*, March 14, 1923 [dated Madrid]. Selfa's views on the nature of scientific theory are interesting, although not greatly relevant to the present discussion. That which is incomprehensible is not a theory; nor are simplistic formulas whose evidence is obvious even to the blind able to effect scientific change. The efficacy of a sci-

One cannot argue with Selfa's observation that Einstein's supporters had appropriated his theory to serve their own ends, just as the allegation of incomprehensibility served the ends of his detractors. It was obviously self-serving for scientists such as Eddington to claim publicly that only a handful of initiates could understand the theory. This kind of posturing made it easy for intellectuals to claim, as Carles Soldevila did, that they were being pressured to praise what they could not understand. This defensive attitude, as Rius noted, was a perfect excuse for not having to deal with the problem at all. Lurking below the surface of such protests, however, was the wounded defensiveness of an intellectual class that had, its members felt, been dislodged as the arbiters of culture. Biezunski notes the defensive posture of the Parisian intellectual *gens du monde*, unable to assume their traditional role of interpreters of high culture. Their protestations over the incomprehensibility of relativity, he concludes, constituted an angry response to their loss of prestige. The same was true in Spain, particularly in regard to the Catalan intellectual bourgeoisie, of whom Soldevila was a representative figure.[45] When Lucanor, in a passage previously cited, expressed despair that José de Echegaray had not lived to explain relativity, he was making a similar kind of statement: Echegaray, winner of the Nobel Prize in literature, as well as conservative politician and mathematician, was more than just a good popularizer of science: he was the quintessential Castilian *gens du monde*. His departure was symbolic of his class's loss of control over a segment of high culture, pure science, whose significance it had only recently begun to appreciate.

There is an interesting literary reflection of the incomprehensibility issue in *Las veleidades de la fortuna*, a 1926 novel by

entific theory, that is, is predicated on its ability to be communicated, and on the level at which it is communicated.

[45] On *gens du monde* as cultural intermediaries, see Biezunski, "La diffusion de la théorie de la relativité en France," doctoral dissertation, University of Paris, 1981, pp. 75, 83. On Soldevila as the personification of a bourgeois "gentleman," see José Tarin Iglesias, "Carles Soldevila, en el contexto de la burguesía barcelonesa," *El Noticiero Universal*, October 2, 1982.

Pío Baroja. In one episode, which takes place in Zurich, the protagonist Larrañaga is discussing psychoanalysis with Dr. Haller and a young physician:

> The doctor from the sanatorium, who was a young humoralist, said that once after reading things about relativity he had dreamed that Euclidean space had become non-Euclidean. During the dream he had felt very happy thinking that now he understood non-Euclidean space, but when he woke up he saw that this was just an illusion.
>
> "The same thing happens reading Einstein," said Haller.
>
> "Don't you believe in relativity?" Larrañaga asked.
>
> "It makes me very suspicious."
>
> Larrañaga was sure that he didn't understand the theories of Einstein; true, they said that to understand them fully you had to know mathematics, but he professed a humble pragmatism—a little in the style of Flaubert's Homais—believing that all of educated Europe could not be wrong.
>
> "I don't believe there are any theories that cannot be expressed in a rational summary," said Haller. "The part of Einstein's theory that is deduced by reasoning contains nothing new. It is the subjectivism of the elemental notions of time and space and causality, something that is already well explained in Kant. The rest, the mathematical, one doesn't understand."
>
> "But Einstein's theory might be exclusively physical-mathematical."
>
> "Without being susceptible to rational explication? It's strange. It's the same as Steiner said, the charlatan of anthroposophy: according to him, you have to know special mathematics in order to understand his doctrine of higher worlds which ended up, in practice, in shelling out money for his time."[46]

[46] Pío Baroja, *Las veleidades de la fortuna* (Madrid: Caro Raggio, 1926), pp. 114–115.

It is wholly consonant with my conception of the scientific middle class that a discussion of relativity should come up among physicians, and that it should be linked to psychoanalysis is not surprising either, inasmuch as revolutionary scientific ideas tended to be linked in peoples's minds, whether there was any underlying logic to such linkages or not. In the above passage, the topic comes up almost by free association. Monsieur Homais, as already noted, was regarded as the prototypical popular consumer of science. Larrañaga espouses the "take it on faith" approach, whereas Haller seems to echo Rafael Selfa's view that one must be able to give a rational account of any theory, if it is indeed rationally valid.

A particularly analytical and thoughtful journalist, Arturo Mori, taking account of the wealth of press comment on Einstein, wondered why so little of it was really serious. Indeed, humorists like Julio Camba (see chapter 9) and Wenceslao Fernández Flores had said the worst that could possibly have been said. Antonio Zozaya, another popular Madrid columnist, told him, by way of explanation, "The era of respect for philosophers has passed here." A few years before, any philosophical novelty created either polemics or wild enthusiasm. But, since people no longer trust philosophers, it was easier to speak in jest than to make transcendental assertions. Einstein deserved better. Why, Mori asked, had Bergson been received with such "emotional solemnity" on his Spanish visit of 1916, when he is far less a figure than Einstein? Because, he answered, Bergson represented continuity, and Einstein was a revolutionary figure.[47] To carry Mori's comment to its logical conclusion, we must return to the notion of a traditional society opposing change when such change appears to cast its world view in doubt. This is most likely the root cause of Spanish defensiveness with regard to Einstein, and it was a reaction that transcended political or ideological differences.

[47] Arturo Mori, "Crónicas de Madrid, La visita de Einstein," El Progreso, March 6, 1923 [dated Madrid, March 1]. On Bergson's Spanish trip, see Juan-Miguel Palacios, ed., "Le voyage espagnol d'Henri Bergson (avril-mai 1916)," Les Etudes Bergsoniennes, 9 (1976), 7–122.

Mori's comment ends on a cynical note, indicating his disbelief in the ability of a traditional culture such as Spain's to grasp the significance of Einstein, the philosopher/scientist: "Einstein is nothing more than a philosopher; if he were a king. . . ." I argue below that traditional Spain's image of science had changed dramatically in the 1920s from the hostile caricature it had been in the nineteenth century. Such images do change over time, and, in the case of Spain, where sectors of a disunified elite holding different images of modern science have alternated in power, the image can change rapidly and dramatically. Under Franco, after two decades of a very idiosyncratic, authoritarian/traditionalist/religious official image of science (with resultant confusion on the popular level), the pendulum began to swing the other way. In 1972 a Madrid cab driver, stalled at a military parade, exclaimed to a reporter: "Parades! And what the devil does Spain need military parades for? Now if it were a parade of scientists, that would be worth going to see!"[48] Mori's point, that Spaniards would only turn out for a king, was contradicted by press reports of Einstein attracting attention as he passed through the streets in Madrid and Toledo.

Why was Einstein acclaimed at all? Numerous answers made the rounds. The vogue of relativity was owing to the disenthronement of absolute time (Schrödinger); doubts about comprehensibility fueled popular curiosity (Rutherford); a war-weary world looked for a new kind of hero (Infeld).[49] Einstein himself noted that "The nonmathematician [is] seized by a mysterious shuddering when he hears of 'four dimensional' things, by a feeling not unlike that awakened by thoughts of the occult. And yet there is no more commonplace statement that the world in which we live is a four-dimensional continuum."[50] All of these reasons, in addition to some culture-spe-

[48] Richard Eder, "Spanish Joke," *New York Times Magazine*, August 27, 1972, p. 40.
[49] Clark, *Einstein· The Life and Times* (New York: World, 1971), pp. 241, 249.
[50] Erik H. Erikson, "Psychoanalytic Reflections on Einstein's Century," in

cific ones, were adduced in the Spanish press. Manuel Aznar observed a lack of proportion in Spanish praise of Einstein, which he thought was due to defensiveness over the culture's supposed inability to produce science. Sick of criticism, Spain wanted to prove in the eyes of the world that the judgment held of it in this regard was wrong.[51] In my view, demystification was perhaps a more potent force than mystification, and I agree therefore with the reviewer in *Nature* that Einstein's principal attraction was that in his cosmology, "The commonest phenomena become organic parts of a great plan."[52] That he made the universe seem less mysterious was a powerful attraction. Blas Cabrera put it better than anyone else:

> The privilege of attracting the curiosity of the crowd is a natural consequence of the ample domain over which these principles extend their influence, because they affect the most basic concepts of knowledge, which transform ideas developed by a secular science, so deeply ingrained in our minds that they seem to belong to that class of ideas which are imposed upon us by our own mental organization. Whoever understands how arduous a task it was to challenge these residues that classical patterns of thought have imprinted upon us . . . cannot help but feel admiration for the inspired mind that knew how to conceptualize them in a time when no one dared doubt the probity of the mechanics of Newton or Galileo, which was contradicted by the marvelous achievements of modern science, their direct heir.[53]

G. Holton and Y. Elkana, eds., *Albert Einstein: Historical and Cultural Perspectives* (Princeton, N.J.: Princeton University Press, 1982), p. 164, citing Lincoln Barnett, *The Universe and Dr. Einstein* (New York: W. Sloane, 1948), p. 67.

[51] Manuel Aznar, "El profesor Einstein en Madrid," *El Diario Español*, April 5, 1923 [dated March 5].

[52] Clark, *Einstein*, p. 247.

[53] Cabrera, in Real Academia de Ciencias Exactas, *Discursos* [en honor de Einstein] (Madrid: 1923), p. 8. Emilio Mira y López made a similar point in comparing the reception of Freudian psychology to that of relativity: "Our

IMAGE OF A GENIUS

Einstein's personal appearance and his unique personality did much to create a special mystique surrounding his scientific achievements. This conclusion is, of course, a commonplace in the Einstein literature. However, somatic images differ from culture to culture, and it is interesting to see how Einstein's persona was perceived by Spaniards:

> He is an affable and simple man, young, about forty-five, tall, robust, dark, with not very long hair and a well trimmed mustache.[54] . . . Einstein is a tall man, impressive, of noble bearing, with quiet manners, dark, lively eyes, with a penetrating look, a thick head of hair, an enigmatic smile and an energetic expression.[55] . . . He is tall, strong, dark, with a profile that proclaims his race.[56] . . . His dark eyes have a melancholy expression; he has a distant look, as if he is accustomed to the infinite . . . his mouth is sensual, very red, rather large; on his lips plays a permanent smile: kindly or ironic? Who could define it? He is tall (perhaps one meter 75 centimeters).[57] . . . His leonine head sits majestically on his robust Teutonic neck and his gray hair parts to reveal his broad forehead, a little too Beethovenian.[58] . . . With the appealing carelessness of his appearance and his beautiful head of wavy gray hair, he looks more like a Latin artist than a German thinker.[59]

---

natural curiosity for everything new is increased when the novelty also embodies boldness and leads to extreme modifications of our convictions or actions; this explains the rapid diffusion of Einstein's theory of relativity, in spite of the insignificance of its practical applications"; *El Psico-análisis* (Barcelona: Monografías Médicas, 1926), p. 7.

[54] *El Noticiero Universal*, February 24, 1923.

[55] *El Imparcial*, March 3, 1923; *La Voz*, March 2, 1923.

[56] *La Voz*, March 2, 1923. The referent to *raza* could be either Germanness or Jewishness.

[57] Révéscz, in *ABC*, March 2, 1923.

[58] Vera, in *El Liberal*, March 4, 1923.

[59] Rafagas, in *El Heraldo de Aragón*, March 14, 1923.

It is noteworthy that Spaniards perceived Einstein as a "Teuton," more than a Jew, and a tall one at that; in Spain, of course, Einstein appeared taller than average. In the photograph at the Academy of Sciences (Fig. 8), Einstein is taller than the scientists but shorter than the king and some military figures. At Cabrera's laboratory (Fig. 7), only two of the thirteen other persons depicted are clearly taller than Einstein. Famous persons are often perceived as being taller than they actually are, a common psychological quirk that lent even more magic to Einstein's persona. In contrast, in Buenos Aires (in 1925) he was seen as short.[60]

Einstein's Jewishness was almost never mentioned in the Spanish press in 1923. By contrast, Corpus Barga, reporting from Paris the year before on Einstein's appearance at the Société de Philosophie, included in his description some nuances that were common in the French press: "Einstein is not likeable, it seems to me, at first glance. Nor is he a typical German professor, as they say. No glasses, no bald head. An intentionally disheviled mien. Olive complexion, somewhat buttonnosed. The shirt collar firmly closed with a tie; white stripe on the jacket. Strong shoulders, a little stooped. Thick and flaccid wrists, with real fists, hovering heavily around the blackboard."[61] Many of the French descriptions of Einstein were hostile, while none of the Spanish ones were. References to his nose and his "olive" complexion were standard in unsympathetic French accounts.[62]

[60] *La Razón*, March 25, 1925: "De estatura mediana, más bien baja. . . ."

[61] Corpus Barga, "El 'colloquium' de Einstein con los sabios franceses," *El Sol*, April 14, 1922.

[62] Biezunski, "Einstein à Paris," *La Recherche*, 13 (1982), p. 506, citing *Le Gaulois*, April 1, 1922: "Sa physionomie est celle d'un Méridional; son teinte *olivâtre* . . . tout en lui dément ses origines germaniques" (my emphasis). One of the few dogmatic anti-Semitic references to Einstein made in Spain in the 1920s was by the priest Manuel Barbado, who implied that the ideas of Freud, Einstein, Bergson, Paul Ehrlich, and other Jewish scientists were promoted in press campaigns through "interests of race"; "Boletín de psicología, II. Tratados, textos y orientaciónes generales," *La ciencia tomista*, 32 (1925), 254.

The quality of Einstein's voice was persuasive. Francisco Vera described it as "mellow, straight, scarcely with inflections."[63] Writing in *El Debate*, Enrique de Benito stated that, although admittedly incompetent in scientific matters, he had been an enthusiastic listener at the University of Madrid, "pressed by the curiosity which the mystery of these beautiful and transcendent things awoke in every eager spirit, even if not everything was comprehensible." He was impressed by Einstein's calm and rationality, and probably also by the quality of his voice: "Listening to Einstein, I am inclined to accept his relativistic theses."[64]

Einstein's problems with the French language have been mentioned before. Romain Rolland had noted in 1915 that Einstein spoke it with difficulty, mixing in German words.[65] By 1923 his French had improved to the point where he appeared fluent, at least to the casual observer: "Einstein's French is clear and exact, but with a certain Germanic coarseness in pronunciation."[66] (In his travel diary in Argentina two years later, Einstein still referred to himself as a "French stutterer.")

Ramón Gómez de la Serna, a witty, epigrammatic writer, made the most astute comment of all: with his theory, Einstein was guilty of "having slandered" time pieces. As a result, said Ramón, "I no longer wind mine, having observed that Einstein is a man who doesn't wear a watch."[67]

Another significant aspect of Einstein's public image, one which endeared him to the entire left and even to alliophile conservatives and centrists, was his moral stature. A Barcelona

---

[63] Vera, "La tercera conferencia del profesor Einstein: Consecuencias relativistas," *El Liberal*, March 8, 1923.

[64] Enrique de Benito, "Las conferencias de Einstein. Notas de un oyente profano," *El Debate*, March 6, 1923.

[65] Clark, *Einstein*, p. 185.

[66] *El Heraldo de Madrid*, March 3, 1923. In secondary school, Einstein was "promoted with protest in French"; see *The Collected Papers of Albert Einstein*, vol. I (Princeton: Princeton University Press, 1987), p. 17.

[67] Ramón Gómez de la Serna, "El birrete de Einstein," *El Sol*, March 8, 1923.

daily noted that Einstein was not only a great scientist but also "a moral giant." He had refused to sign the manifesto and despised pan-Germanism. If Newton's laws were false, so were Germany's, the newspaper editorialized, and to the extent that Europe had a moral unity, it was Einstein who had sustained it.[68] Of course, the syndicalists had invited him not because of his scientific achievements but "bearing his pacifist ideas in mind" and for not having signed the manifesto.[69] Noting this, Wenceslao Fernández Flores opined that the fact he refused to sign the manifesto was of minuscule importance in comparison with what he had really accomplished, which "escapes the understanding of those syndicalists." Those who could not understand Einstein, the conservative humorist implied, praised him for those elements of his persona they could relate to, by way of compensation. Had Einstein been more astute, he *might* have noted that the slayings in Barcelona were insignificant compared to those in Russia, Hungary, or Fascist Italy, and of what importance is all this compared to what really mattered—"that Euclidean theory has been overthrown."[70]

Indeed, commentary on the syndicalist episode linked together a number of conservative ideas on the nature of science which had their roots in nineteenth-century views on the opposition between science and traditional values. "Consider now the workers of Barcelona," Ramiro de Maeztu adjured, "assembled to honor, not a man, but Einstein's theories." He suspected that the Confederación Nacional de Trabajo (CNT) admired Einstein because he symbolized change in what was once believed.[71] Miguel Adellac, commenting on Maeztu's article, noted that the syndicalists, taking on blind faith that which they did not understand, submitted "the burning ques-

[68] "Crónica diaria. Einstein," *El Diluvio*, February 24, 1923.
[69] *El Noticiero Universal*, February 28, 1923.
[70] Wenceslao Fernández Flores, "Einstein y los comunistas," *El Diario Español*, April 7, 1923 [dated March 6].
[71] Ramiro de Maeztu, "Fuera de la cultura," *El Sol*, March 6, 1923.

tion of their conduct to his judgment, as if the saving formula, capable of molding . . . a new morality would flow from his lips."[72] For the anarchists, of course, science was salvation, a fetishism to which they—and the Left in general—had long been prey, Adellac insinuates: "In Spain the popular idolatry of the experimental sciences has, for a long time, had well-known promoters whose activity has been stimulated more by their political ideas than by their enthusiasm for science. They are the ones who advocate the excellences of natural history and chemistry as the leading disciplines of general education and the formation of nuclei of professionals in advanced specialties." Adellac, as it happened, was critical of the limited (in his view) pedagogical value of science in a general education. His reasoning was peculiar: the classificatory sciences—zoology, botany, mineralogy, and all their divisions—are reducible to the concepts of genus and species. Thus a small number of examples is sufficient because the sum of the data "would only repeat the foundations of the method, without being able to discern the characteristics of coelenterates, which is pure specialization and useless for education."

This kind of critique of science as a fetish (of the Left, it is understood), links up easily to the criticism of Catholic ultras, who repeated nineteenth-century verities: science was materialism, and the proof of this was the liberal saw that science could cure all social ills. "Betibat," columnist for the traditionalist, Carlist newspaper *El Siglo Futuro*, welcomed Einstein by noting that "Nowadays one cannot pass as a scholar without having a well-established reputation for incredulity or atheism, nor establish oneself as a man of science without displaying a streak of scepticism or materialism." He cites the *Diario de Lérida* for March 9, 1923, quoting the lieutenant mayor of Barcelona as asserting that neither law, justice, nor religion had been able to achieve the union of all men: "Only science

[72] Miguel Adellac, "La fe en la ciencia," *El Heraldo de Aragón*, March 21, 1923.

has achieved it." For Betibat, this translates into "Only science is popular." In fact, he continues, "Christ is the bond of union between all mankind."[73]

The Madrid press did not dwell excessively on Einstein's past and present political views. *El Liberal* hailed him as "a herald who announces universal peace at the bidding of science."[74] A science writer for the same paper, Mariano Poto, asserted that Einstein should be considered a symbol and conjectured that had he signed the manifesto, his theory might have been blocked (among the victors, one assumes) by a conspiracy of silence.[75] On the other hand, the press of the capital was sensitive to the political ramifications of Einstein's visit. In an editorial, *El Sol* noted that the French now suspected Einstein because of his trip to Spain, whereas previously they had admired him as the hero of German resistance. How could "a nation of illiterates" have received Einstein so enthusiastically? *Le Petit Parisien* had gone to the length of printing a telegram from Geneva stating that Einstein's trip to Spain was pure propaganda, backed by the German authorities.[76]

EINSTEIN AND THE PERCEPTION OF SCIENCE IN SPAIN

Einstein's visit lent itself to a public discussion of the role that science ought to play in Spanish society. To some, his presence was a hopeful sign that Spanish science had come of age; to others, he was only a painful reminder of the inadequacies of the local scientific establishment. Cabrera's statement on the embryonic nature of Spanish science stimulated many defensive analyses. For some, Catholics were to blame: "Because of the clerical rabble Sr. Cabrera has rightly been able to say that

---

[73] Betibat, "Saludando a Einstein. Chispazos racionalistas," *El Siglo Futuro*, March 13, 1923.

[74] "Einstein en España," *El Liberal*, February 25, 1923.

[75] Mariano Poto, "Einstein y su teoría," *El Liberal*, March 1, 1923.

[76] "Después del viaje de Einstein," *El Sol*, March 28, 1923.

Spanish science is embryonic."[77] But Cabrera's statement moved others to hope. In a front-page editorial, *El Heraldo de Madrid* declared that it had been wise for those making public utterances to have been modest and to have refrained from citing lists of "great" Spanish scientists, to have admitted that the best of Spanish mathematicians had only been great popularizers. Today, however, Spanish science had entered a new phase of integration with European science. In this sense, the grants of the Junta para Ampliación de Estudios "to study abroad are the most productive expenditures of the national budget." Moreover,

What is the presence of Einstein himself in Spain but a clear demonstration that we are interested here in all new ideas and that we know how to offer their authors . . . the attention demanded of all hypotheses which rectify principles which seemed authoritative? . . . It is to our credit that we are not indifferent to the accomplishments of research scientists, and that Philistinism is an exception in our country.[78]

For others, Einstein's presence brought forth a defensive reaction. A practical geometrician who attended Einstein's lectures reported (with tongue in cheek) that he had dreamed Einstein was explaining his theory "to an audience composed of Spanish scholars of the mid-nineteenth century, those who believed themselves in possession of absolute truth, those for whom classical geometry and Newtonian fields were incontrovertible truths . . . and we saw Albert Einstein, in a straitjacket, led out between guards who put him in a cab and told the driver: 'to the house of Esquerdo, in Carabanchal Alto' "

[77] Roberto Castrovida, "La triufulca de Santo Tomás," *El Pueblo*, March 11, 1923. Catholic students, protesting the nonappointment of a Catholic professor, rioted at the University of Madrid on one of the days that Einstein lectured there.

[78] "Sobre la ciencia nacional. Confesiones y esperanzas," *El Heraldo de Madrid*, March 5, 1923.

(a noted psychiatric clinic of the day).[79] Of course, it was not simply Spanish professors educated in the tradition of Newtonian physics who were opposed to relativity.

One of the most peculiar defensive statements was made by Josep Escofet, who noted that Spanish detractors of Einstein were so ill-educated that they were unable to offer effective criticism. Spain had demonstrated only "an accidental and intermittent interest" in higher learning. True, there had been some notable geographical and navigational accomplishments related to the discovery of the New World, "but this is the extent of our scientific tradition and if we need to lend tone to solemn occasions we can present some rudimentary knowledge, which is what was trotted out to give Einstein the illusion of Spanish scientism." In Barcelona, as well as in Madrid, Einstein met with the opposition of Spanish scholars, "who have emerged from their obscurity for a moment to remind us that they exist and are also skeptical." These Spanish refuters of Einstein have had to use the same arguments used by their colleagues in other countries, especially France, "supporting their positions on what Professor X, Professor Y, and Professor Z said." Their argumentation is wholly second-hand, "whereby it is proven that they would have found themselves in difficult straits had not these foreign critiques existed." If scientists in France, Italy, or England had wanted to know what Spanish scientists thought about Einstein's theories, "wouldn't they have smiled ironically on noting this backwardness? That much is clear."

This same ignorance, according to Escofet, explained Einstein's popularity in Spain. In other countries, interest in his theories waned once the crowd discovered that familiar views of the physical world had not been altered. But in Spain the crowd included educated—poorly educated—people, so interest remained high, even though there was no understanding. "It hurts to say so, but in truth [relativity] has been discussed

[79] "El profesor Einstein en Madrid," *El Auxiliar de la Ingeniería y Arquitectura*, 3 (1923), 69.

in Spain for pure ostentation, making manifest both great superficiality and greater ignorance. Whenever a Spanish mathematician has attempted to popularize Einstein's theory, his failure has been complete and overwhelming. The public has not understood it; he himself has not understood it." The result was that Einstein had come, although without his knowledge, to demonstrate to Spaniards "our scientific poverty, our mathematical simplicity and our childish petulance." They should be more discrete in the future and put up a sign at the frontier declaring "Let no geometrician enter here!"[80]

Escofet's diatribe is notable for the extreme defensiveness that led him to a number of demonstrably false conclusions concerning Einstein's reception in Spain. The first is that more scientists opposed than favored his theories. The second is that the level of mathematical culture in Spain was very low. Probably he was influenced by conditions peculiar to Barcelona, where there was more opposition to Einstein than in Madrid, where the cultural leadership commonly expressed dismay and anger over its inability to grasp the ideas, and where mathematicians had a sense of being at a disadvantage when compared to the more active nuclei of mathematicians in Madrid.

A third generalization, also erroneous, is Escofet's perception that Spanish scientists (those opposed to Einstein in this case) suffered an information lag with the rest of Europe. We have noted that Spanish relativists suffered no such disadvantage, and we can also presume that opponents such as Comas Solà were well in command of anti-relativity materials emanating from European scientific centers. Escofet's position is all the more strange in view of the fact that Comas was a frequent contributor to his own newspaper, *La Vanguardia*, where the astronomer explicated his original ideas on relativity.

Possibly the most valuable aspect of Einstein's trip was that

[80] Escofet, "Einstein y los matemáticos." Escofet here makes a word play on Plato's famous epigram, which also appears on the title page of Copernicus's *De revolutionibus orbi*, "Let no one ignorant of geometry enter here."

it stimulated an open discussion on the value of pure scientific research, as opposed to that with utilitarian or practical objectives. Science writers were at pains to make the point that science was pure ideation and that the public ought not to be overly concerned with practical results emanating from Einstein's theory, In this light, Mariano Poto drew a comparison with Darwinism: ". . . Darwinian evolution was [once] no more than pure ideation. But with the passage of time we have seen it applied to the transmutation of living species; and nowadays don't researchers seek to acquire in the laboratories a model to demonstrate the route that nature followed in the evolution of the elements? . . . Darwin revolutionized mankind's soul; Einstein has achieved even more."[81] Many made the point that utilitarian-minded persons were unable to perceive what Einstein meant to them.[82] For some, the very absence of utility was praiseworthy. Comparing public interest in relativity with that in King Tut's tomb, José María Salaverría declared that people pursue such knowledge precisely because there was no practical benefit to be had from it. Salaverría had been present at the tea at the palace of the Marqueses de Villavieja. In his view, what united the guests, aristocrats, and intellectuals was exactly "that cult of the useless."[83]

I have already alluded to the unspoken pact that existed in the 1920s between previously and potentially warring sectors of the Spanish elite and permitted open, civil discourse on scientific ideas without the necessity of loading each one with some political or ideological charge. Salaverría's article is significant because it specifies two parties to this contract, the intelligentsia and the aristocracy. His testimony should be taken at face value. Other testimony, such as Fernández Flores's lampoon of the minister Amalio Gimeno as an example of a provincial politician unable to connect with Einstein[84] and *El Sol*'s

[81] Mariano Poto, "Einstein y su teoría," *El Liberal*, March 1, 1923

[82] See, for example, J. M. Sagarra in *La Publicitat*, March 4, 1923.

[83] José María Salaverría, "Las originalidades einsteinianas," *ABC*, March 10, 1923.

[84] Wenceslao Fernández Flores, "Einstein y Gimeno," *ABC*, March 14,

editorial scoring the Minister of Public Instruction, corrobo-
rate Salaverría's observation that politicians per se were excluded from the pact. He neglects to mention that the Church
and the Army, each with some notable exceptions, were also
excluded.

Salaverría's reference to the "cult of the useless" was sardonic, but not derogatory. He went on to point out that Einstein was not the kind of scientist to traffic with industrialists,
in the manner of Nobel or Edison. In fact, "The world is replete with practicality. We have a practical science, a practical
morality, virtually a practical religion (pragmatism). If culture
is not practical, it isn't culture." And this, indeed, was the distinguishing feature of the wisdom of Einstein: it is not practical, it does not produce money and—another of its singular
qualities—"no one understands it. . . . But precisely in this re-
sides the charm and nobility of the phenomenon." Relativity
brings to mind the highest moments of Western civilization,
when the greatest truths of science and religion were forbidden
to the masses.

Many, like "Abecé," writing in *En Patufet*, made the point
that Einstein's "discovery did not convey practical advantages
of any kind, unlike the inventions of Pasteur and Marconi."[85]
For Francisco Vera, the utility of relativity was a purely *scientific* utility—"a pretty theory, useful in the same way as a ray of
sunlight is," in that it opens up new fields of mathematical and
philosophical speculation.[86] The theory's specific utility was
that it provided answers to a host of previously intractable astronomical conundrums. It also, he noted, had awakened an
anxiety—"an anxiety to revise all old concepts."[87] Others also

---

1923. Although portrayed as pompous by the humorist, Gimeno was a distinguished physician and prominent liberal politician who played an instrumental role in establishing the early autonomy of the Junta para Ampliacion de Estudios, of which he was a trustee.

[85] Abecé, "Albert Einstein," *En Patufet*, 1923, p. 182.

[86] Francisco Vera, "La tercera conferencia del Profesor Einstein: Consecuencias relativistas," *El Liberal*, March 8, 1923.

[87] Vera, *Espacio, hiperespacio y tiempo* (Madrid: Paez, 1928), p. 160.

noted the same undercurrent of anxiety but linked it to a specific historical context. J. Menéndez Ormaza noted that the war had made Europeans anxious; they had lost confidence in old concepts and were anxious about new directions. That is why they seized upon the new and mysterious theory with such enthusiasm.[88]

Einstein, of course, was also seen as the representative not only of a new and revolutionary science, but of German science. Eugeni d'Ors noted sarcastically that someone should have presented Einstein with a collection of Spanish alliophile magazines from the years 1916–18, wherein it was declared

> that German science was only a bluff; that a German had never invented anything, nor ever would, impeded as he was by the special and lamentable constitution of his brain; that all the German scholars together over all the centuries taken together have only known how to cook, in the most indigestable sauce of pedantry, the original contributions of the Latins and the English . . . ; and, finally, taking these judgments into account, the most urgent duty for any friend of science . . . was to tear apart the Germans.[89]

Ors goes on to note the irony of Einstein's rejection, as a Jew, by German nationalists, just at the time when he was creating the maximum prestige for German science.

Very few of the literary or journalistic figures whose ideas we have been presenting took the care to analyze their views of science in any depth. One who did, and who in so doing betrayed a profound hostility to it, was Jaime Mariscal de Gante, in a digression from a warm and perceptive appreciation of Einstein's wife Elsa ("perhaps one of the most interesting women we have met"). Elsa's dedication to Einstein's abstract science was unusual:

[88] J. Menéndez Ormaza, "Significacion del éxito de Einstein," *El Imparcial,* March 7, 1923.

[89] Eugenio d'Ors, *Nuevo Glosario,* 3 vols. (Madrid: Aguilar, 1947–49), I, 794–795.

This is even more surprising because it involves a woman who is very sensitive to poetry, and poetry is the antithesis of science, which has no more beauty or sensitivity than scientific aridity, an endless deserted steppe upon which ice freezes and the sun burns, but which lacks the coloring of the oasis, the scent of flowers, the clear waters of a pool, sentimental emotions.

The sensibility of science is that of a barren field, of the rugged mountain, of the hurricane and the storm; it is the sensibility of strong and vigorous spirits, but not poetry—poetry is feminine.[90]

Mariscal's appreciation is better characterized as ambivalent than as hostile. Science is masculine, poetry feminine; science is arid and cold, without emotional content—just the opposite of poetry. Some French commentators of the Einstein phenomenon took the opposite tack, of equating science with poetry.[91] In Spain science was viewed within a wholly positivist framework, and therefore the dichotomy between the scientific and the poetic was a widely held perception.

[90] Jaime Mariscal de Gante, "La mujer del sabio," *La Voz Valenciana*, March 8, 1923 [dated Madrid, March 7].
[91] Cf. Biezunski, "La diffusion de la relativité," p. 149.

EIGHT

Flow and Transformation of Ideas

DISARTICULATION OF SCIENTIFIC IDEAS

The subject of this chapter is a continuation of the last, for here we deal with popular permutations of Einstein's theory, in particular those that reached the man in the street. There are few documented responses at this level, but notions broadly held at the popular level are reckoned to have been reflected in cartoons and in the writing of humorists. Since such perceptions were typically based on elements of Einstein's ideas disarticulated from their scientific context, they link up with similar concepts expressed by members of the intelligentsia. The appropriation of disarticulated ideas is a normal way (perhaps *the* normal way) in which the general public—those with no formal, articulated background in science—assimilate complex scientific ideas. In this context, the popularization of science is only partly explicable in terms of difficulties in mastering a particular set of ideas. A disarticulated idea may have some truth in it but appear as an absurdity when presented out of context or in mixed philosophical and scientific contexts. Biezunski's "semantic slides" can frequently be understood against a specific social or cultural background.

A variety of "popular" responses to Einstein are indicated in Table 4, chapter 7. Many authors freely mixed elements relevant to the theory with other elements appropriated from non-scientific contexts to achieve a kind of impressionistic picture that had little or nothing to do with the physical theory.

To divide the reading public into highbrow, middlebrow, and lowbrow groups based on educational background[1] does

---

[1] This is the scheme adopted by Alvar Ellegard, *Darwin and the General Reader* (Goteborg: Goteborgs Universitets Arsskrift, 1958).

not necessarily resolve the problem of classifying popular reception. The reason for this is, of course, that a "highbrow" reader with no formal scientific education may have as little understanding of a given idea as a lowbrow reader (or non-reader) may have. Consider, for example, a reflection of relativity in Proust. The novelist had been flattered when Jacques Emile Blanche compared him to Einstein, and he was embarrassed when Christian Romestead, president of the Swedish Academy, wrote, "You accelerate and decelerate the rotation of the earth, you are greater than God!" This statement, typical of those which were popularly regarded as having something to do with relativity, must have gotten Proust's mind turning in relativistic ways, for he later explained some apparent anachronisms in *Le côté de Guermantes* by saying that these were due "to the flattened form my characters take owing to their rotation in time."[2] The lesson this anecdote holds for the study of the popularization of relativity is that Proust's comment would be regarded as imbecilic, had not Proust been its author. Because of Proust's tremendous prestige, a reader of this statement might be led to suspect (or assume) that the novelist possessed, at the very least, some intuitive understanding of Einstein's ideas, when in fact he more than likely had none. Likewise, Ortega's article sounded plausible, and its readers were willing to take his word that relativism had something to do with relativity. In this fashion a scientific corpus can pick up nonscientific accretions due to the prestige of their authors.

How does the impressionistic mode work in practice? An explicit answer is given by Salvador Dalí, who has been, in many ways, the prototypical consumer of science in twentieth-century Spain. Dalí has been a voracious reader of science in fields as diverse as psychology, physics, evolutionary biology, and biochemistry, and his writings and public utterances provide an excellent commentary on the scientific atmosphere in Spain of the 1920s and 1930s as refracted through the prism

[2] George D. Painter, *Proust, the Later Years* (Boston: Little, Brown, 1965), p. 336.

of impressionistic (that is, disarticulated) popularization. He has had Darwinian dreams ("tertiary landscapes," he calls them), which he interpreted according to Freudian psychology; and his conscious world was one of discontinuous matter operating according to the rules, as Dalí understood them, of quantum physics.[3] Yet, in spite of the massive amount of verbiage that Dalí has produced on scientific themes, he has never entertained the slightest illusion of his own comprehension of them. His candor is refreshing:

> For myself, le more complicate things for this reason like it le best today; le nuclear microphysics and mathematics because no understand, myself no understand nothing of these. Is *tremendous* attraction for understand something in this way.[4]

> I do nothing more than read books which I don't understand. Scientific books. I cause myself terrible problems. But I make a synthesis of these problems and arrive at valid conclusions, so much so that, at times, in conversations with scientists and Nobel Prize winners, they ask me, "How did you get to know this?" and I reply, "This I have deduced by reading things I don't understand."[5]

One example of the Dalinian scientific synthesis in action will suffice to make my point. Dalí's most interesting and sustained work of scientific popularization is a Freudian interpretation, written in the early 1930s, of Millet's canvas, *The Angelus*. It is an essay that actually occupies a place in the history of modern art because in it Dalí develops his famous "Critical Paranoid Method," a technique whereby he created delirious scenes in his mind and then analyzed and/or painted them. The

[3] Pronouncements on physics are scattered throughout Dalí's works and occur frequently in interviews. See, for example, Alain Bousquet, *Conversations with Dalí* (New York: E. P. Dutton, 1969), pp. 61, 110.

[4] Cited by Jacques Barzun, *Science, the Glorious Entertainment* (London: Secker and Warburg, 1964), p. 228n (emphasis Dalí's).

[5] "Dalí: El equilibrio de la contradiccion," *Destino*, no. 2193 (October 17–23, 1979), 7. Translated from the Spanish.

technique and Dalí's views on the subconscious processes underlying it were worked out while Dalí was in contact with Jacques Lacan, then preparing his doctoral dissertation on paranoia.[6] In his analysis of this painting, Dalí notes that painters frequently compress a temporal sequence of events into a single scene. These sequences can be reconstructed analytically by using Freudian notions of condensation, substitution, and displacement, but the temporal reference system will differ from painting to painting and such differences must be understood relativistically. The solution to the tension between concrete and psychic time ("oneiric time," he calls it, the time-frame of memory) must be resolved dialectically for each painting. The temporal dialectic suggests a more encompassing framework, also relativistic:

> The paranoid phenomenon which, in the field of poetry, makes the very dialectic of the surrealist delirium tangible and objectively recognizable, that paranoid phenomenon—I repeat—which is the true dialectic of the surrealist delirium, I can only now understand (considered in the field of natural sciences) as the poetic equivalent of that "conciliation" of everything irreconcilable, as that translucent clarity, born of the entangling and rapprochement of the most irreducible and distant antagonisms, as the sum of the "concrete dialectic" objectified in that grand theory, whose theoretical heights are only accessible to us intuitively and which is called "the special theory of relativity."[7]

Here Dalí synthesizes a number of themes in the popular reception of relativity. First is the appreciation that Einstein had

---

[6] Patrice Schmitt, "De la psychose paranoiaque dans ses rapports avec Salvador Dalí," in *Salvador Dalí, restrospective* [Catalog] (Paris: Centre Georges Pompidou, 1979), pp. 262–266; Thomas F. Glick, "The Naked Science: The Reception of Psychoanalysis in Spain, 1914–1948," *Comparative Studies in Society and History*, 24 (1982), 565–566.

[7] Salvador Dalí, *El mito trágico del "Angelus" de Millet* (Barcelona: Tusquets, 1978), pp. 126–127, 152.

resolved important contradictions in physical theory and that the significance of this achievement could be appreciated without alluding directly to the content of the ideas involved. Second, he stands the intuition argument on its head by stating that the Special Theory is accessible only through intuition; Emilio Herrera and others argued the opposite. Finally, he makes a common, and erroneous, deduction from the Special Theory that "mental phenomena in a human observer have thereby been introduced into the very definitions of physical science."[8] Given Dalí's psychoanalytical approach to painting, the attractiveness of such an idea is obvious.

In their "popular" reception, then, scientific ideas acquire subjective meanings that are impossible to objectify. It is not even useful to attempt to identify "objective" elements in such cases because even those are imbedded in idea sets—frequently ideological—that have nothing to do with science per se and certainly have no physical meaning. With Dalí, for example, defeating "Maxwellian science" was a point of honor—that was why he supported quantum physics and relativity. But the new physics was important to him only as an analogue to the overthrow of classical perspective and to the ideological aims of the surrealist movement.[9] In a 1935 article, Dalí elaborated the significance of special relativity as he traced the history of conceptions of space from Euclid, through Descartes and Newton, to Maxwell and Faraday. "But in that period [the second half of the nineteenth century] scientists thought it so absurd to concede to space the function of physical states, that

---

[8] As characterized by Gerald Holton, "Einstein and the Shaping of our Imagination," in Holton and Yehuda Elkana, eds., *Albert Einstein: Historical and Cultural Perspectives* (Princeton, N.J.: Princeton University Press, 1982), p. xxiv.

[9] Cf. André Breton, "Limits, not Frontiers for Surrealism" [February 1937], in *From the N.R.F.*, Justin O'Brien, ed. (New York: Farrar, Straus and Cudahy, 1958), p. 87: "Just recently I sought to show that to an *open rationalism*, which defines the present position of scientists (as a consequence of the concept of non-Euclidean geometry, followed by generalized geometry, of non-Newtonian physics, of non-Maxwellian physics, etc.) there corresponds an *open realism* or *Surrealism*, which brings down the Cartesian-Kantian structure and overthrows sensibility."

the thousand tentative subtleties of the aether were needed in order to reach the modern theory of relativity in which space has been transformed into something so important, into a true and material thing, that it has even acquired a fourth dimension, time, which is the surrealist and delirious dimension par excellence."[10] Dalí is an example of a nonscientist whose personal cosmology was changed through the popularization of special relativity.

We have noted that philosophers like Unamuno or d'Ors were given to dropping Einsteinian epigrams, based on disarticulated bits of information, for example, the paradox of relativistic time. Other intellectuals did the same. In *Juan de Mairena*, his collection of the wit and wisdom of his apocryphal philosophy teacher (who supposedly had died in 1910), Antonio Machado observed: "Upon reading what is currently being written about the modern theory of relativity, Mairena would have said: 'What an elegant way to stop the clock of the Divinity himself.' The truth is that God—unlike the god of my mentor—is not omniscient. What terrible blunders we commit when we judge the order of events."[11] Mairena's concept of an immanent, abstract, and rational divinity seems broadly consonant with at least the spirit of Einstein's public statements on the God of Spinoza who does not throw dice. Gaziel, to cite another passing reference to Einstein in the same vein, recounted that Einstein "in my presence, assured Esteve Terradas . . . that what is most incomprehensible about the universe is its comprehensibility."[12] Squibs of Einsteinian wisdom were, in this fashion, passed from person to person and be-

[10] "Aparíciones aerodinámicas de los seres-objeto," *Minotaure*, no. 6 (1935), reprinted in *Sí* (Barcelona: Aríel, 1977), p. 67.

[11] Manuel and Antonio Machado, *Obras completas* (Madrid: Plenitud, 1967), p. 1166. The paragraph is entitled "Sobre el tiempo local." There is no doubt that Machado was well informed scientifically. Freudianism appears in his play, *Las Adelfas*, coauthored with his brother, and elsewhere he shows his familiarity with Rocasolano's biocolloidal theory of aging (see *Obras*, p. 1105, on the "mineralization of our cells," and p. 1073 for a reflection of Marañón's theory of the climacteric).

[12] Gaziel, *Memòries, Historia d'un destí (1893–1914)*, 4th ed. (Barcelona: Aedos, 1967), p. 13.

came fare for more popular consumption, in cartoons or in utterances by the man in the street.

CIRCLES OF AFFINITY: TERTULIAS

Circles of affinity are informal groups of friends and acquaintances which serve as media for the diffusion of (in this case) scientific ideas beyond the strict confines of the scientific discipline. Such groups may include, around the same table, scientists, philosophers, writers, poets, bullfighters, and others of the most diverse backgrounds. Ortega's famous *tertulia* was, in the 1920s and '30s, a focus of discussion both of Freudian psychology and of the new physics. In his memoirs, one of the participants, the novelist Francisco Ayala, described the gatherings as "a scientific or philosophical seminar, although lacking the pedantic formality . . . which usually accompanies similiar academic rituals." According to Ayala, Manuel G. Morente and Xavier Zubiri, both of whom were philosophers who commented on relativity, were conspicuous members.[13]

Another influential *tertulia*, that of Ramón Gómez de la Serna, which held forth in the Café Pombo in Madrid, was more literary and less philosophical than was Ortega's but still numbered among its participants scientists from the main circles of Freudian and Einsteinian discourse (for example, J. M. Sacristán and G. R. Lafora, psychiatrists, and Manuel Martínez Risco, physicist). Unlike Ortega, Gómez de la Serna left a detailed, if chaotic, record of how his *tertulia* functioned, including a description of how science was discussed. The episode in question was the visit of Georg Nicolai, the German physiologist who had joined with Einstein in opposing the manifesto of the German intellectuals and who soon after took up a position at the university of Córdoba, Argentina (where

[13] Francisco Ayala, *Recuerdos y olvidos* (Madrid: Alianza, 1982), pp. 95–99; Glick, "The Naked Science," p. 543; see also Horacio Bentaból's reference to the discussion of relativity in *tertulias* on the premises of the Athenaeum of Madrid in *Observaciones contradictorias* (Madrid: Imprenta Velasco, 1925), p. 12.

in 1925 he not only greeted Einstein but was a major popularizer of relativity in Argentina). Nicolai, during a visit to Madrid in the 1920s, was accompanied to the Pombo by Juan Negrín, and there he replied to questions on the nature of life, issuing a general invitation to attend his lecture the following day on the subject "What is life?" Gómez de la Serna's comments on how little he understood reflect his experience both at the lecture as well as in the more informal setting of the *tertulia*: "After applauding the science, the keenness, the Andalusian expressiveness of this distinguished German, we all returned to our daily routines without knowing any more than we knew before going, confused about everything, with the same unanswered questions."[14] *Mutatis mutandis*, the same experience was true, truer by far, of relativity. Many of the comments we have recorded may also have reflected a cleavage between the intelligentsia and scientific culture generally.

POPULAR ACCESS TO SCIENTIFIC IDEAS

What distinguishes the popular reception from that of the scientific middle class is that, in the former, there is no organized or formal, articulated access to scientific ideas. Such access requires, first, a specific level of scientific education necessary to place the new facts in perspective and, second, some institutional supports whereby the ex-science student can relate to authority figures (for example, former professors) in order to validate or invalidate the new ideas. I have in mind engineering journals, public lectures, newspaper articles or interviews, and so forth. Free-form reading of popularizing literature, on the other hand, is bound to lead, in the vast majority of cases, to a subjective, impressionistic, disarticulated, and incomplete assimilation of the idea in question.

When discussing the incidence of relativity and Einstein in humorous articles, jokes, and cartoons, one would like to

[14] Ramón Gómez de la Serna, *La Sagrada Cripto de Pombo*, (Madrid: G. Hernández, n.d.), II, 449–451 (list of participants), 414–416 (visit of Nicolai).

know if humorists or cartoonists were representing popular themes—attitudes held by the "man in the street"—or whether such attitudes were simply so attributed by the writer or artist. What is the sociological evidence that popularization actually occurred, that articles were really read and relativity really discussed? There is every indication that popularizations were widely read and relativity was discussed by all social classes in large cities, and by middle class individuals in small towns. The beginning of the phase of intensive popularization antedated by a year Einstein's trip to Spain, probably stimulated by his trip to France in April 1922. In that year, Gregorio Marañón recalled, traveling by rail to Spain "on my way from Paris, I observed that, of the five people who shared my compartment, three were reading books about relativity which were on sale on the railway platforms, alongside the daily papers. Relativity is discussed in serious articles, in theatrical works, in cartoon captions and now, finally, there have appeared critiques, some fair, others exaggerated, which oppose the hyperbolic diffusion of a doctrine whose seriousness is incompatible with the man on the street."[15] Marañón mentioned this phenomenon in a book on the endocrinological doctrine of internal secretions in order to support his view that the receptions of important scientific ideas follow pretty much the same pattern. First, there is a period of latency, when a new idea is known only to a few; then, an explosive period of discussion when publications multiply at a rapid rate; then, a hyperbolic period, "in which new ideas which are excessively diffused and which lose their original scientific structure and seriousness, changing into all-purpose answers, applied to everything, explaining everything." The hyperbolic period is followed by a movement of reaction, "also exaggerated," wherein the pendulum of opinion swings the other way, and critics of the idea are able to score points by criticizing the

[15] Marañón, *Problemas actuales de la doctrina de las secreciones internas* (Madrid: Ruiz, 1922), p. 6.

overgeneralizations of the previous phase.[16] For Marañón relativity had entered its hyperbolic phase in 1922. Einstein's visit the next year simply added fuel to the fire.

In Madrid during Einstein's visit, Francisco Vera reported, "it has been a week now since we have heard the word 'relativity' everywhere, in cafés, in offices, and on the street," and, Eugeni d'Ors added, "It was inevitable that Einstein should make a brief appearance in after-dinner conversation."[17] Outside the first lecture one writer noted an "a large street crowd" reminiscent of the "bullfight fever" of several years earlier.[18] Another warned that what Einstein would say in his lectures "had absolutely nothing in common with what was said on streetcars and in cafés."[19] Popularizers originating in the scientific middle class were trying to reach this level of reader, those who spoke of Einstein on trams and in bars. Mariano Moreno-Caracciolo introduced a popularizing article by supposing that his readers did not know absolute differential calculus. Neither did he: "This is an unpretentious chat, among simple folk, about Einstein and his theories, as if talking about Rafael el Gallo [a bullfighter] or García Prieto [a politician]."[20]

In Zaragoza, of course, the intellectual atmosphere was provincial, perceptibly different from that in Madrid. There, informal conversation was identified as *the* major source of diffusion of relativistic ideas: "Now people in cafés, *tertulias*, and cliques (*corrillos*) talk about Einstein and his immense work. Even the local wits have had to submit to this irresistible wave . . . and, laughing all the way, they have been the most successful popularizers of the scientist's name and the great im-

[16] Ibid., pp. 2–5.

[17] Francisco Vera, "La tercera conferencia del profesor Einstein," *El Liberal*, March 8, 1923; Eugenio d'Ors, *Nuevo glosario*, 3 vols. (Madrid: Aguilar, 1947–49), I, 943.

[18] Miguel de Castro, "Einstein y los madrileños," *Las Provincias*, March 11, 1923.

[19] Rigel, "Einstein y la relatividad," *El Heraldo de Madrid*, March 1, 1923.

[20] Mariano Moreno-Caracciolo, "La teoría de la relatividad," *El Sol*, October 8, 1920. El Gallo was a bullfighter, and Manuel García Prieto a politician.

portance of his work."[21] The sense of *corrillo*, suggesting a knot of impromptu conversationalists in some public square, is indicative of the currency of Einstein. Such gatherings, in addition to more typical ones in cafés and *tertulias*, are the wellsprings of journalistic humor. And it was, therefore, to the humorists that fell a fundamental burden of disseminating, more than Einstein's name, his persona; and, more than the contents of his work, a sense of its enormity, its hyperbolic nature. Another Zaragozan columnist, C. Sánchez Peguero, embroidered the same theme:

> Virtually no one understands his famous theories; but everyone is talking about relativity and the artistic physicist. The coffee-house strategists . . . move lumps of sugar around to give an idea of what an "inertial" system is. The press fills column after absurd column, horribly distorting the relativistic harmonies. Cartoonists and humorists squander their charm and good humor on Einsteinian curves and space. The role of the generalizer, never lacking in any latitude, tends to make relativity the panacea of universal knowledge which, at the same time as it resolves problems of the interatomic world, poses the gravest questions about bullfighting. Even in village *tertulias*, where the vicar, the doctor and the mayor play cards in the back room of a grubby pharmacy, they speak of its portent, one believing it a corroboration of some theologian's conjecture, another holding it to be no more than a hoax based on the application of useless calculations to what a forgotten and insignificant book by an old professor of physics said on such and such a page of some chapter.[22]

For Sánchez Peguero most of the talk of relativity was on the level of M. Homais—so beautifully typified in the *tertulia* at the

---

[21] Rafagas, "Lecciones de humildad," *El Heraldo de Aragón*, March 14, 1923.
[22] C. Sánchez Peguero, "Un aspecto mínimal de la relatividad," *El Noticiero*, March 13, 1923. On the apothecary's *tertulia*, see José Luis Urreiztieta, *Las tertulias de rebotica en España, Siglos XVIII–XX* (Madrid: Ediciones Alonso, 1985).

village pharmacy, where the vicar, the doctor and the mayor spouted sententious opinions on relativity based on vaguely relevant readings, dimly remembered. The locus of such discussions is significant, moreover, because it represents the geographical and social limits of the diffusion of relativity in Spain as of 1923.

When Einstein departed the scene, so did discussions of relativity, at least on this level. In 1924 Miquel Masriera observed: "Since Einstein's trip to Spain almost no one has spoken of Einstein again. He has lost topicality. Or, better said, our topicality—the province of humorists and soccer fans—has lost Einstein." We noted in chapter 2 that intellectual or scientific fads tend to decline very rapidly after reaching their peak. Because Einstein's trip was a "media event" that pumped up the incidence of discussion much higher than if no trip had occurred, the decline was more dramatic and would have so impressed a scientist like Masriera who was interested in relativity. In losing interest in relativity, the general public was also taking a cue from the scientific community: from a high point in 1922, scientific publications fell off sharply from 1925–26 on. Theoretical physicists, preoccupied with the study of the infinitely small (quanta), abandoned gravitation and the universe to cosmologists and did not begin the process of integrating general relativity into the theoretical corpus until the 1950s.[23]

EINSTEIN AND SPANISH HUMORISTS

Literary humorists typically produced jokes that were plays on the incomprehensibility of mathematics, and on space, time, gravity, or other specific elements in relativity theory. An example of the first is a parody of mathematical reasoning that appeared in the Catalan humor magazine, L'Esquella de la Torratxa:

[23] Miguel Masriera, "El estado actual de las doctrinas de Einstein," *La Vanguardia*, October 25, 1924. For a quantitative and qualitative analysis of the decline of interest in relativity, see Eisenstaedt, "La relativité générale à l'étiage," *Archive for History of Exact Sciences*, 35 (1986), 178–185.

CHAPTER 8

The passage through Spain of our colleague in physics and mathematics, the studious Dr. Einstein with his theory of relativity, has caused so much discussion among us scholars who devote ourselves to the field of exact sciences—or almost exact, at least in good part exact—that it has cast into doubt all the fundamentals which we held, as well as some we never held.

This confusion, more than anything, exists because his theory has not been clearly explained, and that is what we are going to do!

We already know what A and B are, and light and space. Thus space with respect to light is like light with respect to A, and A with respect to B. Light conserves mass and mass conserves energy while energy is conserved.

$$X + 2i = n2$$

This is the trajectory of things which have motion or, better said, acceleration, and acceleration is more or less rapid according to its degree of speed, and thence to dynamics, and so much from i to X that when we multiply it by the coefficient we get the result H, in a relative manner.

If we use the equations discovered by Zorenz and distinguish between bodies and forms, whether physical or mathematical, and we are assured of Newton's dynamics, it cannot be true, and we accept the simultaneity of luminous rays, as well as opaque ones, and if we further accept phenomena and reject sensations, we will have the relativistic statement that two is, in fact, two, albeit in a relative way.

Mass and energy are insubstantial, but if there is peace, and having both mass and energy, we have

$$Y. \quad f()$$

which is all, and all is not nothing if nothing isn't everything.

I believe the reader will have understood Einstein's science, since I have explained the theory concisely, clearly and without rhetoric. If not, go to some of his lectures, because you already know what they'll be like.[24]

Of course, the "clear, concise and nonrhetorical" exposition was not unlike much of what passed for popularization, Einstein's own little book included. Either the intrusion of algebraic formulas added nothing to the explanation, or, from the layman's point of view, they defeated the objective of popularization by recourse to what amounted to a language as foreign as Sanskrit. The article also faithfully mimics the hyperconcision of many popularizations.

There were plays on the relativity of time, which typically made the point that clocks no longer work as they should: Corpus Barga, riding the Paris-Rome express after having attended Einstein's colloquium at the Collège de France in April 1922, noted that his watch, set to the time of the Paris station, should now be slower with respect to it and, as a result, he will be late for a meeting in Rome.[25] During Einstein's visit the joke was repeated often. The satirical poet "Nyic," writing in the Catalan humorous weekly En Patufet, noted that Einstein was the inventor of relativity

> according to which it can't be said,
> nor even less insisted
> that the time shown on the clock's face
> is really the correct time,
> nor that a meter always has
> exactly one hundred centimeters;
> that is, the notions of time and space
> are relative, he says.

However, one need not be a physicist, nor have spent years "making algebraic theorems and binomials," because simple observation leads to the same conclusion. For example,

[24] "Relativitzant," L'Esquella de la Torratxa, 46 (1923), 176 [March 16].
[25] Corpus Barga, "Cronica einsteiniana. El 'colloquium' de Einstein con los sabios franceses," El Sol, April 14, 1922.

And so, of course, I own a clock,
perfect and guaranteed;
its brand name is impeccable
but still this doesn't mean
that I don't have to have it repaired
every other week,
and when it says it's four o'clock,
it's really only three
and when it says it's almost twelve,
it's only ten-fifteen.

Another example is the lady who goes to buy several meters of cloth, only to find that each meter lacks three-quarters of a palm. A final example illustrates the relativity of space: those who ride the streetcars at the luncheon hour know that a platform designed for ten or twelve persons holds twenty.[26] The central theme of this poem is not so much that commonsensical notions of time and space have gone awry, but that common sense had already discovered these absurdities. The same point was made epigrammatically by a Basque humorist, commenting on press reports of Einstein's lectures, which "a few, very few, take seriously. The rest take it as a good joke." Newsmen, indeed, have long known that there was no true time or space, because they habitually use the phrase, "due to lack of space and time."[27]

Another tack was to use Einsteinian references to make points about current politics, the preferred strategy of *El Sol*'s humorist, Julio Camba (1882–1962). He recounted having read a book about Einstein's theories and learning that light had weight. If a ray of light suffers a detour of 1.75 degrees when passing the sun, he observed, a speech of Juan La Cierva's, whose speed is much less, would fall into the sun like a "a crushing mass, neither more nor less than had it fallen on the Congress of Deputies." The literary universe, he concluded,

[26] "Pel·licula de la setmana," *En Patufet*, 1923, p. 192.
[27] Fulano de Tal, "La teoría de la relatividad es muy vieja," *El Noticiero Bilbaino*, March 11, 1923.

was still a Euclidean one; it needed Einstein's help.[28] In 1923 Camba welcomed Einstein to Madrid with another well-worn conceit: that Spaniards were innately relativistic. Scholars in London and Paris had opposed him, Camba noted, but that will not happen in Spain:

> Spain, Sr. Einstein, has never been Euclidean, except officially. In fact, it has existed beyond any defined space, and the system of measurements with which it derives its values has nothing to do with the geometries of the rest of the world.
>
> Time does not exist in itself, or in absolute terms? We already knew that, Sr. Einstein! Time doesn't exist because we Spaniards have killed it. . . . Our politicians are eternal and our academicians boast the title of immortals.
>
> If you aspire to a minor government job, I cannot say that our scholars will shower your path with roses. But if you limit your activities to turning the universe upside down, Official Spain will receive you with open arms.[29]

The image of a non-Euclidean Spain was often repeated. In our discussion of Einstein's visit to Zaragoza, we noted Marcial del Coso's glib statement that relativity was more widespread there than the sugar beet. "Lotario" urged the anti-relativist Bentaból to dog Einstein's steps, wherever he might speak, to destroy his theories, but predicted that Bentaból would not succeed in Madrid because "Madrid is a relativistic town. At least, its largest social class—that of government functionaries—is. What could be more relative that 4000 pesetas, with a discount!" (the latter, a snide reference to Einstein's honorarium).[30] The ribaldry was upsetting to some commentators. Arturo Mori's view was that "From Antón de Olmet to Julio Camba, passing through Fernández Flores, our

[28] Julio Camba, "Las teorías de Einstein y el universo literato," *El Sol*, July 8, 1922.

[29] Camba, "Bienvenido a Einstein," *El Sol*, March 1, 1923.

[30] Lotario, "Relativismo y Bentabolismo," *El Heraldo de Madrid*, March 15, 1923.

CHAPTER 8

newspapers have said the worst that could be said of a philosopher who has agreed to be our guest: that he is a joker. . . .
With his customary comic flourish, Camba says that the idea
of erasing time, killing time, is something that enchants us
Spaniards." And the public laughs.[31]

"Pop" Relativity

The humorous uses of relativity become even clearer when
one examines newspaper cartoons. Following George Basalla's guidelines, we note the stripping down of scientific ideas
at the most "popular" level to a small number of baseline
clichés with varying degrees of fidelity to the original idea.[32]
In this regard, relativity was somewhat distinct from other
significant theories of contemporary science. In the cases,
for example, of Darwinism and psychoanalysis, the baseline,
stripped-down, key-word, "pop" definitions, although frequently misused, bore some relationship to the theories in
question (for example, the struggle for existence, survival of
the fittest, humans descended from anthropoid apes; dreams
have unconscious meaning, sexual problems cause neurosis,
and so forth). Relativity was too complex and abstruse for
similar derivations. For this reason, the "pop" representation
of Einstein was somewhat broader than that of the other two
cases. The analog of the typical Freud joke (sexual innuendos
of various kinds) or of Darwinian monkey jokes was the incomprehensibility joke: relativity was the epitome of all that
was obscure, complex, or, *mutatis mutandis*, that which, in
their simple but profound wisdom, the common folk had already grasped. Some relativity jokes related to some aspect of

[31] Arturo Mori, "Crónicas de Madrid," *El Progreso*, March 6, 1923. For an
appreciation of Camba's black humor, see Francisco Umbral, "Los ajos," *El
Pais*, April 4, 1980.
[32] George Basalla, "Pop Science: The Depiction of Science in Popular Culture," in G. Holton and W. Blampied, eds., *Science and Its Public: The Changing
Relationship* (Dordrecht: D. Reidel, 1976), pp. 261–278.

292

the theory (for example, that light has weight); others merely built on Einsteinian-sounding terms. Indeed, *anything* geometrical could be made into a relativity joke: curves, for example, seem to have been inherently funny (because women are endowed with them), even if the curvature of space was too esoteric an idea for a one-panel cartoon.

The following is a sample of fifteen cartoons appearing in the Spanish press during Einstein's visit. Of the fifteen, five deal with the incomprehensibility issue, four with "relativism," two with general geometrical jargon, two with the notion that light has weight, one with the relativity of time and space, and one with snobbism.

Snobbery

1. Two well-to-do men are conversing:
   "Einstein left . . ."
   "Sure, but Basallo's here!"
The cartoon is entitled "The Hero of the Week," and Basallo was a hero of the war in Morocco who was on the Spanish banquet circuit during Einstein's visit.[33] The implication here is that Einstein's visit had no social meaning other than its notoriety, which was reason enough to attend a banquet because the upper classes had to have cultural heroes.

Incomprehensibility

2. In one of a number of cartoons inspired by Einstein drawn by *El Sol*'s cartoonist Lluis Bagaria (1882–1940), a child addresses his father:
   "Tell me, Dad, is there anyone smarter than Einstein?"
   "Yes, son."
   "Who?"
   "The person who understands him."[34]

[33] *El Heraldo de Madrid*, March 16, 1923.
[34] Bagaria, "El tema de actualidad," *El Sol*, March 8, 1923.

3. Two priests converse:
"What do you think of Einstein's theory of relativity?"
"Frankly, Don Zenón, I don't understand it . . ."
"But, for God's sake, it's clearer than the light of the Electric Company, clearer than the water of the Canal . . ."
"Golly, Don Zenón, well . . . that's some kind of clarity!"[35]

4. "Absolute Curvilineal Movement and Absolute Rectilineal Movement." A *chulo* (typical figure of the Madrid underclass), reading the newspaper, exclaims:
"Gosh! This theory of Einstein's is clearer than water!"[36]

5. Two *chulos* are conversing near a gate of the Plaza Mayor:
"I envy Einstein, because as smart as he is, no one 'unnerstans' him."
"On the other hand, nobody understands you, a total goon, either!"[37]

6. Under the title, "Einstein's Audience," Bagaria presents two panels: in the first, women are listening to relativity lectures; in the second, men listen to Einstein playing the violin.[38]

The notion of clarity doubtless originated in statements made by scientists praising the clarity of Einstein's exposition. If this was so, the cartoonists imply, then the concept of clarity is in need of revision. The men listening to the violin and the two priests make the point that the class of persons usually counted on to understand new scientific theories had failed in its role as intermediary.

Relativism

7. A university professor and a top-hatted bigwig are conversing (Fig. 14):

[35] *El Heraldo de Aragón*, March 14, 1923.
[36] *Heraldo de Madrid*, March 7, 1923.
[37] *Heraldo de Madrid*, March 13, 1923.
[38] *El Sol*, March 9, 1923.

EINSTEIN EN MADRID

—¡SI, SEÑOR GEDEON; NO ME DEJAN EN PAZ NI UNA DIEZMILLONÉSIMA DE SE-
GUNDO. ME TRAEN Y ME LLEVAN EN UN MOVIMIENTO ABSOLUTO RECTILÍNEO, CURVI-
LÍNEO, UNIFORME Y ACELERADO...!

14, 15. Newspaper cartoons on Einstein's Madrid lectures.

"And you, Calinez, have you understood the theory of relativity?"

"Friend, the truth is: I have understood it . . . very relatively!"[39]

The ceremonial dress of the two figures leaves no doubt that this is a comment on the incomprehension of the elite and its inability to popularize the theory.

8. "Einstein in Spain." Two men are conversing in a classroom, with a blackboard filled with formulas in the background:

"And have you seen him?"

"Yes. He's relatively young . . . relatively tall . . . relatively stout."

"Sure, sure, I can imagine . . . relatively."[40]

This joke makes the same point, while parodying the stock descriptions of Einstein included in all newspaper accounts of his visit.

9. "Scientists." At the market, a vegetable lady, bargaining with a customer, says:

"Come on, lady, fourteen small leeks. Don't come back at me with the theory of relativity!"[41]

Relativity, therefore, should not constitute an excuse for altering normally accepted quantities. Putting this sentiment in the mouth of an uneducated person adds to the joke, while convincing us that "relativity" was discussed at all social levels, at least in Madrid.

Finally, the conjunction of Einstein's visit and civil strife in Barcelona produced this cartoon:

10. "Relativity and Terrorism." Two professors are conversing in front of a "Real Acade[mia]" (evidently that of Barcelona, during the labor unrest concurrent with Einstein's visit):

---

[39] *ABC*, March 3, 1923.

[40] *El Mercantil Valenciano*, March 7, 1923.

[41] *El Debate*, March 3, 1923.

"The authorities assure us that relative peace reigns in Barcelona."

"But, who is the civil governor of Barcelona, Raventós or Einstein?"[42]

Scientific Allusions

References to various aspects of Einstein's theories range from the very general to the specific. Any reference to geometry could be stretched to suggest a relationship with Einstein, but curves (because of the curvature of space) were the most obvious:

11. In a cartoon entitled "Einstein's Theories," Bagaria depicts men observing a woman and exclaiming, "What curves! Long live Einstein!"[43]

In another drawing, Einstein expresses the weariness that has resulted from his fame, using jargon that the cartoonist has borrowed indiscriminately from Einstein's first two Madrid lectures (Fig. 15):

12. "Einstein in Madrid." Einstein addresses a man:

"Yes, Sr. Gedeón, they won't leave me in peace. Not a ten-millionth of a second. They take me here and there in an absolute, rectilineal, curvilineal, uniform, and accelerated motion!"[44]

The fact that familiar concepts of time and space had been altered (the former more severely than the latter) also lent itself to the theme of relativity as an excuse for certain actions. Note that a usual distortion of Einstein's ideas was to omit modifiers like "absolute" or "infinite" from space and time, thereby challenging the authority of physicists to define their own concepts.[45]

[42] *ABC*, March 18, 1923.
[43] *El Sol*, March 5, 1923.
[44] *ABC*, March 8, 1923.
[45] Cf. Biezunski, "Einstein à Paris," *La Recherche*, 13 (1982), 507, and "La

13. "Einstein's Theories." One man says to another:

"I can't take it any more. Look at the time you've owed me this bill!"

"But, man, don't you know Einstein says that time doesn't exist?"

In the second panel, the creditor chases the debtor with a stick, exclaiming:

"Is that so, you bounder! Time doesn't exist but, on the other hand, space sure does!"[46]

Finally, there is the appreciation that "light had weight" was the only "hard" fact that cartoonists were able to sift out of the mass of available information that lent itself easily to jokes:[47]

14. "Light has weight, according to Einstein." Einstein speaks:

"These Spaniards are going to ruin my theories, because I see that here they produce 'light' [that is, news, learning, and so forth] without any weight."[48]

15. Two wealthy men converse:

"The gravity of light is owing to Einstein!"

"I beg your pardon! Before he discovered that, it already 'weighed' on me to the tune of 50 pesetas a month!"[49]

This last joke is of particular interest because it can be traced

---

diffusion de la théorie de la relativité en France," doctoral dissertation, University of Paris, 1981, p. 174.

[46] *El Sol*, March 7, 1923.

[47] It is not at all surprising that people referred to the weight of light. Of course, according to general relativity what happens is that light rays follow geodesics, or the shortest lines, of a Riemannian—that is, a "curved"—space. However, as Hans Reichenbach pointed out many years ago, it is in principle possible to think, in Newtonian terms, that light rays move on a flat space and that, however, they follow curved paths because they are attracted (gravitationally, and this means, hence, that they have weight) by massive bodies such as the sun. This is an easy concept for people not sophisticated enough to transcend Newtonian physics. With the displacement of the perihelion of Mercury and gravitational red shift, however, there is no such immediate Newtonian image to substitute for a relativistic explanation. This is doubtless why the "weight of light" issue was so frequently raised in Spain.

[48] *ABC*, March 6, 1923.

[49] *ABC*, March 4, 1923.

to a specific source, namely, the polemic between Herrera and Eddington in the pages of *El Sol*, decribed in chapter 6. Such jokes in particular, as Biezunski notes, reinforced doubts about the scientific reality of Einstein's theories and underscored their inaccessibility.[50] The distribution of themes in cartoons confirms that the incomprehensibility issue was basic to Spanish popular appreciation of Einstein, that relativity was confused with relativism, but that there was also popular awareness of the basic subject matter of relativity: time, space, and gravitation. Such allusions, as Biezunski observes, do not in themselves constitute the popularization of scientific ideas. They are simply markers to identify for the public the issues at stake in the scientific debate.[51]

Einstein's notoriety, when combined with the abstract nature of his ideas, gave rise to numerous absurdities when the public confronted him in person. In Madrid a chestnut vendor, seeing the famed figure (whom she recognized by his distinctive hair) getting out of a car, cried out: "Long live the inventor of the automobile!"[52] Two years later a fellow passenger on the *Cap Polonio*, the boat carrying Einstein to Buenos Aires, had no difficulty in identifying the savant as "the inventor of relativity glands."[53] Such episodes were only to be expected. Inventors were popularly regarded as men of "science," and the rejuvenation craze had vulgarized the relative nature of physiological time.

THE MYTH RECONSIDERED

The Einstein myth can be usefully compared to the distinctive myth that arose in the United States in reaction to the feats of the inventor Thomas A. Edison. In his innovative study of the

[50] Biezunski, "La diffusion de la relativité," p. 133.

[51] Biezunski, "Einstein's Reception in Paris in 1922," in Thomas F. Glick, ed., *The Comparative Reception of Relativity* (Dordrecht: D. Reidel, 1987).

[52] Federico Bravo Morata, *De la semana trágica al golpe de estado* (Madrid: Fenicia, 1973), pp. 338–339.

[53] *La Razón*, March 25, 1925.

Edison myth, Wynn Wachhorst notes that Edison was viewed as a wizard who displayed the cerebral control of nature common to both magic and science and who was able, thereby, to control nature by means of simple devices. Because of the dramatic and magical qualities of inventions such as the electric light and the phonograph, Edison was accused of refabricating the laws of nature.[54] Einstein, of course, had done just that: his formulation of relativity had upset Newton's world and redone the laws of nature themselves. Accordingly, he too was invested with the qualities of wizardry and magic, because he was able to perceive secrets of nature that others of equal brilliance had been unable to see. The "magic" of Einstein's achievement was felt by many scientists who expressed their astonishment at a theory that, although not empirically founded, had nonetheless been confirmed by experience. Their sense of awe was transmitted to the public.[55] Then, too, part of the public's ambivalence toward Einstein, as Wachhorst notes with regard to Edison, was owing to "the inference of evil in the control of nature through science and technology, just as through magic."[56] (Recall Emilio Herrera's animadversion to the control of nature implied by a physics overly based on mathematical constants.) The assumption of godlike powers by both men was viewed with hostility by many, and the opposition of certain scientists to Einstein can certainly be traced to their perception of him as presumptuous. The German paleontologist Hugo Obermaier, who spent much of his career in Spain, made the following remark, "with malice," to Pío Baroja apropos of the current discussions over relativity: "We cannot understand nature, any more than a mosquito can understand the geographical structure of the Alps or the Himalayas."[57]

What was presumption to some was plain magic to others.

[54] Wynn Wachhorst, *Thomas Alva Edison, An American Myth* (Cambridge, Mass.: MIT Press, 1981), pp. 23, 25.

[55] Eisenstaedt, "La relativité générale à l'étiage," p. 154.

[56] Wachhorst, *Edison*, p. 30.

[57] Baroja, *Memorias* (Madrid: Minotauro, 1955), p. 765.

That indeed is how Ortega understood Einstein's hold on the public mind. When Einstein asked him how such abstract ideas could be of interest to the masses, the philosopher replied:

> I believe . . . that it is very understandable, Sr. Einstein. It could have been predicted that, given the condition of the human spirit, peoples' enthusiasm would burst forth if any great invention of high and pure science had appeared. There have been happy wars. . . . Because even though war is always a crude and cruel task, at times it has produced an intoxication with hope. But the last war was a sad one. Men fought for things which did not inspire hope, as if they were already exhausted. The economy and political organization of Europe had lost its attractiveness for the very same people who were fighting for them.
>
> The faith of men was empty, therefore. In such a circumstance there appeared your work, in which laws are promulgated for the stars, which obey them. The human masses have always perceived astronomical phenomena as religious. In them, science is conjoined with mythology and the scientific genius who masters them acquires a magical halo. You, Sr. Einstein, are the new magus, the confidant of the stars.[58]

Einstein, by challenging the commonplace notions of time and space, had challenged the metaphysics of the common man, a feat both presumptuous and magical at the same time.

[58] Ortega, "Con Einstein en Toledo," in *El tema de nuestro tiempo*, 18th ed. (Madrid: Revista de Occidente, 1976), pp. 196–197.

NINE

After Einstein's Visit

EINSTEIN AND THE SECOND REPUBLIC

The discussion of relativity in Spain before, during, and just after Einstein's visit demonstrated that Spaniards of virtually all ideological persuasions understood both that the physicist had raised important questions about the cosmos and that these issues could be debated publicly without doing violence to one's political or social ideology. But that consensus did not last long. On January 28, 1930, the dictator Primo de Rivera resigned and fled the country in the face of popular unrest. Subsequently, the strong showing of Republican and Socialist candidates in the municipal elections of April 12, 1931, revealed the great strength of sentiment favoring political change. King Alfonso XIII fled the country and the liberal Second Republic was declared on April 14. In the course of the next five years the civil discourse in matters of science that had played so prominent a role in the recuperation of Spanish science in the previous two decades and had provided a positive medium for the reception of new ideas like relativity and psychoanalysis was increasingly imperiled by the rising pace of political polarization that pitted right against left. Finally, the political struggle evolved into a military one when, on July 18, 1936, rightist generals rose against the Republic, initiating the Spanish Civil War. Here we examine changing views of Einstein and his work in the 1930s and 1940s, first as political polarization increased and, second, as a rightist regime eminently hostile to civil discourse came to power in 1939.

Einstein had few contacts with Spaniards after his visit. In 1932 Rafael Campalans, recently elected a deputy to the constituent Cortes of the Republic, wrote again to report on his political activities, notably those related to pacificism. An address Einstein made to the War Resisters League had been publicized in Spain, and Campalans wished to have a group to which he belonged designated as the Catalan affiliate.[1] None of Einstein's other friends from 1923 appears to have corresponded with him during this period. But the depth of Spanish interest in him was revealed in the spring of 1933 when Einstein left Hitler's Germany and numerous universities bid to acquire his affiliation. The physicist had previously made arrangements to be a visiting scholar at the California Institute of Technology, Oxford, and Princeton, and in April 1933 he accepted offers from the universities of Paris and Madrid as well. Although he most likely accepted the latter two positions simply to show solidarity with two prominent anti-Fascist countries, the protagonists of the Spanish offer certainly entertained hopes, at least at first, that he might elect to reside permanently in Madrid. The offer of an "Extraordinary Chair" was negotiated by the novelist Ramón Pérez de Ayala, the Republic's ambassador in London; the Hebraist A. S. Yahuda (Fig. 16), formerly professor of Hebrew at the University of Madrid and an ally of Einstein in the internal politics of Hebrew University; and, for the cabinet, the Socialist minister of public instruction, Fernando de los Ríos. Having once accepted the Republic's offer, Einstein spent the next three years (from Princeton) trying to escape from his obligation. As negotiations evolved, the Spanish government agreed to establish an Einstein institute for relativistic physics in Madrid, to be directed by a refugee scholar named by Einstein, who would head the institute in name only. Although many names were considered, including those of Einstein's collaborator Walter Meyer, Leopold Infeld, and several of Max Born's assistants, the position was never filled.[2]

[1] Campalans to Einstein, September 2, 1932, Einstein Duplicate Archive, Seeley Mudd Library, Princeton University.

[2] See J. M. Sanchez Ron and Thomas F. Glick, *La España posible de la segunda Republica: La oferta a Einstein de una cátedra extraordinaria en la Universidad Central (Madrid 1933)* (Madrid: Universidad Complutense, 1983).

16. Ramón Pérez de Ayala, Einstein, and A. S. Yahuda in Belgium, April 1933.

The press debate over the Republic's offer to Einstein revealed striking contrasts to the uniformly laudatory comment of ten years before and also demonstrates the extent to which Einstein and relativity alike had now become ideologically charged symbols. Liberal newspapers of 1933 attached positive value to Einstein's Jewishness, while conservative comment centered on the contrast between the Republic's generosity to Einstein and the harsh treatment it had accorded to certain conservative Catholic scientists. Indeed, the liberal press declared that the offer to Einstein was a belated rectification of the intolerance shown the Jews by Ferdinand and Isabella, who had expelled them from Spain in 1492.[3]

In contrast, the Catholic daily *El Debate*, whose coverage of

[3] *El Liberal*, April 9, 1933; R. Blanco Fombona, "Einstein en España: La República reivindica a los judíos," *El Sol*, April 16, 1933.

Einstein in 1923 had in general been laudatory and objective, bitterly attacked the government that had acquiesced in Terradas's loss of chair in Madrid (ostensibly because of his conservative politics) and had, of course, borne complicity in the Jesuit Pérez del Pulgar's expulsion from the country with the rest of his order.[4] A columnist in the same issue as the editorial just cited contrasted the voluntary exile of Einstein from Germany with the forced exile of conservative professors like Pérez del Pulgar: "The socialist Ministry has rushed to offer him protection. Judaism and Marxism are identified and confused with one another. A Jew gave life to Marxism and Jews are its most competent leaders all over Europe. Thence, that gesture which hypocrisy disguises with a veneer of comprehension and expressions of human brotherhood." If Einstein had been a Jesuit, the columnist concludes, he would not have been invited to Spain.

Einstein, formerly viewed as a "Teuton" by the Spanish right, was now fully a Jew, with all of the burden of associations that stigma implied. In particular, a Jew was automatically under suspicion of Marxism. Inasmuch as Spanish conservatives had little or no contact with real Jews—the Jewish community in Spain was practically nonexistent at this time—anti-Semitism had, in this context, a purely symbolic value and was pressed into use to stigmatize the hated center-left government.

By the early 1930s, relativity was no longer a burning issue in physics. In Spain, as elsewhere, attention was focused on quantum mechanics and, especially for those with philosophical interests, on indeterminacy. It is noteworthy that many of those who had choked on relativity as destructive of classical concepts and intuitive science, had no such qualms about quantum's approach to causality or even about indeterminacy (although it was not uncommon for conservatives to point out the doleful philosophical implications of the latter, no matter how valid a physical concept it might be).

[4] "Todo es relativo," *El Debate*, April 12, 1933.

A Catholic philosopher who commented on the new physics in this period was Xavier Zubiri (1898–1983). Zubiri had
studied both physics and mathematics and frequented the
Mathematical Laboratory in the early 1920s. There, Rodríguez Bachiller imparted higher mathematics to him in return
for lessons in philosophy.[5] One can presume they spoke of relativity, among other topics. Rey Pastor came to regard Zubiri
as one of his own circle: "The philosopher Zubiri, whom I
consider almost my student, has been in contact with [the
Mathematical Seminar], because he went there in search of ori-
entation. After all, Zubiri has very intensive mathematical formation, more than what is necessary for a philosopher."[6] Indeed, Zubiri's interpretation of the new physics was in large
part based upon his understanding of the role of mathematics
in it, which he, among Spanish philosophers, was perhaps
uniquely qualified to judge. In classical physics, he noted in a
1934 article in the Catholic intellectual journal *Cruz y Raya*,
mathematical equations are the formal expression of what
really happens in the system measured, without any reference
to the observer. The structure of the equations is the structure
of reality. Given certain conditions, we can use these equations
to predict future measures of the same object or phenomenon.
In the new physics, however, mathematical symbols are
merely "operators" (*operadores*), symbols of operations to
carry out on other symbols, designated "observables." Mathematics is now a theory of operations, and laws cease to be real
laws and have become statistical ones. Turning to uncertainty,
Zubiri calms his presumedly conservative readers by asserting
that Heisenberg gives no cause for scandal, and that his principle has nothing to do with subjectivity or objectivity in human knowledge. (This is the same kind of statement made a
decade earlier by Catholic apologists for relativity.) Nor is it a
principle of general ontology. It belongs, rather, to regional

[5] Tomás Rodríguez Bachiller, personal communication, Madrid, April 10,
1980.
[6] Ramiro Ledesma Ramos, "El matemático Rey Pastor," *La Gaceta Literaria*,
II, 30 (March 15, 1928).

ontology, which defines the primary sense of nature and the natural.

Zubiri had introduced his discussion of the role of mathematics in the new physics with a discussion of the nature of observation, based on implicitly relativistic principles. In classical physics, the observer was foreign to the content of what he observed, whereas in the new physics he modifies the nature of what is observed through the act of observation itself. Near the end of the essay, Zubiri counterposes Weyl, who proposed the total geometrization of physics (by eliminating all reference to the real movement of bodies, replacing this with the simple variation of the field in which they are found) and Einstein, for whom physical nature still implied real measurability. In this sense, Einsteinian physics was the crown of classical physics, seeking to provide a valid description of the universe for any observer and any point of view.[7] Einstein's 1923 protestations that he was not a revolutionary had, in general, been lost on the right. Conservative relativists like Terradas, Rafael, and Urbano were concerned to defend the theory theologically, while anti-relativists were obsessed by defending Newtonianism. The real work of philosophical assimilation, Ortega's essay aside, began with Zubiri, who concluded the essay under discussion with a reconstruction of a conversation he had with Einstein in Berlin in 1930:

> There are among physicists [Einstein told him] those who believe that all that science consists of is weighing and measuring in a laboratory, and who think that everything else (relativity, unification of fields, etc.) is non-scientific work. They are the *Realpolitiker* of science. But there is no science with numbers alone. A certain religiosity is required of it. An electrical phenomenon has an associated value of probability. Fine, but a probability which is somewhat subject to Coulomb's law. And this law? Also,

[7] Xavier Zubiri, "La idea de la naturaleza: La nueva física" [1934], in *Naturaleza, historia, Dios*, 3rd ed. (Madrid: Editora Nación al, 1955), pp. 255–259, 262–265, 271.

in turn, a probability. It is conceivable that God could have created a different kind of world. But to think that God plays dice with all the electrons of the universe at every instant, this, frankly, is too much atheism.[8]

Zubiri's inclusion of this statement is significant because he himself appears to have had considerably less difficulty in coming to grips with the philosophical consequences of quantum mechanics than did Einstein. The presentation of Einstein to Catholic intellectuals as conservative, religious, even a bit passé, helps to explain why Einstein's prestige, on the whole, remained high among conservative intellectuals.

The Breakdown of Civil Discourse

Other conservatives, though anti-relativists, may have perceived in the increasingly polarized social ambience of the years just preceding the Civil War an opportunity to promote views for which there had previously existed only a scant audience. Here we will consider two cases: an anti-relativist physicist (Eduard Alcobé) who remained silent for two decades only to become openly anti-Einsteinian after 1930, and a physician and educator (Ricardo Royo-Villanova) who had been favorable to Einstein in 1923 but saw fit to change his position later.

Since at least 1917, Alcobé had been ridiculing the Lorentz contraction in his undergraduate classes at the University of Barcelona.[9] A physicist of the old school who held firmly to a mechanistic view of the Cosmos, Alcobé believed that the the the concept of the aether had been altered to suit "the ideological requirements of certain minds—privileged ones, no doubt," and attempted a defense of physical absolutes by asserting that "it is not reasonable to subordinate the presence or absence of a vital force to the presence or absence of an arbitrary system of coordinates, completely detached from matter." Although

[8] Ibid., p. 275.
[9] Miquel Masriera, personal communication, Barcelona, February 1980.

his opposition to the Special Theory was tied, therefore, to specific physical concepts, he opposed the General Theory because it was metaphysical and because of its excessive mathematical complexity: "For there goes a flood of differential equations, and among the new algebraic symbols, covariants and countervariants are mixed together in addition to pseudocurvatures in captious hyperspaces, with all of which Einstein achieves the most resounding success for his famous theory of general relativity." For Alcobé, "physics is experimental or it isn't physics," and "the fantastic mathematical scaffolding" of the General Theory was seemingly beyond the capability of the experimentalist to grasp.[10] Alcobé's position, that relativity (the General Theory especially) was alien to the experimental approach and was more a mathematical than a physical theory was rarely articulated by Spanish physicists of the epoch, although there must have been some who shared his views but remained silent. Anti-relativists in other countries frequently railed against Einstein's mathematization of physics and the consequent abstruseness of the theory, but in Spain the presence of so many mathematicians favorable to relativity surely acted to dissuade dissenting scientists to take this line openly.[11] The loud complaints of scientific quacks like Horacio Bentaból only served to further silence Einstein's detractors.

Yet Alcobé's views were significant, particularly from 1935 through 1941, when he wrote the articles on physics for the supplemental volumes of the influential Espasa-Calpe encyclopedia. In the 1934 supplement he summarized the current state of relativity, repeating all the usual doubts: whether c is really a constant, whether aether has really been disproved, and so forth. He noted that he and Cabrera had attended the International Physical Congress at Como in 1927, where rela-

---

[10] Eduardo Alcobé, "Sobre un concepto dinámico," *Memorias de la Academia de Ciencias y Artes de Barcelona*, 22 (1931), 363–371; "Física," *Enciclopedia Ilustrada Universal, Suplemento 1936–39*, p. 1003.

[11] On France, see Biezunski, "La diffusion de la théorie de la relativité en France," doctoral dissertation, University of Paris (1981), pp. 177–182, where he discusses opposition to relativity as (1) metaphysical and (2) mathematical.

tivity did not have the "preponderance which one might have assumed." The article's bibliography is replete with the major anti-relativist tracts of the 1920s, and he cites approvingly Giuseppe Gianfranceschi's opinion that a hyperspace of n dimensions was "pure scientific juggling." Although Einstein had some, even excessive, support, "complexity is always suspicious. Some see things another way, though, and assert that Einstein's theory conceals a childish simplicity."[12] This kind of innuendo was designed to discredit relativity without dealing with the issues directly. In the supplement appearing directly after the Civil War, Alcobé simply notes that supporters of the theory are now "strongly diminished."[13] The Espasa's characteristic way of dealing with controversial scientific articles was to allot some articles to proponents of a given idea (evolution, relativity, and so on) and others to opponents, thus satisfying everyone. It is amusing to note Alcobé introducing cross-references to Terradas's article in volume 50 for those wishing further details. In spite of his problems with relativity, Alcobé was the Spanish translator of Theodor Wulf's *Tratado de Física*, a standard text used in Catholic universities. Wulf presents a balanced and favorable discussion of relativity and other contemporary theories (Einstein is the third most cited author, after Newton and Maxwell). Although Alcobé had the free use of footnotes to express his conservative opinions, he limited himself to identifying Heisenberg's uncertainty principle as "a metaphysical principle applied to physics," a procedure with which he did not agree.[14]

If some of the preeminent stalwarts of the Right did not lose their admiration for Einstein, others were led to associate the overthrow of classical science with the revolutionary Left's attack upon traditional social values. An example is Ricardo

[12] Alcobé, "Física," *Enciclopedia Universal Ilustrada, Suplemento 1934*, pp. 361–365.

[13] Ibid., p. 1003.

[14] Teodoro Wulf, S.J., *Tratado de Física*, trans. from the 2nd German edition of 1929 (Barcelona: Tipografía Católica Casals, n.d.), pp. 515–521 (relativity), 536n (uncertainty principle).

Royo-Villanova, who, as rector of the University of Zara-goza, had ordered the blackboard used by Einstein there in 1923 preserved for posterity. By 1935 Royo had changed his mind about Einstein's greatness, asserting in a newspaper ar-ticle that astronomical observations by John S. Plaskett, the experiments of Dayton Miller, and the theoretical formula-tions of Comas Solà and Willem De Sitter had discredited rel-ativity, and that Einstein himself (in his Willard Gibbs Lecture in Pittsburg the year before) had cast doubt on his own cos-mological ideas.[15] The following year, Royo published a long essay embodying (and emending) a series of shorter articles on the crisis of science in which he articulated, in darkly apoca-lyptic terms, his lack of confidence in the new scientific theo-ries, which had "a sinister significance and carry within them-selves a destructive seed which seems to approach a type of nihilism." The crisis of science, moreover, was related to social crisis, although the latter involved lower-order principles. Royo not only regretted the loss of classical principles, but also of the "blind faith in science professed by yesterday's schol-ars," now replaced by "a desolating agnosticism" that had in-vaded all the precincts of Galilean positivist experimental thought.[16]

In this context, Royo-Villanova was more categorical about the moral consequences of relativity, in particular the General Theory, which had replaced Euclidean space with that of Rie-mann: by extension, the world is not only insufficient, but rel-ative, "tending to destroy mercilessly our self-confidence, our capacity for work, and even our individual substance and structure." Relativity was "an intellectualist creation which lacks all absolute scientific value" and which "leads us again to the deep conviction that absolute time and space were consti-tuted, nothing more nor nothing less, by the infinite presence

[15] Ricardo Royo-Villanova Morales, "La teoría de Einstein," *El Norte de Castilla*, June 12, 1935.

[16] Royo-Villanova, "La crisis de la ciencia," *El Siglo Médico*, 98 (1936), 58–59.

of God."[17] Still, Royo's short popularization displays more than a passing acquaintance with the General Theory, and he clearly followed the experimental results closely. Considering his position of 1923 (to which he does not here allude), he seems to be saying that he had been deluded and had lost sight of the real locus of truth, which is a matter of faith, not science. He then discusses quantum theory and indeterminacy, whose physical results he seems to accept, while pointing out the demise of classical notions of causality. He accepts Zubiri's admonition that uncertainty overturned the old notion of causality in physics only and does not suppose "a renunciation of the absolute idea of cause."[18] Yet, taken as a whole, the panorama of modern physics suggested nothing less than anarchy, because "before the excess of mechanics we have lost control of mechanisms." This crisis was a necessity, he admits. (This explains why he is accepting of uncertainty and of quantum mechanics, which, following Comas, he believed had proven relativity wrong.) But it has left us, because of its skepticism, possessed of "the distressing terror of not knowing." Nietzsche had declared that the culture of his time merited its own demise. The same held true, in Royo's view, for contemporary science:

> Today's science has violated human tradition, human precepts and commandments, and it ought thereby to suffer the punishment of divine justice, the sanctions of earthly justice, the hatred of mankind, the contempt of history. Our science has lost its soul or else it has already hardened; it wants to pass for good, but at base it is terribly bad; it is hypocritical and drags culture, the individual and even the masses after it, to make them serve its crude, false needs, the more shameful and unspeakable they are,

[17] Ibid., pp. 61–62. In his 1935 article (see 15) Royo had mentioned the Special Theory, which he clearly did not understand, claiming that it was based on the "principio de la *variedad* de la velocidad de la luz" (my emphasis). In the 1936 version (n. 16) there is no reference to special relativity.

[18] Ibid., p. 64.

the more peremptory they seem. Our science is extraor-
dinarily excited, lively, animated, which are the watchwords of an epoch that is enervated, overly civilized, too
mature, which is beginning to fail, which begins its de-
composition and will soon rot. . . . False values and
empty words: this is the panorama of contemporary
science.[19]

The solution, according to Royo, was to revise the overly restricted concept presently given to the concept of science, to
give it a more general significance so that the criteria of human
reason and sense perceptions could be applied to it. In science,
there was no true knowledge, only "a constant posing of problems." True knowledge, of course, reposed only in the inscru-
table power of God.

It is interesting to note that, whatever relation there may
have been between the mounting political polarization and the
association by conservative scientists of the "anarchy" in physics with social upheaval, Einstein's prestige still remained high
among figures of the political Right, and Spanish fascists'
views of Einstein seem not have been influenced at all by German anti-Semitism. We have already noted Ramiro Ledesma's
support of Einstein against his philosophical opponents. There
are other examples of right-wing testimonials to Einstein. In
his *Memorias*, originally scheduled for publication in February
1932, Emilio Mola, one of the "Four Generals" of the 1936 re-
bellion, reversed the typical thrust of the comprehensibility issue in order to criticize the Republican constitution: "For me,"
he wrote, "it is easier to digest the Lorentz contraction, the experiments of Eötvös, or Minkowski's time-space continuum,
so intimately linked to the complicated philosophical physics
of Albert Einstein, than the present interpretation of Spanish
legislation."[20] A more personal testimony is attributed to Juan
March, the Mallorcan financier who provided capital for the
rebel cause. "Once," March's biographer writes, "he had the

[19] Ibid., pp. 68–69.
[20] Emilio Mola Vidal, *Memorias* (Barcelona: Planeta, 1977), p. 316.

opportunity to meet Albert Einstein, and when he spoke of sages he always said that the author of the law of relativity was the one whom he considered the ideal type of scientist."[21] During the 1930s, attempts to associate relativity with Marxism were confined to tracts of the extreme right wing.[22]

The "Unity of Science" Restored

Royo's essay was published on July 18, 1936—the day the Fascist uprising began—a fitting introit to the conception of science that the victorious Franco crusade was to instill by virtue of the Law of November 24, 1939. That law declared the Junta para Ampliación de Estudios dissolved and vested control of its assets and the centers formerly under its supervision in the newly created Consejo Superior de Investigaciones Científicas. The spirit informing the new institution, however, was totally distinct from that of its predecessor, inasmuch as its founders saw in Franco's victory an opportune moment to realize at long last the ideal of the nineteenth-century neo-Catholics and to restore to knowledge the "Catholic unity" that had been lost to rationalism and liberalism. The law's preamble held that the renovation of Spain's glorious scientific tradition would be based on

> the restoration of the classical and Christian unity of the sciences, destroyed in the eighteenth century. In its place, one must repair the divorce and discord between the speculative and the experimental sciences, and promote the harmonious growth and evolution of the whole tree of science, avoiding the monstrous development of some of its branches and the wasting away of others. There must be created a counterweight to the exaggerated and solitary specialization of our epoch, returning to the sciences their regime of sociability, which supposes a frank and certain

[21] Ramon Garriga, *Juan March y su tiempo* (Barcelona: Planeta, 1976), p. 37.
[22] René Llanas de Niubo, *El judaísmo* (Barcelona: José Vilamila, 1935; Biblioteca Las Sectas, 14), pp. 130–131.

return to the imperatives of coordination and hierarchy. One must impose upon culture, in sum, the essential ideas which have inspired our Glorious Movement, in which the purest lessons of universal Catholic tradition are conjoined with the demands of modernity.[23]

The results of the regime's efforts to turn back the clock in science were uneven. Catholic unity was not, and could not be, restored. On the Consejo's "tree of science" (the Lullian *arbor scientiae*, which it adopted as its logo), philosophy and theology held status equal to the natural sciences; but as one Catholic critic complained, the experimentalists, particularly those in applied fields, were favored in the allocation of funds. In experimental physics, perhaps unwittingly, the Consejo's program was strongly operationalized, inasmuch as optics—a seventeenth-century topic par excellence—became a central focus of research in the 1940s.[24] The Franco regime had better luck in impressing its ideological stamp on science through censorship: Darwin—again—and Freud were virtually proscribed. Biology was taught as though evolutionary theory did not exist, and psychological theory was taken from second-line followers of Kraepelin, since avowal of any Freudian tenets was discouraged. The effect on relativity was not so pronounced, however, because so many conservative engineers favorable to Einstein survived the war with their status intact. Terradas and Lucini both published volumes on relativity in the 1940s. As a counterpoint, the Falange republished Félix Apraiz's anti-relativist treatise, which had originally appeared in the 1921.[25] Ecclesiastical journals that had been anti-

[23] *Estructura y normas del Consejo Superior de Investigaciones Científicas* (Madrid, 1943), pp. 7–8.

[24] Catholic critique: Antonio Fontán, *Los católicos en la universidad española actual* (Madrid: Rialp, 1961), esp. pp. 94–97. On optics in the 1940s, see Carlos López Fernández and Manuel Valera Candel, "Estudio bibliométrico-multivariante de los artículos de física publicados en los Anales de la Real Sociedad Española de Física y Química durante el periodo franquista (1940–1975)," *Llull*, 6 (1983), 54.

[25] Manuel Lucini, *Principios fundamentales de las nuevas mecánicas (relativista,*

relativist before the Civil War continued in the same line afterward. Thus *Razón y Fe* published a long article by Luis Prieto in 1941 confusing relativity with relativism, reiterating and extending Pérez del Pulgar's strictures regarding the velocity of light, and insisting that relativity was simply an artifact of measurement, an apparent effect, totally divorced from physical reality.[26] The following year Juan García published a geometrical theory of space which denounced Einstein's concept of space as a function of matter. That this article appeared in the journal of the Academy of Exact Sciences was an apt sign that the criteria by which physical science was judged had, to say the least, changed.[27]

Views of Einstein in Franco's Spain were probably less conditioned by relativity than by Einstein's antipathy to the regime. Although the physicist had not been unusually active or outspoken in his support of the Republic, his fame ensured the notoriety of his few efforts and attracted the wrath of the Republic's enemies, both Spanish and American. In July 1937 Einstein sent a message of support for the Republic to the International Writers' Congress in Spain. Writing from Paris in December 1937, Ortega y Gasset, his liberal days behind him, declared: "Some days ago, Albert Einstein believed himself to have the 'right' to give his opinion on the Spanish Civil War and to take a position on it. Nevertheless, Einstein displays a radical ignorance over what has taken place in Spain now, before, and always. The spirit that informs this insolent meddling is the same which for some time now has been causing the universal discredit of intellectuals who, bereft of spiritual force, cast the world adrift."[28]

---

ondulatoria, cuántica) (Barcelona: Labor, 1966); Esteban Terradas and Ramón Ortiz Formiguera, *Relatividad* (Buenos Aires: Espasa-Calpe, 1952); Félix Apráiz, *Las seis dimensiones del espacio físico. Errores de la física actual y naturaleza de la electricidad y el éter* (Las Palmas: Tip. Falange, 1945).

[26] Luis Prieto, "Relatividad y racionalidad," *Razón y Fe*, 123 (1941), 279–294; 124 (1941), 81–108.

[27] Juan García, "Sobre la teoría del espacio," *Revista de la Real Academia de Ciencias Exactas*, 36 (1942), 263–295, esp. p. 295.

[28] José Ortega y Gasset, "En cuanto al pacifismo," *Obras completas*, IV, 307.

When in 1938 Einstein signed a petition asking the United States government to lift the arms embargo against the Republic, Franco's supporters were outraged. The *Brooklyn Tablet*, a Catholic weekly, urged in an editorial that Einstein be sent back to Europe, since by appealing for an end to the embargo he was supporting the persecution of Spanish Christians.[29] Attacks on Einstein continued for another decade even though he rarely addressed Spanish politics directly. In October 1945 Einstein refused permission for the American Committee for Spanish Freedom, which advocated that the United States break diplomatic and commercial relations with the Franco regime, to send out a fund-appeal letter using his name. Einstein wrote the committee that, because he lacked pertinent information, he did not feel justified in making himself "a public spokesman for the Spanish cause, although I am convinced that justice is on the side of the Spanish loyalists and that the survival of Fascist Spain seriously endangers international security." The Committee used Einstein's name anyway, and, as a result, the conservative congressman John Rankin attacked the scientist in the House of Representatives as a "foreign-born agitator" promoting the Communist cause throughout the world.[30] Einstein also received hate mail stimulated by his Loyalist sympathies. Earlier the same year a man wrote him: "It is with profound disgust that I learn of your activity in behalf of Communist Spanish forces. . . . Genuine Christian Americans resent your arrogance and reactionary European ideas."[31]

Einstein's death stimulated a certain amount of hostile commentary in the Spanish press. Agusto Assía, writing in *La Vanguardia*, noted that Einstein had been a friend of all anti-Spanish organizations. This, Assía implied, made Einstein an ingrate, because Pérez de Ayala had lodged him at the Spanish

---

On the Writers' Congress, see Hugh Thomas, *The Spanish Civil War*, rev. ed. (New York: Harper & Row, 1977), pp. 698–700.

[29] "Send Einstein Back," *Brooklyn Tablet*, May 14, 1938.

[30] Otto Nathan and Hans Norden, eds., *Einstein on Peace* (New York: Schocken, 1968), p. 344,

[31] Einstein Duplicate Archive, 31–1032 (January 3, 1945).

embassy in London in 1933. "I knew Einstein quite well," remarked Assía, who had been a correspondent in Berlin, and "attended one his courses in the Kaiser Wilhelm Gesellschaft." He goes on to misinterpret the exchange between Einstein and Zubiri, which he had witnessed personally: "One of his last disciples in Europe was Professor Xavier Zubiri, and I was present at a discussion they had at the Kaiser Wilhelm Gesellschaft in which, if I understood it correctly, Einstein held that mathematics precluded the existence of God and the Spanish philosopher [held] the opposite." The last time Assía encountered Einstein had been in London, where he spotted the physicist walking about "like a gnome."[32] This article provoked an angry complaint from members of the University of Barcelona professoriate who regarded Assía's remarks as scurrilous.

Anti-relativist authors found Spain of the 1950s and '60s to offer an atmosphere favorable to publication of their counter-theories. José Casares Roldán introduced an aether-drag theory in order to revive absolute time.[33] A more serious challenge, because of the great prestige of its author, came from the physicist Julio Palacios (1891–1970). Palacios had been an early Spanish supporter of the new physics; he had taught relativity, and his discourse of reception in the Academy of Sciences (in 1932) on indeterminacy was one of the major defenses of the new physics in Spain of the 1930s. At that time, Palacios's conservative politics had not yet interfered with his acceptance of the new physics (although it is interesting to note, among quotations beginning each section of the speech, one by Charles Maurras, leader of the French political ultras, stating, "It seems that, more and more, the head loses the direction of things").[34]

Between 1955 and 1959 Palacios wrote a series of both learned and popular articles attacking Einstein's concept of si-

[32] Agusto Assia, "De hijo de un obrero a mago de la física," *La Vanguardia*, April 19, 1955.

[33] José Casares Roldán, *Refutación de los fundamentos de la Relatividad* (Granada: Urania, [1952]).

[34] Julio Palacios, *Discurso leído en el acto de su recepción por . . .* (Toledo: Academia de Ciencias Exactas, 1932), p. 63.

multaneity, using the famous paradox of the clocks as the focal point of his argument. This is the discussion (sometimes also referred to as the "Twin Paradox") that Paul Langevin popularized by asserting that an astronaut returning to earth after traveling at high speeds would find his twin to have aged more than he, since at high speeds time runs slower. Palacios formalized his objections in his 1960 book, *Relatividad, una nueva teoría*. His argument against relativity was similar to those of Herbert Dingle and others who rejected the relativity of time. This he achieved in a novel way, by altering the construction of clocks to eliminate the contraction of time.[35] Palacios in essence recapitulated the objections of many of Einstein's Spanish critics of the 1920s—that relativity, the Special Theory in particular, was "unscientific and contrary to common sense." During this period (according to Tomás Rodríguez Bachiller) Palacios was stating publicly, although not in print, that the lack of scientific response to his countertheory was a product of the "Judeo-Masonic conspiracy."[36] Finally, in 1962, when he read of some experiments conducted by W. Cantor which ostensibly disproved Einstein's theory, Palacios announced the results to his class on relativity at the Faculty of Sciences in Madrid and said: "Go to the Secretariate and have them return the registration fee for this course. Relativity is dead and these classes are devoid of meaning."[37]

We have noted the common inability of many scientists to overcome their suspicions of the Special Theory and particularly to abandon "common sense" for Einstein's conceptuali-

---

[35] *Relatividad, una nueva teoría* (Madrid: Espasa-Calpe, 1960). See the discussion of Palacios's theory by Henri Arzeliès, *Relativistic Kinematics* (Oxford: Pergamon, 1966), pp. 190–191, asserting that Palacios's abolition of the postulate of relativity was arbitrary and "sentimental," and also Manuel A. Sellés, "La teoría de la Relatividad de Julio Palacios," *Actas, II Congreso de la Sociedad Española de Historia de las Ciencias*, 3 vols. (Zaragoza, 1984), II, 427–452.

[36] J. Aguilar Peris, D. *Julio Palacios y el lenguaje de la física* (Santander: Universidad de Santander, 1981), pp. 36–37. Tomás Rodríguez Bachiller, interview, Madrid, April 1980. For a chronicle of Palacios's anti-relativity period, see *Homenaje a Julio Palacios: Vida y obra de un científico* (Santillana de Mar: Fundación Santillana, 1982), pp. 8–9.

[37] Aguilar Peris, D. *Julio Palacios*, p. 37

zation of simultaneity and the relativity of time. Emilio Herrera, we recall, wondered whether our very notions of common sense would have to be adjusted to accommodate the new Einsteinian viewpoint. Hererra, who had been a general in the Republican air force during the Civil War, followed Palacios's attacks on Einstein from exile in Paris. Around June 1966, Herrera wrote to his nephew, Juan Aguilera, to sum up his objections to Palacios:

> I have been, and continue to be, a good friend of Professor Palacios, to whom I gave the chair of physics at the School of Aeronautics which I directed because I believed him to be the most qualified man in Spain in this discipline, with the exception of Blas Cabrera who, owing to his age and position, could not accept this post. But the admiration which I felt for Professor Palacios is not compatible with his present attempt to ridicule and destroy Einstein's theory of relativity which he has undertaken in the Spanish press. . . .
>
> He has invented a personage called Don Ingénuo who, leaving Madrid in a train with the velocity v, with his watch synchronized with that of the station, would have to see his watch slowed down with respect to the other stations he would pass (according to the idea which he has of the theory of relativity), at the same time that the stationmasters would see that their watches had slowed down with relation to that of Don Ingénuo. Since this is impossible, Einstein's theory is wrong.
>
> In order that [Palacios] with this article in which he demonstrates that he hasn't understood relativity . . . might not continue to put Spanish science in ridicule abroad, I have written him a number of letters trying to disabuse him of his error, but without success. He sent some of his calculations to scholars and Academies of Science across the world, but no one answers him. He says that a conspiracy of silence has been established against him, but what really has happened is that scientists and

Academies don't want to waste their time discussing things which are arch-established.

Herrera then presented to his nephew a simple and clear explanation of special relativity, relaying the examples he had addressed to Palacios to disabuse him of his error. Herrera, inventive as always, proposed that if linear clocks (tapes on which hours had been marked) were used in the train experiment, it would be impossible to confirm Palacios's negation of the relativity of time because for the passenger, Don Ingénuo, the tape would have stretched out in equal proportion to his train's velocity. For Herrera, opposition to Einstein had, by this time, acquired ineluctable social and intellectual consequences:

> These articles of Professor Palacios are discrediting our science to the point where Spain is mentioned as being totally uncultivated in these matters. In a number of the American *Atomic Journal*, Professor [Hans] Bethe . . . published an article saying that the flight to Russia of Klaus Fuchs, an atomic scientist who had worked on the atomic bomb, was a serious matter, because there are fine scientists in Russia, as well as great laboratories and factories where Fuchs's knowledge could be of great value. But if he had gone to Spain instead of Russia, the matter would be of scant importance *because there is no one there who understands these matters.*[38]

It is significant that Herrera, a prominent doubter of special relativity during the 1920s because of its "anti-intuitive" nature, should, a year before his death, make its defense into a kind of scientific testament. Only several years earlier, he had published a final statement of his notions of a mechanical Cartesian universe based on rotating vortices comprising a hi-

[38] Text reproduced by permission of Juan Aguilera Sánchez. The complete text is reproduced in T. F. Glick, "Emilio Herrera and Spanish Technology," in Emilio Herrera, *Flying: The Memoirs of a Spanish Aeronaut* (Albuquerque: University of New Mexico Press, 1984), pp. 206–208.

erarchical series of universes of higher dimensions, that of three dimensions (in which Einstein's rules hold sway) enclosed in superior ones of four, five and higher dimensions.[39] As Herrera made clear, opposition to Einstein had become an unacceptable symbol of retrograde and gratuitous resistance to scientific change.

THE RECEPTION of relativity in Spain, which is impossible to disentangle from Einstein's personal impact, illustrates the constraints upon civil discourse attaining in that country in the first half of the twentieth century. So long as the unwritten pact between right and left was in force, only irredentist clerical bigots and their secular epigones saw fit to include Einstein and relativity on the list of familiar "liberal" horrors. Ataulfo Huertas's assertion that relativity has not been politicized was an accurate reflection of the positive operation of civil discourse in the 1920s. As the political cleavage deepened in the early 1930s, both relativity and Einstein's image fell victim to the breakdown of civil discourse. Einstein's Jewish identity, for example, which had not been an issue in the 1920s, became, in the following decade, an index of the deepening political cleavage. Still, it was difficult to turn the clock back completely, given the continuance of political democracy. Even after Franco's victory and the deliberate attempt to shut down civil discourse, pro-Einstein conservatives could not be dissuaded from their relativistic bias. By the time of Einstein's death in 1955, most of the original conservative relativists had passed from the scene, and rehabilitation of Einstein had to wait, as did that of Darwin and Freud, until the transition to democracy was completed.

[39] Emilio Herrera, "L'Univers de Descartes," *Le Genie Civil*, 140 (1963), 280–282, 308–311, 327–329, 352–355, 477–478. Herrera summarized his theory in a letter to Einstein dated February 23, 1952, but without much hope that Einstein would take it seriously; see the text in Herrera, *Flying*, p. 205. Einstein intervened on Herrera's behalf to secure the exiled general a position with UNESCO.

APPENDIXES

APPENDIX 1

Einstein's Travel Diary for
Spain, 1923[1]

February 22–28. Stay in Barcelona. A great deal of strain but kind people (Terradas, Campalans, Lana, Tirpitz's daughter), folk songs, dancing, Refectorium. It was lovely!

March 2. Arrival in Madrid. Departure from Barcelona, warm farewell. Terradas, German consul and Tirpitz's daughter, etc.

March 3. First lecture at the University.

March 4. A drive with the Kocherthalers. Wrote an answer to Cabrera's Academy address. In the afternoon a meeting of the Academy with the King as chairman. Wonderful address by the president of the Academy. Afterward tea with an aristocratic lady companion. Evening at home, however, totally catholic.[2]

March 5. Afternoon meeting of the Society for Mathematics. Honorary member. Discussion about general relativity. Dinner with Kuno [Kocherthaler], visit with Cajal, wonderful old man. Seriously ill. Invited to supper by Herr Vogel. Kind, humorous pessimist.

March 6. Trip to Toledo camouflaged by many lies. One of the most beautiful days of my life. Radiant sky. Toledo was like a fairy tale. An enthusiastic old man, who has supposedly produced some significant writing on Greco, guides us; the streets and marketplace, view of the city, Tajo with stone bridges; stone-covered hills, lovely plain, cathedral, synagogue; sunset with flowing colors on our way home. A little garden with a view near

---

[1] Einstein Duplicate Archive, Seeley Mudd Library, Princeton University. Identifications in brackets were supplied by Einstein's secretary, Helen Dukas.

[2] *ganz katholisch*. Dukas adds: "The latter expression should not be taken literally. A. E. uses it to characterize a confusing state resulting from a great deal of exertion, especially of a social nature."

the synagogue. A magnificent picture of Greco in a small church (burial of a nobleman) among the most profound things that I saw. A marvelous day.

March 7. 12 o'clock. Audience with the King and the Queen Mother. The latter reveals her knowledge of science. One sees that no one tells her what he is thinking. The King, simple and dignified, I admired him in his way. During the afternoon the third University lecture. Attentive audience, which surely understood almost nothing because of the difficult problems being treated. Ambassador and family, splendid, straightforward people. Party, painful as usual.

March 8. Honorary doctor. Real Spanish speeches with accompanying Bengal fire. The German ambassador spoke on the subject of German/Spanish relations—long speech but the content was good, German through and through. Nothing rhetorical. Then a visit with technical students. Talking and nothing but talking, but well intentioned. In the evening, a lecture. Then an evening of music at Kuno's home. An artist (director of the Conservatory), Bordas, played the violin splendidly.

March 9. Trip to the mountains and Escorial. A marvelous day. In the evening a reception in the Residencia with talks by Ortega and me.

March 10. Prado (mainly works by Velázquez and Greco). Farewell visits. Dinner with the German ambassador. Spent the evening with Lina [Kocherthaler] and the Ullmanns in a primitive, small dance hall. Merry evening.[3]

March 11. Prado (magnificent works of Goya, Raphael, Fra Angelico).

March 12. Trip to Zaragoza.

[3] Dukas adds: "Lina Kocherthaler, cousin and close friend of the family."

German Consular Reports of
Einstein's Trip to Spain

1. BARCELONA[1]

Barcelona, March 2, 1923

German General Consulate for Spain
K. no. 23
Follow-up to cable dated 26th
of previous month, no. 5

This will confirm my aforementioned cable with text as follows:

"On the 24th Einstein held his first lecture which was warmly received by the press. Guest of the city. Receptions planned today and tomorrow by Mancomunidad and Municipal administration. Discretionary press representation. Hassell."

Professor Einstein, in the company of his wife, arrived here on his return trip from Japan on the 22nd of last month. It had been agreed well in advance (refer to my report of March 6 of last year, J. No. 1674) that he would visit Spain, but his firm commitment was not received until a day before his arrival in Barcelona, and even then the exact date was left open. It was, therefore, not possible to prepare an immediate reception for him, especially in the press, as would have been in the best German interests. At the invitation of the Mancomunidad, Einstein held three lectures at the Institut d'Estudis Catalans and an additional lecture in the Real Academia de Ciencias. Tickets for the lectures, delivered in French and dealing with Einstein's theories, were limited to those with a scientific background and official representatives. Notwithstanding that, there was a large audience present, which listened attentively to every word, extended him an extremely warm welcome, and thanked him with enthusiastic ap-

[1] Zentrales Staatsarchiv, Merseburg, Rep. 76 Vc Sekt. 1 Tit. 11 Teil Vc Nr. 55, Bl. 150–151.

plause. Professor Einstein was the object of numerous honors from all sides. During his visit here, he was the guest of the city of Barcelona; the Ayuntamiento arranged a formal reception in the city hall which was attended by representatives of numerous governmental departments and cultural organizations. The Deputy Alcalde delivered the welcome speech in Catalan. Einstein's reply in German contained many significant comments that were immediately translated into Catalan. The president of the Mancomunidad together with the Institut d'Estudis Catalans held an intimate dinner party for Einstein and his wife which was attended by my wife and me, the Deputy Alcalde, as well as some professors. Numerous groups and private individuals have sent him tokens of gratitude.

The press published detailed reports of his visit here and his lectures, emphasizing his significance and paying tribute to his scientific investigations. Most newspapers adopted a muted political tone, perhaps not without help from some local political circles, referring to Einstein's pacifist views while mentioning the fact that he had not signed the well-known manifesto of the 93 German scholars. A political overtone, which probably would have been best avoided, was sounded on another occasion. The radical "Sindicato Unico," which has gained a bad reputation as a result of its terror tactics, welcomed Einstein as a fellow supporter, so to speak, in the hotel and extended an invitation to him. At the instance of the socialist professor Campalans, who is a fervent supporter of cultural relations with Germany—as is well known from my reports—and who has been invaluable in organizing Einstein's lecture tour, Einstein accepted this invitation. The well-known socialist leader Angel Pestaña welcomed him, using the opportunity to mention the social struggles in Barcelona. According to newspaper accounts, Professor Einstein replied, among other things, that he, too, was a revolutionary, albeit only on the intellectual plane. Einstein went on to say that he considered the suppression of social struggle by force as more a sign of ignorance than of evil. Subsequently, Einstein himself told an interviewer that this account was incorrect—in fact the opposite was true. Einstein maintained that he was not a revolutionary in the field of science but rather an advocate of evolution and that he rejected the program of the Communists.

All in all, Einstein's visit here, during which time he always presented himself as a German and not a Swiss, can be characterized as a complete success for him and for the reputation of German science as well as for German/Spanish cultural relations, thanks primarily to the personality of the savant which inspired the keenest affection from all quarters.

[signed]
Hassell

2. MADRID[2]

Report of the German Embassy in Madrid
to the Foreign Office

Madrid, March 19, 1923.

Professor Einstein, traveling from Barcelona in the company of his wife, arrived in Madrid on March 1. Those who gathered at the railway station to greet him included a number of German relations residing here as well as various members of the local German community, myself, and a commission of Spanish professors from the University of Madrid. A large crowd of curious onlookers was also present. Needless to say, photographers and reporters were also there. A representative of *ABC*, who had boarded the train a few stops before Madrid, had already obtained an interview before the train arrived here.

Professor Einstein gave three lectures in the Madrid University lecture hall for physics. In spite of the fact that tickets had been issued primarily to individuals with a purely scientific interest and corresponding requisite background, there were extraordinary crowds for all three lectures. Additional lectures took place in the Athenaeum, the Residencia de Estudiantes, the Academia de Ciencias, and in the Asociación de Alumnos de Ingenieros. There was great enthusiasm everywhere, even though only very few individuals were able to follow Einstein's ideas. All manner of honors were lavished on Professor Einstein and one can say, without any exaggeration, that since time immemorial no foreign guest has ever enjoyed such a spirited and extraordinary reception in the Spanish capital.

Every day the newspapers printed reports, several columns long, concerning his activities. Science reporters of the major newspapers

[2] Zentrales Staatsarchiv, Merseburg, Rep. 76 Vc Sekt. 1 Tit. 11 Teil Vc Nr. 55, Bl. 155–156; reproduced in Christa Kirsten and Hans-Jürgen Treder, eds., *Albert Einstein in Berlin*, 2 vols. (Berlin: Akademie-Verlag, 1979), I, 232–233 (Doc. 154).

discussed his theory of relativity in long articles. In their reports on Einstein's lectures, journalists tried to explain, in terms that could be readily understood by the average layman, how "Einstein's discoveries had shed light" on the great problems in physics. Newspaper photographs captured Einstein in a variety of poses and also those in attendance at the festivities in his honor. Caricaturists attempted to reproduce his head with its great intellectual power, and the watchwords Einstein and "relativity" could even be found in popular humor magazines.

The sole discordant note could be heard in the clerical *Debate*, where there was a subdued and carefully formulated opposition to the "Einstein mania." There was a warning against overestimating Einstein's accomplishments and it was suggested that only time will really tell how accurate "Einstein's theory" is. In addition to the lectures, the following events took place:

There was a welcoming address by the mayor, and Professor Einstein showed his gratitude by making a personal visit to the Ayuntamiento. Following that, the College of Doctors of Madrid gave a banquet in his honor at the Palace Hotel, during which the chairman, Dr. Bauer, and the rector of the University, Dr. Carracido, praised the accomplishments of the German scholar. On March 4 there was a special session of the Academy of Sciences, chaired by His Majesty, the King, and Professor Einstein was awarded the diploma of a corresponding foreign member. At this occasion, the Spanish minister for Arts and Sciences gave a speech, and he concluded by offering the scholar the hospitality of Spanish soil and the financial support of the government should present conditions in Germany render the continuation of Einstein's research temporarily impossible. On the same day, the Marqués de Torrevieja [*sic*] held a reception in honor of Herr and Frau Einstein and on the evening of March 7 there was a reception for 110 persons at the embassy. Numerous local professors and other scientists as well as members of society and the colony participated so that Professor Einstein had an opportunity for stimulating discussions with a wide variety of personalities.

In a special session at the University on March 8, Einstein was awarded an honorary doctorate with all traditional pomp and ceremony. In addition to the Rector, a professor, and a student, I gave a speech in accordance with local custom. My speech was in Spanish. The subject I chose dealt with the historical relations between Germany and Spain. Excerpts from the speech appeared in almost all the Madrid newspapers and seem to have been generally well received.

Prior to his departure, Professor Einstein was received by the King and Queen Mother; the Queen herself was spending some time with her mother in Algeciras.

On March 10, Professor Einstein and his wife departed Madrid for Zaragoza, where he had agreed to give three lectures prior to returning home. Due to a lack of time, Professor Einstein had to decline an invitation from the University of Valencia.

In summary, we can say that Einstein's visit was a complete and unequivocal success. As a result of his Spanish visit, Professor Einstein has rendered an invaluable service to the prestige of German science and its achievements, a fact also noted in the English press. The unpretentious and genial character of the scholar contributed substantially to this success.

I take the liberty of enclosing some newspaper clippings.

> [signed]
> Langwerth

3. ZARAGOZA[3]

March 21, 1923

1638

In reference to my report 2nd of this
    month—K. No. 22

Permit me to submit the enclosed newspaper clippings dealing with Professor Einstein's visit and lectures in Zaragoza. The German Consul in Zaragoza informed me that upon his arrival in Zaragoza, Professor Einstein was received by representatives of all authorities who also attended his first lecture. In addition, I was advised that Professor Einstein was made an honorary member of the Academy of Sciences in Zaragoza and that during his visit the Ateneo Escolar created a fund for needy German students. I was further informed that Einstein expressed his appreciation for the warm reception in Zaragoza. On his way back to Germany, Einstein spent one day in Barcelona, although his presence there was not widely known and no special events took place.

> [signed]
> Pilger

Embassy Madrid

[3] Zentrales Staatsarchiv, Merseburg, Rep. 76 Vc Sekt. 1 Tit. 11 Teil Vc Nr. 55, Bl. 154.

APPENDIX 3

Einstein's Madrid Lectures on Relativity

ON March 3, 5, and 7, 1923, Einstein delivered three lectures on relativity under the auspices of the Royal Academy of Exact Sciences. The following resumés were taken at the lectures by Tomás Rodríguez Bachiller (1899–1980), then a graduate student in mathematics, and published in the Catholic daily *El Debate*, whose editor Angel Herrera was a close friend of J. M. Plans. After each lecture Bachiller, accompanied by a fellow mathematics student Teófilo Martín Escobar, repaired to the Café Vinces, a popular meeting place for journalists near the offices of *El Sol*, to prepare the following day's copy. Bachiller later wrote Einstein to seek his opinion of the transcripts. If any errors were found, they would be corrected before appearing in a special number of the *Revista Matemática Hispanoamericana*. The articles were never reprinted, however.[1] Mindful of the multiple linguistic and cognitive screens involved in the note-taking process, I have not attempted, in my translation, to correct Bachiller's notes in the frequent places where he was confused or wrong; in the third lecture especially he frequently loses the thread of Einstein's narrative. His constant shifts between direct and indirect discourse further add to the confusion. Equations, which the newspaper's typesetter formed with numerous improvised substitutions for symbols he lacked, have been reconstructed from the context, when necessary.

[1] Bachiller to Einstein, July 26, 1923. Einstein Duplicate Archive, Seeley Mudd Library, Princeton University. In the same letter Bachiller, in the name of Plans, asks Einstein to sign a photograph to appear in the same number and even suggests the wording of a message in German!

With the hall filled by the numerous public avid to hear from the lips of its discoverer the theory that had so profoundly revolutionized the bases of physical science, Professor Einstein was greeted upon entering by a great salvo of enthusiastic applause. As long as these signs of affection lasted, we noted many personalities entering and taking their seats in representation of the governing, aristocratic, and scientific classes. Among others, whom we regret not to mention, there were Sres. Maura, Salvatella, Gimeno, the Count of Grove, Gascón y Marín, the dean of the Faculty of Sciences Octavio de Toledo, and Señores Cabrera, Plans, and P. Enrique de Rafael, well known for their relativistic publications. At a desk, conveniently situated, we saw taking notes for the publication of the lectures on behalf of the Faculty of Sciences, Professors Carrasco, Palacios, Lorente de Nó, translator of Einstein's book, and Dr. T. Rodríguez Bachiller.

There fell a solemn silence in which we heard the anxious respiration of impatient chests. The professor of Mathematical Physics, Sr. Carrasco, makes the presentation of Professor Einstein, a brief but expressive speech, and Einstein begins his lecture. With a relaxed mien, sure intonation, and slow diction, which eases the comprehension of his French for little-accustomed ears, he faces the audience, while fixing them with a glance from his noble head, as if he wished to imprint upon them the ideas that inspired him, and said: "In the first place I must excuse myself for two reasons: first, because I cannot address you in your beautiful language, and second because I do not speak well the one in which I must address you."

He outlines the three lectures, proposing to devote the first to special relativity, the second to general, and the third to problems recently arisen, noting that the first two can be followed with basic mathematical knowledge, while the third would be harder to understand, inasmuch as he would have to use a more complicated mathematical algorithm, absolute differential calculus.

333

"Now I will say something about the general method of special relativity. The theory of relativity is a deductive theory based on two experimental facts, being in this sense analogous to thermodynamics, which is also based on two experimental facts, and its whole edifice is constructed without any additional empirical data."

He pointed out that the theory had arisen from research on the optics of moving bodies and from the application to such phenomena of the Maxwell-Lorentz equations, equations which he says should be considered the most likely expression of the physical content of our knowledge, and which show that light must have a constant velocity in empty space independent of the motion of bodies or of its color. Such a simple and completely general law seems as if it ought not to cause any difficulties, and, yet, it has produced those of such a nature that, in order to resolve them, one had to recur to the general conception that informs the entire theory of relativity, to the relativity of motion.

"There is," he says, "a lack of understanding. One hears that everything is relative, without noting that physics deals with things that are absolute in a certain sense, because it deals with objective things. It is totally false to suppose that this theory has anything to do with philosophical relativism."

He recalls that the relativity of motion was already known in antiquity, since it is impossible in ordinary life to conceive of motions that were not relative, at least with respect to the earth. In physics one cannot speak of anything except relative motions. This is not the object of the theory of relativity. The basis of classical mechanics is Galileo's principle of inertia, according to which a body that is not subject to external forces must move rectilineally and uniformly. If one states it [i.e., the principle] in this way, one ends up considering motion as absolute, and any motion would be rectilinear because there is no reference element.

The principle must be completed as follows: there exists a reference body relative to which the body is in uniform motion: the system that the reference body defines we will call an inertial system; without it the statement lacks meaning. Taking into consideration the content of this general principle and that of classical mechanics, we arrive at a general law that we can call that of special relativity. Let us suppose that the law [i.e., Galileo's principle] is correct for an inertial system k; if we take another system, k', in uniform motion without rotation with respect to k, it is easy to see that Galileo's principle is also valid with respect to k'. This means that there is an infinity of inertial systems in uniform translatory motion with respect to one another.

This is the principle of special relativity valid in mechanics; it asserts that there is no privileged state of motion because the inertial systems are equivalent, preserving all the laws of motion.

In essence, "it is true that from the concept of motion one cannot deduce a priori that there is a privileged system; however, from the point of view of physics, it seems a priori possible that its laws will be simpler with respect to a privileged system." We have seen that this does not occur for mechanics.

Thus, the principle of special relativity is valid in classical mechanics; we say special because this equivalence refers only to systems in relative uniform translatory motion.

Now, having the principle of relativity and preserving the principle of the constancy of the velocity of light in empty space, contradictions arise in optics which make the two principles appear irreconcilable: applying the theorems of classical kinematics, it seems that the velocity of propagation of light must depend on its direction.

Bearing in mind that a terrestrial laboratory fulfills the conditions for an inertial system (thanks to the movement of the earth), many experiments have been performed to search for this dependence. The most famous was that of Michelson which, like all the others, yielded a negative result.

Nature appears to show that the principle of special relativity is valid and we thus find ourselves in a difficult situation, because experience contradicts logic; and it is therefore necessary to see whether the chain of reasoning that results in the dependence of the velocity of propagation of light on its direction has any weak point.

It is necessary to analyze the concepts of space and time. "At this point I want to say that in general it is not possible for the human mind to know the origin of its concepts, and it frequently finds itself tempted to believe that these concepts [that is, of space and time] are drawn from experience and that, owing to their importance, they appear to have a higher dignity. As for time, it appears that the following is a fact to our senses: Everything that happens is called an event. Physical time, then, is a certain order of these events that seem given to our senses in an immediate way. This is Kant's way of thinking, and we could say that all those events are such that, taken two at a time, either one is earlier and one later, or else they are simultaneous. It appears to follow from this that time really exists in Nature without any restriction. Why do we believe this? We fix what we see with our senses, which give a temporal order that we apply univocally to events. One must distinguish between sensations and events. It is enough to say that one believes the temporal order given by the nature of the immediate sensation, but it is easy to see that this is not certain. For the physicist it is necessary that there be a means of knowing if the statement is true or not.

Are the twinklings of two stars simultaneous? We don't know how to proceed in order to demonstrate it, because the fact that a moving observer sees them as simultaneous does not imply that they are so." After some reasoning, he poses the following definition: "Two events, for example, two lightning bolts, must be considered simultaneous when an observer located in a system of reference equidistant from the places in which they occur sees them at the same time."

Afterward he says that there is no simultaneity when one takes another inertial system in uniform motion with respect to the first, that is, simultaneity is not absolute, and therefore it is meaningless if the system of reference is not specified. This, conjoined with the relativity of distance, leads us to think that the difficulty is not a serious one and that the contradiction is only apparent; Einstein resolves it by attributing to each system its own time.

In this fashion four numbers are assigned to each event which define it with respect to a system k, numbers that will be the three coordinates x, y, z and time t; to this same event in another inertial system k' four other numbers, x', y', z', and t', are assigned, and, to complete our study of the phenomenon we must now only establish the relationship between the sets of variables x, y, z, t and x', y', z' and t'.

It is easy to demonstrate that if we require that the law of constancy of the velocity of light in space be valid in the systems k and k', there will be a one-to-one transformation and it will coincide with the so-called Lorentz transformation.

From the study of this transformation we can immediately deduce several physical consequences, especially those relative to the behavior of solid bodies and clocks located at rest with respect to system k' and viewed from k: we find that solid bodies undergo a contraction in the direction of their movement, a contraction that increases with velocity, and that clocks are slowed down.

This transformation, together with the principle of relativity, provides a means to obtain natural laws. I must note here an interesting analogy with thermodynamics: in it one seeks the form of natural laws in such a way that a "perpetuum mobile" not be possible. We find something similar in relativity: natural laws must be such that their expression not vary from a system k to another k'.

If for k and k' the natural law is not the same, then the principle of relativity has been violated and one could establish a privileged system or state of motion that could be determined by the Lorentz transformation. Express mathematically a natural law in a system k using the four coordinates x, y, z, t, applying the Lorentz transformation, and by an analytical process of elimination one obtains the mathematical expression of the law in a system k' with the variables x', y', z', t'. If the two expressions for k and k' are not identical, the principle of relativity is not fulfilled and the expression of the law is not acceptable. On the contrary, if the two expressions in the systems k and k' are identical, the law is well formulated.

Minkowski gave a very elegant method for finding the form of the laws eliminating the transformation, the method he employed being very similar to the vectorial one used in classical physics. In the classical realm, one tries to find equations that do not vary as the position of the system of coordinates changes. This condition only affects space, not time.

One can easily explain why one speaks of a space of four dimensions in relativistic physics, while this is not done in classical physics. In classical physics, even when phenomena are defined by four variables, these are of a very different nature, since three refer to space and the other to time, the last being common to all systems of coordinates. In relativistic physics, the four variables have an identical meaning insofar as they are dependent on the systems of coordinates. Important physical consequences of special relativity: the electrodynamics of Maxwell-Lorentz preserves the form of its equations because they satisfy the conditions expressed above. On the other hand, Newton's dynamics are not preserved and must be modified in the sense that there be no body whose velocity is superior to that of light in space. The resistance that a body offers to receiving ever higher velocities grows as these velocities increase and it becomes infinite when the velocity equals that of light, or, what is the same, the velocity of the propagation of light in space is an unreachable upper limit. This result has been confirmed experimentally in the theory of electrons in which enormous velocities are found.

The most important consequence refers to the principle of the conservation of energy, which is found to be valid for all systems. The inertia of a body increases with the energy communicated to it, and in this way it comes about that the inert mass of a body is energy, thus fusing the two classical principles of the conservation of mass and the conservation of energy into one principle only of conservation of energy.

This principle is very important for the General Theory.

At this point Professor Einstein summed up his beautiful

and attractive discourse with the following sentence: "All we have said today is no more than an immediate consequence of two principles: special relativity and the constancy of the velocity of light in space."

As we said before, in this first lecture Einstein remains on elementary terrain, and one can complete the ideas expounded by reading his popularizing volume, translated into almost all European languages.

El Debate will give the most complete extract possible of the subsequent lectures, more difficult to understand, as the lecturer indicated at the start, always attempting to remain on a level acceptable to the majority of readers.

GENERAL RELATIVITY

With the hall brimming with people, Professor Einstein punctually begins his second lecture, a natural continuation of the first and filled with interesting ideas, as the reader may judge.

Today I begin with the theory of general relativity.[2] We have seen that the main idea of the special theory was the restricted principle of relativity, that is, the equivalence of all systems of coordinates in uniform motion with respect to one another, that is, of all inertial systems. This can be expressed thusly: there is no privileged state of motion. But it is necessary to note that it is true that there is no privileged state, but an infinitude: all inertial systems. It is thus natural to ask: Is it possible that Nature is such that it does not know such privileged states of motion? Are all systems, whatever might be their state of motion, equivalent for the expression of natural laws? At first glance this seems impossible, because the first law of mechan-

---

[2] Einstein's Madrid lecture on general relativity reflects his latest thoughts on the subject and may be fruitfully compared with his article, "Zur allgemeinen Relativitätstheorie," *Sitzungsberichte der Preussischen Akademie der Wissenschaften*, February 15, 1923, pp. 32–38, written in January on the *Haruna Maru*, the ship that carried the Einsteins from Japan to Europe, and published just before his arrival in Spain. On the basis of this article, I have corrected some of the equations in this lecture.

ics is not valid for all systems. Let k be a system relative to which the law of inertia is valid. That is, in k a material point free of all exterior action will have rectilineal and uniform motion, but if we take a system k' in rotation, the movement of the point is no longer rectilineal but follows a curve. Then we can say that the law of inertia prefers a certain system and does not accept others. We will see that this argument has no value, for the reason that the ordinary definition of an inertial system has a weak point. It is said that if there are no exterior forces the point describes a straight line with uniform motion, but the inexistence of exterior forces is known by the point's describing a straight line. There is a vicious circle here, then.

On the other hand, there is an argument for the validity of the principle of general relativity which seems to me of such force that I am convinced of the necessity of establishing such a principle. It is known to everyone: the equivalence between gravitational and inertial masses.

In mechanics there are two definitions of mass: one as a measure of inertia, that is, of the resistance that a material point offers to a variation in velocity; the other, as a measure of weight, that is, the force acting on a body in a gravitational field. It is almost a miracle for classical mechanics that these two measures, which are independent, coincide. There are measurements whose precision reaches one part in ten million, demonstrating their equivalence. It is easy to see that if one formulates the hypothesis of general relativity, all states of motion are equivalent for the enunciation of natural laws, the equivalence of masses appears to be a natural thing. The law of the equivalence of gravitational and inertial masses can be stated thusly: acceleration in a gravitational field does not depend on the physical or chemical state of a body, but only on the ratio of inertial and gravitational masses.

Now it is easy to see that this law can also be stated saying that an inertial action can be considered the same as a gravitational one. To picture it, let us imagine a region of space without a gravitational field, that is, an inertial system k. Let us consider a coordinate system k' in accelerated movement with

respect to k, and consider a free material point that is accelerated with respect to k. But now, who can say that the movement of k' is accelerated? In the same way we can say that k is
in accelerated movement with respect to k', and thus the point
would be in accelerated movement with respect to k'.

If, relative to a system, all bodies have the same accelerated
motion, we will say there is a gravitational field with respect to
the system, and nothing prohibits us from saying that the bodies are at rest and the system is accelerated. On this point of
view the equality of both masses is completely natural. We will
present an example. Let us suppose two systems k and k' and
in the former a mass suspended from a wire whose other end
is fixed in k'. The wire stretches out. Why? There are two ways
to view the question. First, let us suppose k' accelerated. The
acceleration of the system is transmitted to the mass by the
wire, and then tension measures the inertial mass. Second, let
us consider that k' is at rest. It is necessary to suppose that a
gravitational action acts on the body and the tension of the
wire will measure the gravitational mass. It thus results that
that same action can be interpreted as inertia or as gravity, and
by the general postulate of relativity the equality of the two
masses is absolutely necessary. It is natural to accept this law as
valid for any other accelerated motion.

Why is this interesting? Because we can produce a gravitational field by merely choosing an arbitrary state of motion,
and in this manner we are in possession of a purely theoretical
method for discovering the characteristics of a gravitational
field.

Let us suppose a space with no gravitational field referred to
a certain system. Referred to another system with known motion there is then a gravitational field that can be determined by
calculation. It is now necessary to see whether the laws thus
obtained for the field coincide with those of the first system,
and this is not as easy as it seems, because one must be aware
that the fields obtained by this procedure are not the most general ones, that is to say not all fields can be obtained by a change
of coordinates, or, what amounts to the same thing, one can-

not always make gravitation disappear by changing the system of reference. So, to find a general law of fields a generalization that we believe to be possible seems necessary. The simplest law of gravitation is that of Newton, which for spaces exterior to matter corresponds to the mathematical condition expressed by the Laplace-Poisson equation (vanishing of the Laplacian). But there is a great difficulty: the theory of special relativity has changed somewhat the nature of physical time, but it has preserved the geometry. Likewise it has conserved the theory of time in its basic characteristic, that is, in the manner of measuring it by identical clocks distributed throughout space, and it was sufficient to give a rule to regulate them.

In the General Theory this does not work. One can easily see that if the postulate of general relativity is true, neither Euclidean geometry nor the measure of time by identical clocks can be preserved. Let us take a system k with respect to which there is no gravitational field, that is, let k be an inertial system. Let us now take a disk in rotation which constitutes a new system k'. It is easy to see that, relative to the disk, Euclidean geometry does not properly describe the localization of solid bodies. We have seen that the Special Theory requires of bodies in motion a contraction in the direction of the motion. Let us imagine some operators on the disk who try to find the ratio of its circumference to its diameter. Toward this end they take identical rulers and apply them to the circumference and on a diameter. They read the rulers in both cases and obtain accordingly two numbers that, divided according to Euclid, should yield as a quotient the number *pi*; but they will find it to be greater because they are ignorant of the influence of the gravitational field (produced by the rotation) on the rulers. There are centrifugal forces relative to the disk which can be considered a gravitational field, whose action on the rulers is neither known nor required by our line of argument. We know that, when the disk is still, the measurements are made in the inertial system k and we obtain the results of Euclidean geometry, because the laws of special relativity are valid for a system with no field. One understands that the previous measurement must be taken for a fixed time t of system k. When one puts the disk in motion, the rulers placed on its circumference contract and, therefore, the number of them has to be greater, while those located on the diameter undergo no alteration. In this way the experimenter finds that, with respect to the disk, the laws of Euclidean geometry do not work.

The same occurs with clocks. If we have two identical ones, one in the center and another on any point on the disk, the latter will run slower than the former. This means that the field acts on the clocks and, therefore, time cannot be defined by means of identical clocks.

If, therefore, Euclidean geometry is not valid, a Cartesian system of coordinates cannot be used, and we find ourselves with the difficulty that on the disk and in any gravitational field, an immediate sense cannot be ascribed to the coordinates of space and time. At first glance it seems impossible for physics to make a description of Nature if we do not give, a priori, the meaning of the coordinates and of time, a seemingly insoluble problem, but of such a nature that Gauss had already resolved mathematically a very similar problem, that is, that of the geometry of a surface located in a Euclidean space. What is the geometry of a surface? Let us begin with the plane. The geometry of the plane is Euclidean plane geometry, according to which we locate flat solids on a plane. Let us begin by defining the straight line as a line between two points such that, locating rules on it beginning with one point until we reach the other, the number of necessary rules be the minimum. To resolve this question there is no need to leave the plane, and everything that interests us happens on it. Let us now try to do the same on a surface, imposing on ourselves the condition of not using points exterior to it, because the surface should suffice to give the law of localization upon it. In general, it is not done this way. For example, if one seeks a curvature, exterior points are used, and the problem resolved by Gauss was that of doing away with such exterior points.

Euclidean geometry does not in general work in locating points on surfaces, and, as an example, we can take a hemisphere and a small square that we can subdivide into smaller ones, thus obtaining a lattice that cannot be accommodated on the hemisphere as it would on a plane.

In view of this, Gauss stated the problem in the following way. It is necessary to give a method to enumerate the points of a surface, and he resolved it by taking two entirely arbitrary systems of curves, numbering each one of them in this fashion: through each point on the surface passes a curve of each system, and to each point there thus correspond two numbers called its curvilinear coordinates which lack any physical meaning, because it is necessary to know the curves. Nevertheless, all the properties can be described with these numbers. In spite of the fact that Gauss's coordinates lack physical meaning, they give the relations among real things when they are eliminated.

The entire geometry of the surface is known if, for points infinitely close to one another, we know their distance in any region of the surface. We are going to see how this condition can be translated into formulas.

Taking a portion of an infinitely small surface, one can consider it a plane and apply Euclidean geometry to it, in particular the Pythagorean theorem. In local coordinates the distance between two nearby points would be

$$ds^2 = dX_1^2 + dX_2^2$$

If we take Gauss's coordinates x_1 and x_2, the local coordinates will, for each point, be functions of those of Gauss, and the distance will take the general form

$$ds^2 = g_{ik}\, dx_i\, dx_k\ (i, k = 1, 2),$$

where the g's are functions of the x's and depend on the form of the surface and of the network of coordinate curves selected. The g's characterize the surface and cannot be taken as functions given beforehand.

In the general relativistic problem something completely analogous happens. We have seen, first, that one cannot derive, a priori, any physical meaning from the coordinates. Second, for a small part of our universe the Euclidean coordinates are valid, for example, on the earth for a freely falling observer. Free bodies have no acceleration and therefore there is no gravitation. For an infinitely small portion of our universe one can find a system of coordinates such that special relativity is valid with respect to it. In special relativity the magnitude

$$ds^2 = dx^2 + dy^2 + dz^2 - c^2dt^2$$

referring to two events infinitely close together in time and space does not depend on the inertial system of coordinates. This statement is equivalent to the Lorentz transformation. ds^2 turns out to be an invariant for the transformations of the Lorentz group and is measurable by means of rulers and clocks, inasmuch as the dx's etc. are directly observable differences.

We have seen that in our space of four dimensions we can take small parts without gravity, and then, for these, invariance persists for the local coordinates. If we take Gauss's coordinates (four arbitrary numbers), which fulfill only the condition of continuity, we can describe the universe in the same way that Gauss describes the surface. The coordinates alone mean nothing and give to ds^2 the form we mentioned before. The g's, of which there are now ten, represent directly measurable magnitudes and give a metric relationship or also define a gravitational field. In effect, let there be an inertial system without solid bodies; the g's assume the values 1, 1, 1, −1, which are constant; this is the only special case known up to now; if there is a gravitational field the g's are no longer constant. If we change the coordinate system, for example, by taking a rotating one, the g's appear as variable coefficients and it can be said that the variability of the g's gives the gravitational field. It turns out that not only the gravitational field but also the metric turn out to be given by the g's, which are functions of the coordinates. These functions are arbitrary, since they are x's. If one has understood well what has just been discussed, which is the essential part of the theory, the rest consists merely of formal difficulties.

The first question to face is that of finding a law of pure

gravitation, that is, in a field where there is only gravitation. We will follow a method analogous to that of special relativity, for we will look for laws that do not change in form when the system of reference is changed arbitrarily, as long as the transformation be a continuous one. This condition is very exacting and few laws satisfy it.

In order to obtain the law one can put a limit on the number of derivatives that must figure in it. In classical mechanics and in Laplace's equation the maximum number of derivatives is two and the equation is linear. We can admit the same condition for general relativity: thus the equations are determined and one finds the law of gravitation.

Next we need a law of motion analogous to Galileo's, and this we find in the condition that a free point describe a geodesic, a direct translation of Galileo's law.

It is frequently said that the theory of relativity is a revolution. This is not true. Rather it is, in a certain way, a translation: the language varies but at base the idea is the same. It is not a simple translation, but it adds the condition of the covariance of natural laws.

If there is matter in the field, the equation of Laplace-Poisson does not have a null second member, but the density of mass figures in it. In general relativity the corresponding equation also has a second member, which is also a density—not a scalar but a tensorial one, with ten components. The form of the equations is

$$R_{lk} - \tfrac{1}{2} g_{lk} R = -f_{lk}.$$

To obtain it the principle of the conservation of energy has been applied.

The method for obtaining the equations is formal. We have employed the postulate that the maximum number of derivatives figuring in them is two. As a result, the method says nothing about physical meaning—even a priori it may be doubted whether it implies anything similar to the laws of Newton, because instead of one potential there are ten magnitudes. Therefore, one must see whether, as a first approximation, Newton's theory is found, and, in effect, it so turns out with such precision that no phenomenon has been found for which there is any divergence.

This case shows how speculative methods are necessary for science. The equations are of such enormous complexity (as anyone who has dealt with any particular problem knows well) that it is impossible to obtain them by the inductive method, there being required a general principle as a deductive base. Naturally all general principles come to us from experience in a more or less direct way.

The Consequences

The theoretical consequences of the theory that lead to measurable magnitudes are three:

First. The ellipses of planets are not absolutely fixed, but each in its plane has a movement of rotation in the direction of the planet's revolution. This movement is unobservable for the planets, with the exception of Mercury, for which it reaches a value of 40" per century, approximately. Leverrier knew this, but classical mechanics could not explain it.

Second. Curvature of a light ray in a gravitational field. For the solar field the corresponding deflection should be $1''7$, as was confirmed in 1919.

Third. We have seen that the relative rate of two clocks depends on their position in the field. The light emitted by an element in the sun should differ from that emitted on earth by the same element, and this difference has to be noted by a shift in the spectral lines. Owing to the smallness of the difference, this phenomenon is difficult to observe. Most physicists believe it exists and has the magnitude predicted by the theory. Nevertheless, doubts remain with respect to its magnitude.

The most essential part remains to be said. Geometry, that is, the law for locating bodies, is not a law given a priori, but depends on the gravitational field, and this depends on the bodies. This means that geometry depends on the position of bodies and on all other physical phenomena and, as a result, it cannot be taken as the basis of physics.

In my next lecture I will speak of problems of purely speculative interest, for which I must make use of mathematics.

As OUR readers will observe, *El Debate* has given a most exact account of the first two lectures and will attempt to do the same with the third, within the limits imposed by the nature of the material.

LATEST RESULTS OF RELATIVITY

Before a public as numerous as attended the two previous lectures, Einstein delivered his third, devoted to the latest results of the theory.

He began by saying:

"We have seen that the special theory was imperfect from the philosophical viewpoint. Now we want to discuss the points of imperfection in the General Theory and the attempts made to avoid such weak points."

It is clear that all theories leave something to be desired. They are incomplete by any criterion.

In what pertains to the General Theory, it is necessary, in the first place, to consider Mach's point of view, which has achieved considerable importance in the theory of relativity. Mach was not satisfied with his analysis of classical mechanics. From a descriptive standpoint, motion is only relative, and one cannot give an exact sense to the concept of absolute motion. Like Newton, Mach saw very well that in ordinary mechanics everything is based on acceleration relative to space, that is, to something absolute. To avoid this he conceived that inertia, rather than working against absolute acceleration, worked against relative acceleration. If we have two masses and we accelerate one of them, its inertia or resistance reacts against relative acceleration with respect to the second mass, and in general he thinks that inertia is a resistance to the acceleration of the mass with respect to all the rest of the masses in the world.

He sought an analytic expression for this but could not find it because *actio in distans* appeared and physics of the late nineteenth century had already done away with such a concept, which the quality of the phenomena did not permit either. At that time he thought that inertia is only relative.

In the General Theory of relativity the physical qualities of space depend on matter, for geometry, which gives the rules for situating bodies, is influenced by the gravitational field, which, in turn, depends on matter. Therefore, absolute properties of space do not exist. Then, if this theory of the relativity or reciprocity of causes really exists, the thing is such that the qualities of space, metrics, of inertia or of gravitation, depend on bodies. That is, space does not exist; there is only matter.

Now then, the "disagreeable point" in Mach also exists in the theory of relativity, and is prohibited by its very spirit. In the first place, space would have properties influenced by matter, and this introduces an unacceptable dualism. In the second place, when the laws of gravity are calculated with greater accuracy, one should find the influence of masses in motion on others, finding things in agreement with the ideas of Mach. For example, if we accelerate all the masses of the universe except for one, this one should also turn out to be accelerated. If we only accelerate a portion, there would appear a kind of force of induction in the same direction.

Besides, if all masses are in rotation, the whole system should have rotation in order to avoid relative accelerations, and thus it is necessary to believe that if we seek a system of local coordinates that behave in a Galilean way, the system should also take on a slight rotation. For example, a gyroscope should be influenced by the mass of the other bodies. All these consequences of Mach's ideas are confirmed by the gravitational equations, only that these actions are too weak to be observed.

Thus the hypothesis that the world be almost Euclidean in the infinite is unnatural, and it turns out to be much more logical to admit that all the laws of space are a consequence of matter. There is a more direct cause that impedes us from believing that space is Euclidean in the infinite. We can form an idea of the structure of the universe by imagining that there is matter everywhere whose mean density is not null—something we could not know from experience inasmuch as it is very hard to believe that our stellar system constitutes a kind of island in space. The first hypothesis contradicts both Newton's theory and that of relativity. If we admit that there are stars everywhere and that space is not totally empty in totality, it turns out there is an imperfection in the law of gravity, and it is easy to see that the postulate of relativity permits us to modify the form of the equations a bit, adding to its second member a term that will not change the character of the equations, and thus one sees that a world with a mean density that is not null exists which, insofar as space is concerned, is closed, and Mach's postulate is satisfied because if the additional term is not null, one cannot have a static empty space. The magnitude of space depends on that term, and if the mean density is known one can calculate such a magnitude. Since we do not know it, we can only say that it should be approximately that of our own neighborhood.

The basic thing is that the properties of space are given by matter, and the reciprocity is complete, since space gives the law of localization and matter gives the properties of space. This is more natural than to hold that the laws of space are given a priori (for example, the laws of the motion of bodies before they exist); we would thus have something both absolute and without cause: that is to say, we would have an incomplete causality.

There is another point on which the theory of relativity is incomplete. In the gravitational equations there is a first member (see the summary of the second lecture) that represents the gravitational potential, which can be obtained exactly through purely theoretical procedures, while the second contains a density obtained empirically. The electromagnetic field has been constructed by inductive means from Coulomb to Maxwell. On the basis of things already known, it is certain that the mathematical nature of this field is that of an antisymmetric tensor f_{ik}, and in the Special Theory of relativity it appears thus on writing Maxwell's equations with covariant notation. These equations have been written in a variational form that is very pretty from a mathematical standpoint. The magnitudes to vary are the g_{ik} and the f_{ik}; nevertheless, there are things that do not satisfy me here. There is a problem that already existed in Maxwell's electrodynamics. We know what protons and electrons are: the two elemental bodies must be solutions of field equations "and they aren't." There is no way to equilibrate the force of repulsion of the particles forming electrons. Poincaré introduced a pressure, but this is not natural; and it only constitutes a means of calculation. If we do away with the electron and occupy ourselves only with the field, we find ourselves with the gravitational and electromagnetic fields, that is, with two things totally independent of one another from the logical point of view. We thus face a dualist system, something which one perceives recurring to the variational form

$$\text{var} \{ \textstyle\int H \, dv \} = 0 \, .$$

If this integral is invariant, the equations obtained satisfy the principle of general relativity. The function H is known for gravitational phenomena, and, according to Maxwell, it can be calculated for electromagnetic phenomena, but one thus has two logically independent things whose sum defines space. It would be preferable to have a theory in which the function from which all things are derived might have a single structure and not be composed of the sum of two independent things. This is now a mathematical problem. It is a matter of constructing a function that is not composed of logically independent members, and what we say for H we must say for the f_{ik} and the g_{ik} that express the properties of space.

The problem turns out to be very difficult because we attempt to satisfy a desire for unity without having any physical facts on which to base ourselves, the question being reduced to a problem of mathematical simplicity. If it is resolved, it would have to begin by obtaining observable consequences.

Let us see what mathematical ideas have guided this endeavor. The theory of general relativity is based on the metric of space, that is, on the invariance of the lineal element ds^2. We have seen that the laws of gravity were obtained by seeking tensorial magnitudes derived from the g's. There have been attempts to simplify this theory, which, in its mathematical part, was already given by Riemann, who calculated the tensorial magnitudes that could be derived from the g's. Levi-Civita and Weyl have achieved a very great simplification in the obtainment of a second-order tensor, yielding at the same time a very practical concept: that of parallelism in any kind of manifold. In Euclidean geometry, given two points and a vector, one can trace another vector parallel to the first; that is, a parallel translation of the vector can be made. It cannot be done thus in non-Euclidean continua, for example, on a sphere. But if we take a small part of the continuum, it can be supposed Euclidean, and in it made the parallel translation of the vector, repeating the operation as many times as is necessary; the translation can be made (in a certain sense parallel) all along a curve. If one takes another curve with the same ends one obtains another vector, and one arrives at a general formula for the parallel translation:

$$dA^m = -\Gamma_{lk}{}^m A^l dx_k .$$

The first member represents the variation undergone by the components of the vector as a consequence of the variations d x_k of the coordinates of its origin; in these formulas intervene forty Γ which define the law of parallel translation, that is, the affine structure of space. If we have a space of given metric, we find that the Γ are determined by the g's. It is sufficient to write the condition that the modulus of the vector not vary, a natural thing since it involves a repeated Euclidean process; that is, one must write that

$$A^2 = g_{lk} A^l A^k$$

is invariant, which gives the condition which the Γ must satisfy and one finds that they are precisely Christoffel's symbols with three indices. It is easy to see that the basic tensor of Riemann can be obtained with the parallel translation.

In a closed cycle, the vector returns with a different value, and Christoffel's symbols permit the calculation of the difference. We can say that Levi-Civita and Weyl have generalized geometry with respect to Riemann, for we can imagine geometries without metrics, but which have a given law of parallel transport, that is, an affine law more general than the metric.

In order to modify the theory as little as possible, he reasons in this way, the ds^2 can be measured by solids and clocks that are not immediate things. He believes that one cannot depart from the basic facts. There is a basic fact, the propagation of light given by $ds = 0$; in this equation the ratios of the g's intervene; it turns out thus that the g's do not, in themselves, have any physical meaning other than their ratios, and we are still left with an overall factor devoid of physical meaning. One finds a new geometry, for if two vectors are taken, the ratio of their magnitudes is fixed, but for one vector only there is no absolute magnitude. The same occurs in the translation, and only the angle remains as an absolute; in this way congruence is replaced by similarity. He attempts to find what can be said about the Γ's and he finds that they can be calculated and they depend on four functions f, a very interesting thing, inasmuch as the electromagnetic field is also defined by four functions.

In Weyl's geometry the g's and f's give the theory a complete sense, and it seems natural to say that the ten g's and the four f's define the gravitational and electromagnetic fields. One must bear in mind that this solution is not natural, because ds^2 is an objective thing; Nature shows us that ds^2 has a definite value, inasmuch as we possess natural clocks in the vibrating atoms that yield spectral series with well-defined spectral lines. For Weyl, in his analysis, the f's exist only as results of calculation. Moreover, in order to find the laws of pure gravitation and of electromagnetism, it was necessary to define a function H, composed of two absolutely independent parts. We thus had a dualism, in order to avoid that which he arrived at the solution mentioned. But the ds^2 has physical meaning, and, as a result, this solution is not satisfactory either.

He sought another ingenious route, which consisted in starting with the Γ's and finding the metric, deduced from the parallel translation, that is, from the affine structure of space. One must construct all the concepts on the basis of the Γ's. In the tensor of Riemann the R's can be expressed by the Γ's, which, since they are more general than the g's, give a much more general result, and one arrives at the form

$$\lambda\, g_{lk} = R_{lk}$$

that is, one obtains a tensor that defines the metric, and one finds some equations that contain the cosmological term.

According to Eddington, this is of no use either because one would have to find forty conditions in order to determine the Γ's, and he did not know how to determine them. If there is only a gravitational field, the matter offers no difficulty. In general, the Γ's are not symmetrical, but rather are composed of a symmetrical part and an asymmetrical one, which corresponds well to the existence in Nature of metric and electromagnetic fields.

Only the conditions required to have the Γ's remain to be found. I have lately found a natural way. I follow a variational method, providing the integral

$$\int \sqrt{|R_{lk}|}\; dv$$

be invariant. This integral is analogous to the integral of ordinary geometry and is the simplest invariant possible; upon taking the variation of the same which ought to be null, it is necessary to make the Γ's vary independently.

I will indicate the results obtained:

If there is no electromagnetic force, that is to say, if the f_{lk} do not exist, one obtains gravitational equations as in the old theory and with the cosmological term.

In first approximation, if the electromagnetic field is weak, one obtains Maxwell's equations, which is "almost a miracle."

The procedure is, in any case, arbitrary, because we have

taken the Γ's to have basic physical meaning, and then we take the simple expression as a tensor in order to deduce by invariance the laws of gravitation and electromagnetism. Nevertheless, we avoid Weyl's weak point.

Up to now, the calculations I have made with respect to gravitational and electromagnetic fields have given the results already known. Insofar as the structure of electrons is concerned, the calculations are so complicated that I have not been able to obtain anything definitive up to the present.

WITH A THUNDERING salvo of applause, Professor Einstein ends his third lecture and, with it, the series he gave at the University.

Bibliography

1. Newspapers

ABC, Madrid
Avui, Barcelona
El Correo Catalán, Barcelona
El Debate, Madrid
El Diario de Barcelona, Barcelona
El Diario Español, Buenos Aires
El Diluvio, Barcelona
La Epoca, Madrid
El Heraldo, Madrid
El Heraldo de Aragón, Zaragoza
El Imparcial, Madrid
El Liberal, Madrid
El Luchador, Alicante
El Mercantil Valenciano, Valencia
La Nación, Buenos Aires
El Norte de Castilla, Valladolid

El Noticiero, Zaragoza
El Noticiero Bilbaino, Bilbao
El Noticiero Universal, Barcelona
El País, Madrid
La Prensa, Buenos Aires
Las Provincias, Valencia
La Publicitat, Barcelona
El Pueblo, Valencia
La Razón, Buenos Aires
El Siglo Futuro, Madrid
El Sol, Madrid
La Vanguardia, Barcelona
La Veu de Catalunya, Barcelona
La Voz, Madrid
La Voz Valenciana, Valencia
Ya, Madrid

2. Relativity and Physical Sciences in Spain (Select Bibliography)

Abecé [pseudonym]. "Albert Einstein." *En Patufet*, 1923, p. 182.

Adellac, Miguel. "La fe en la ciencia." *El Heraldo de Aragón*, March 21, 1923.

Aguilar Peris, J. D. *Julio Palacios y el lenguaje de la física*. Santander: Universidad de Santander, 1981 (Aula de Cultura Científica, 2).

Alcobé, Eduardo. "Sobre un concepto dinámico." *Memorias de la Academia de Ciencias y Artes de Barcelona*, 3rd epoch, 22 (1931–32), 363–371.

———. "Física." In *Enciclopedia Universal Ilustrada*, suplemento 1934, pp. 361–386.

———. "Causalidad e indeterminación en la física contemporánea."

Memories de l'Acadèmia de Ciències i Arts de Barcelona, 3rd epoch, 25 (1935), 85–108.

Andreu Tormo, José. *La relatividad descifrada*. Valencia: ECIR, 1978.

Aparici, Rafael. "La gravitación ondulatoria o sea una opinión más acerca de las teorías de Einstein." *Madrid Científico*, 32 (1925), 22; also in *Estudios* (Buenos Aires), 14 (1925), 483–484.

Apráiz, Félix. *Une réponse aux interprétations égarées du principe de Relativité. L'Ether existe et les phénomènes électromagnétiques sont purement mécaniques*. Paris: Gauthiers Villars, 1921.

————. *Las seis dimensiones del espacio físico. Errores de la física actual y naturaleza de la electricidad y el éter*. Las Palmas: Tip. Falange, 1945.

Araquistáin, Luis. "Einstein o la razón estremecida." In *El arca de Noé*. Valencia: Sempere, 1926. (Originally published in *El Sol*.)

Arraras, J[oaquín]. "Una lección de Einstein." *El Debate*, March 2, 1923.

Assia, Agusto [Felipe Fernández García y Armesto]. "De hijo de un obrero a mago de la física." *La Vanguardia*, April 19, 1955.

Aznar, Manuel. "El profesor Einstein en Madrid." *El Diario Español*, April 5, 1923.

Baeza, Ricardo. "Delante del profesor Einstein." *El Sol*, July 3, 1923.

Banús, Carlos. "La vieja y la nueva física." *Madrid Científico*, 31 (1924), 321–322.

Barga, Corpus. "Crónica einsteiniana. El 'colloquium' de Einstein con los sabios franceses." *El Sol*, April 14, 1922.

Bentaból y Ureta, Horacio. "Relatividad—Un concurso patriótico." *Revista de la Sociedad Astronómica de España y América*, 13 (1923), 29–30.

————. *Observaciones contradictorias a la teoría de la relatividad del profesor Alberto Einstein*. Madrid: Imprenta Velasco, 1925.

Betibat [pseudonym]. "Saludando a Einstein. Chispazos racionalistas." *El Siglo Futuro*, March 13, 1923.

Blanco Fombona, R[ufino]. "Einstein en España. La República reivindica a los judíos." *El Sol*, April 16, 1933.

Borras, Tomás. *Ramiro Ledesma Ramos*. Madrid: Editora Nacional, 1971.

Buen, Odón de. "Un idea. Antes de que marche Einstein." *La Voz*, March 9, 1923.

Burgada y Juliá, Juan. "Einstein en España." *El Diario de Barcelona*, March 10, 1923.

Burgaleta, Vicente. "Fundamentos de la dinámica," *Anales del Instituto Católico de Artes e Industria*, 1 (1922), 392–398.

———. "Una paradoja relativista." *Madrid Científico*, 30 (1923), 66–68.

Cabot, José Tomás. "Ortega y el antifranquismo de Einstein." *La Vanguardia*, March 20, 1979.

Cabrera, Blas. "Principios fundamentales de análisis vectorial en el espacio de tres dimensiones y en el universo de Minkowski." *Revista de la Real Academia de Ciencias Exactas* (Madrid), 11 (1912–13), 326–344ff.

———. "Conferencias de Física Matemática en Madrid." *Ibérica*, 5 (1916), 46–47, 175.

———. *¿Qué es la electricidad?* Madrid: Residencia de Estudiantes, 1917.

———. "Conferencias sobre la relatividad en la Universidad de Madrid." *Ibérica*, 16 (1921), 306–307, 324, 356, 371–373, 387–389.

———. "Las fronteras del conocimiento en la filosofía natural." *Revista de Filosofía* (Buenos Aires), 14 (1921), 152–160. (Originally printed in *Verbum*, vol. 14, no. 55.)

———. *Momento actual de la física.* Madrid: Real Academia de Ciencias Exactas, 1921.

———. *La teoría de la relatividad.* San Sebastián: Sociedad de Oceanografía de Guipuzcoa, 1921.

———. *Principio de relatividad.* Madrid: Residencia de Estudiantes, 1922.

———. "La obra de Einstein fuera de la teoría de la relatividad." *Revista Matemática Hispano-Americana*, 5 (1923), 142–152.

———. "Principio de relatividad." *Ingeniería y Construcción*, 1 (1923), 113ff.

Camba, Julio. "Las teorías de Einstein y el universo literato." *El Sol*, July 8, 1922.

———. "Bienvenido a Einstein." *El Sol*, March 1, 1923.

Cantera y Villamil, Federico. "*Aviación y relatividad. Problemas del vuelo sin motor.* Madrid: Gráficas Reunidas, 1923.

———. "De relatividad." *Madrid Científico*, 30 (1923), 150.

Carrasco Garrorena, Pedro. "Teoría de la relatividad." In Ateneo de Madrid, *Estado actual, métodos y problemas de las ciencias.* Madrid, 1916, pp. 145–165.

———. "Nuevo método para determinar la velocidad de la luz." *Re-*

vista de la Real Academia de Ciencias Exactas, 17 (1918–19), 201–216, 340–357.

———. "Método para determinar la velocidad de la luz." *Anales de la Sociedad Española de Física y Química*, 17 (1919), 296–306, 316–330.

———. "Estado presente de la teoría de la relatividad." *Anales de la Sociedad Española de Física y Química*, 18 (1920), pt. 2, 67–73, 93–100.

———. "Desviación de la luz por el Sol." *Ibérica*, 7 (1920), pt. 1, 373–374.

———. *Filosofía de la Mecánica*. Madrid: Páez, [1926].

Casares Roldán, José. *Refutación de los fundamentos de la Relatividad*. Granada: Urania, 1952.

Castro, Miguel de. "Einstein y los madrileños o la derrota de los pedantes." *Las Provincias*, March 11, 1923.

Clarover [pseudonym]. "La visita de Einstein." *El Siglo Futuro*, March 2, 1923.

Colominas Maseras, Juan. "Einstein en Barcelona." *El Pueblo*, March 2, 1923.

Coll Gilabert, A. "Einstein: El desconocido era un gran genio." *Diario de Barcelona*, March 11, 1979.

Comas Solá, José. "Consideraciones sobre la aberración de la luz." *Boletín del Observatorio Fabra, Sección Astronómica*, 1 (1919–27), 25–38. [May-June 1919].

———. "Consideraciones sobre el principio de la relatividad y la teoría emisivo-ondulatoria de la energía radiante," *Boletín del Observatorio Fabra, Sección Astronómica*, 1 (1919–27), 62–68 [September-October 1920].

———. "Interpretación de algunos fenómenos ópticos en la teoría emisivo-ondulatoria de la luz." *Boletín del Observatorio Fabra, Sección Astronómica*, 1 (1919–27), 111–113 [March-June 1920].

———. "El fenómeno de la difracción en la teoría emisivo-ondulatoria de la luz." *Boletín del Observatorio Fabra, Sección Astronómica*, 1 (1919–27), 113–114.

———. "Comentarios sobre la teoría de la relatividad." *Boletín del Observatorio Fabra, Sección Astronómica*, 1 (1919–27), 122–129.

———. "Consideraciones sobre la frecuencia ondulatoria de la energía radiante." *Boletín del Observatorio Fabra, Sección Astronómica*, 1 (1919–27), 147–148 [December 15, 1921].

———. "Procedimiento astronómico para revelar el movimiento absoluto de nuestro sistema solar, en el caso de existir este movimiento absoluto," *Boletín del Observatorio Fabra, Sección Astronómica*, 1 (1919–27), 150–153.

———. "Otro método para obtener el valor del movimiento absoluto de la tierra, en el caso de existir este movimiento absoluto." *Boletín del Observatorio Fabra, Sección Astronómica*, 1 (1919–27), 154.

———. "Evolución científica." *Revista de la Sociedad Astronómica de España y América*, 11 (1921), 66–67.

———. "Las conferencias del profesor Einstein." *Revista de la Sociedad Astronómica de España y América*, 13 (1923), 20–21. (Originally published in *La Vanguardia*, March 14, 1923.)

———. "Mas sobre la teoría emisivo-ondulatoria de la energía radiante." *Boletín del Observatorio Fabra, Sección Astronómica*, 1 (1919–27), 177–181 [April 1923].

———. "Consideraciones sobre la energía radiante." *Revista del Observatorio Fabra, Sección Astronómica*, 1 (1919–27), 320–323.

———. "Nueva teoría emisiva de la luz y la energía radiante en general." *Scientia*, 36 (1924), 375–382.

———. "Una experiencia notable." *Revista de la Sociedad Astronómica de España y América*, 15 (1925), 56–58.

———. "Consideraciones sobre la relatividad." *Revista de la Sociedad Astronómica de España y América*, 15 (1925), 87–88.

———. "Teoría emisiva de la radiación." *Boletín del Observatorio Fabra, Sección Astronómica*, 1 (1919–27), 337–343. [January 1926].

———. "Sobre el principi de Doppler en les seves relacions amb l'anomenat moviment absolut o moviment relatiu dels cossos respecte l'èter." *Ciència. Revista Catalana de Ciència i Tecnologia*, 1 (1926), 91.

———. "Teoría corpuscular-ondulatoria de la radiación." *Boletín del Observatorio Fabra, Sección Astronómica*, 2 (1930–32), 25–37.

Corbella Alvarez, Salvador. *La teoría de la relatividad de Einstein al alcance de todos*. Barcelona: España en Africa, 1921.

Coso, Marcial del. "Varios ejemplos clarísimos: Todos los aspectos de la vida son lecciones de relatividad." *El Imparcial*, March 17, 1923.

Durán, Miguel-Emilio. "Einstein en Barcelona: La teoría de la relatividad y la música." *Las Provincias*, March 6, 1923.

Eddington, Arthur. *Espacio, tiempo, gravitación*. J. M. Plans y Freyre, trans. Madrid: Calpe, 1922.

Edwards Bello, Joaquín. "Einstein en Barcelona." *La Vanguardia*, April 29, 1955.

Einstein, Albert. "Sobre la teoría de la relatividad especial y general." *Revista Matemática Hispano-Americana*, 3 (1921), 194–199ff.

———. *La teoría de la relatividad al alcance de todos*. Fernando Lorente de Nó, trans. 3rd Spanish edition. Madrid: Biblioteca Scientia, 1925.

———. "Nuevos experimentos sobre la influencia del movimiento terrestre en la velocidad de la luz con relación a la tierra." *Investigación y Progreso*, 1 (1927), 4–5.

Escofet, José. "Crónicas catalanas: Einstein y los matemáticos." *Las Provincias*, March 18, 1923.

Fabra Rivas, A. "Una visita a Einstein." *El Sol*, March 27, 1930.

Fernández Barreto, Antonio. *Palabras pronunciadas por el Excmo. Sr. Gobernador Militar de Sevilla D. ——*. Seville, Camara de Comercio, Industria y Navegación, 1929.

Fernández Flores, Wenceslao. "Einstein y Gimeno." *ABC*, March 14, 1923. Reprinted in *Obras completas*, 9. Madrid: Aguilar, 1964.

———. "Einstein y los comunistas." *El Diario Español*, April 7, 1923.

Flammarion, Camille. "La doctrina de Newton y las teorías de Einstein." *Madrid Científico*, 27 (1920), 154–155.

Fojo, Eugenio. "Alberto Einstein y la teoría de la relatividad." *El Noticiero Bilbaino*, March 4, 1923.

Freundlich, Erwin. *Los fundamentos de la teoría de la gravitación de Einstein*. J. M. Plans y Freyre, trans. Madrid-Barcelona: Calpe, 1920.

Gallego-Díaz, José. "Alberto Einstein, símbolo de nuestro tiempo." *ABC*, April 19, 1955.

García, Juan. "Sobre la teoría del espacio." *Revista de la Real Academia de Ciencias Exactas*, 36 (1942), 263–295.

García Doncel, Manuel, et al. *Tres conferencias sobre Albert Einstein*. Memorias de la Real Academia de Ciencias y Artes de Barcelona, 3rd epoch, no. 808, vol. 45, no. 4. Barcelona, 1981.

García Moreno, Enrique. "La ciencia y la verdad." *Electricidad y Mecánica* (Barcelona), 19 (February 1923), 5–7.

García Morente, Manuel. "Einstein." *Arbor*, 104 (1979), 35–39.

———. *Sobre la teoría de la Relatividad*. Madrid: Encuentro, 1984.

———. *See* Schlick, M.

Glick, Thomas F. *Einstein y los españoles*. Madrid: Alianza, 1986.

———. "Einstein y los españoles: Aspectos de la recepción de la relatividad." *Llull: Boletín de la Sociedad Española de Historia de las Ciencias*, vol. 2, no. 4 (December 1979), 3–22.

———. "Einstein a Barcelona." *Ciència*, 3 (October 1980), 10–18.

———. "Tomás Rodríguez Bachiller (1899–1980): In Memoriam." *Dynamis*, 2 (1982), 403–409.

———. "Emilio Herrera and Spanish Technology." In Herrera, *Flying: The Memoirs of a Spanish Aeronaut*, pp. 173–215. Albuquerque: University of New Mexico Press, 1984.

———. "Marginalia einsteiniana." *Ciència*, 4 (May 1984), 224–227.

———. "Einstein, Rey Pastor y la promoción de la ciencia en España." In *Actas, I Simposio sobre Julio Rey Pastor*, Luis Español, ed., pp. 79–90. Logrono, 1985.

———. "Huellas de Einstein y Freud in México," *Tezcatlipoca: Anuario de Historia de la Ciencia y la Tecnología*, 2 (in press).

Goicoechea y Alzuarán, José María. "Las teorías de Einstein sin matemáticas," *Revista Calasancia*, 11 (1923), 468–489.

———. "Crítica de las teorías de Einstein." *Revista Calasancia*, 11 (1923), 563–585.

Gómez de la Serna, Ramón. "El birrete de Einstein." *El Sol*, March 8, 1923.

———. "Einstein y Ortega." *El Sol*, March 10, 1923.

Gómez de Nicolás, Tomás. "La relatividad de los valores: Alegrémonos de no ser sabios." *El Imparcial*, March 10, 1923.

González Blanco, Edmundo. *El universo invisible*. Madrid: Mundo Latino, 1929.

Graña, Manuel. "Aspectos de la relatividad." *El Debate*, March 14, 1923.

———. "Polonia a Copérnico." *El Debate*, March 17, 1923.

Gregorio Rocasolano, Antonio de. *Estudios químico-físicos sobre la materia viva*. Anales de la Universidad de Zaragoza, vol. 1. Zaragoza, 1917.

Güell y López, Eusebio. *Espacio, relación y posición*. 2nd edition. Madrid: Nuevas Gráficas, 1942 [1st edition, 1924].

Herrera, Emilio. "Relación de la hipergeometría con la mecánica celeste." *Memorial de Ingenieros del Ejército*, 5th epoch, 33 (1916), 371–388; 34 (1917), 221–235.

———. "Estrambote luminoso: Una carta del profesor Eddington." *El Sol*, March 12, 1920.

———. "La cuarta dimensión: El tiempo." *El Sol*, October 15, 1920.

Herrera, Emilio. "La cuarta dimensión: El hiperespacio." *El Sol*, October 22, 1920.

————. "El experimento de Morley y Michelson." *El Sol*, November 5, 1920.

————. "Sobre la cuarta dimensión." *El Sol*, November 12, 1920.

————. "Más sobre la cuarta dimensión." *El Sol*, December 3, 1920.

————. "Divulgaciones aerodinámicas." *Madrid Científico*, 27 (1920), 393–394.

————. *Algunas consideraciones sobre la teoría de la relatividad de Einstein.* Madrid: Memorial de Ingenieros, 1922.

————. "La intuición y la ciencia." *Madrid Científico*, 30 (1923), 17–19.

————. "Intuición, ciencia y conocimiento." *Madrid Científico*, 30 (1923), 102–103.

————. "Una ojeada al universo." *Madrid Científico*, 33 (1926), 133–136.

————. *Discurso leído en el acto de su recepción por D.*—— *y contestación del excmo. señor D. José Marva y Mayer el día 19 de abril de 1933.* Madrid: Academia de Ciencias Exactas, Físicas y Naturales, 1933.

————. "El universo y la hiperdinámica." *Anales de la Sociedad Española de Física y Química*, 32 (1934), 109–127.

————. "La vitesse de la lumière par rapport aux corps en mouvement." *Le Génie Civil*, 140 (1963), 262–264.

————. "L'Univers de Descartes." *Le Génie Civil*, 140 (1963), 280–282ff.

————. *Flying: The Memoirs of a Spanish Aeronaut.* Elizabeth Ladd, trans. Albuquerque: University of New Mexico Press, 1984.

Huertas, Ataulfo. "La relatividad de Einstein." *Revista Calasancia*, 11 (1923), 241–254, 290–309, 369–384.

————. Review of Teodoro Rodríguez, *Relatividad, modernismo y matematicismo* (Barcelona, 1924), *Revista Calasancia*, 12 (1924), 871–886.

Hunolt, Emilio. "Sobre las teorías de Einstein." *El Sol*, July 6, 1922.

Ibeas, Bruno. *Las teorías de la relatividad de A. Einstein.* Madrid: N. del Amo, 1922.

————. "El einsteinianismo y la venida de Einstein." *El Debate*, March 7, 1923.

Jiménez, Enrique. "Las matemáticas y la ingeniería." *Electricidad y Mecánica* (Valencia), 10, no. 4 (April 1914), 4–10.

Jiménez Osuna, José M. *Algunas aplicaciones matemáticas a las ciencias naturales.* Toledo: J. Peláez, 1921.

Kirchenberger, Paul. *¿Qué puede comprenderse sin matemáticas de la teoría de la relatividad?* J. de la Puente, trans. Barcelona: Juan Ruiz Romero, 1923.

Lafuente Garcia, Antonio. "Apuntes sobre la relatividad en España." *Llull: Boletín de la Sociedad Española de Historia de las Ciencias,* no. 1 (1977), 35–39.

———. *Introducción de la relatividad especial en España.* Memoria de Licenciatura, Universidad de Barcelona, 1978.

———. "La hipótesis del éter en España." *Llull: Boletín de la Sociedad Española de Historia de las Ciencias,* no. 3 (February 1979), 15–28.

———. "La relatividad y Einstein en España." *Mundo Científico,* 2 (1982), 584–591.

La Llave, Joaquín de la. "La cuarta dimensión." *Madrid Científico,* 37 (1926), 72–74.

Lallemand, Ch[arles]. "La teoría de la relatividad." *Revista de la Sociedad Astronómica de España y América,* 16 (1926), 9–10.

Lammel, Rodolfo. *Facil acceso a la teoría de la Relatividad.* Fernando G. Vela, trans. Madrid: Tiempo Nuevo, 1923.

Lamo, Regina. "Interpretaciones sentimentales. Einstein el precursor." *El Diluvio,* March 2, 1923.

Ledesma Ramos, Ramiro. "El Matemático Rey Pastor." *La Gaceta Literaria,* March 15, 1928.

———. "Hans Driesch y las teorías de Einstein." *La Gaceta Literaria,* October 15, 1928. Reprinted in *La filosofía, disciplina imperial,* pp. 91–96. Madrid: Tecnos, 1983.

López Arroyo, M. "La espectroscopía en el Observatorio Astronómico de Madrid." *Boletín Astronómico de Madrid,* 8, no. 2 (1972), 3–15.

López Campillo, E[velyne]. *La Revista de Occidente y la formación de minorías (1923-1936).* Madrid: Taurus, 1972.

López de Gomara, Juan. "Efímeros triunfos de la relatividad. Audaces observaciones de la ignorancia." *El Diario Español,* April 17, 1923.

Lotario [pseudonym]. "Relativismo y bentabolismo." *El Heraldo de Madrid,* March 15, 1923.

Lucanor [pseudonym]. "Después de oír a Einstein." *La Epoca,* March 16, 1923.

Lucia [Ordóñez], Pedro José. "La intuición y el conocimiento." *Madrid Científico*, 30 (1923), 86–88.

———. "Valor y significación de las leyes científicas." *Revista de Obras Públicas*, 71 (1923), 160–164.

Lucini, Manuel. "El profesor Einstein." *Madrid Científico*, 30 (1923), 65–66.

———. "Sobre una paradoja relativista." *Madrid Científico*, 30 (1923), 52.

Lusa Monforte, Guillermo. "Evolución histórica de la enseñanza de las matemáticas en las Escuelas Técnicas Superiores de Ingenieros Industriales," pp. 1–93. Universidad de Santiago, Reunión de Departamentos de Matemáticas de Escuelas Técnicas Superiores de Ingenieros Industriales. Vigo, 1982.

Maeztu, Ramiro de. "Fuera de la cultura." *El Sol*, March 6, 1923.

Mañas y Bonví, J. *Optica aplicada.* Barcelona, 1914.

Mariscal de Gante, Jaime. "La doctrina de la relatividad." *La Voz Valenciana*, March 6, 1923.

———. "La mujer del sabio." *La Voz Valenciana*, March 8, 1923.

Martínez Risco, Manuel. *Oeuvres scientifiques.* Paris: Presses Universitaires de France, 1976.

Masriera Rubio, Miguel. "El estado actual de las doctrinas de Einstein." *La Vanguardia*, October 25, 1924.

———. "El antirrelativismo psicológico." *La Vanguardia*, January 7, 1925.

———. "La verdad sobre Einstein." *La Vanguardia*, January 17, 1925.

———. "El valor del relativismo." *La Vanguardia*, February 4, 1925.

———. "De Einstein para mis lectores." *La Vanguardia*, October 29, 1925.

———. "La otra cara de Einstein." *La Vanguardia*, March 9, 1965.

———. "Spinoza y Einstein." *La Vanguardia*, February 2, 1977.

———. "La polémica con Bergson." *La Vanguardia*, March 14, 1979.

Mataix Aracil, Carlos. *Primeras nociones de mecánica relativista.* Madrid: Koehler, 1923.

Medio, Pedro Nolasco de. "Un nuevo paladín del relativismo." *España y América*, 23 (1925), 97–112.

———. *Relatividad y energía, espacio y tiempo.* Oviedo, 1925.

———. "Resumen de los principales inconvenientes del relativismo." *España y América*, 24 (1926), 15–27.

Menéndez, Jaime. "Intimidades y rasgos característicos de Einstein." *El Sol*, April 12, 1933.

Menéndez Ormaza, J. "Significación del éxito de Einstein." *El Imparcial*, March 7, 1923.

Moreno-Caracciolo, M[ariano]. "Las matemáticas del ingeniero." *Madrid Científico*, 27 (1920), 237–238.

———. "La teoría de la relatividad." *El Sol*, October 8, 1920.

Mori, Arturo. "Crónicas de Madrid. La visita de Einstein." *El Progreso*, March 6, 1923.

Muñoz, D. "Después de la visita de Einstein." *El Noticiero Bilbaino*, March 16, 1923; also in *El Noticiero Universal*, March 20, 1923.

Navarro, Benjamín. "De relatividad." *Revista Calasancia*, 10 (1922), 38–47.

Nordmann, Charles. "Una revolución en la ciencia. Teorías de Einstein." *Redención* (Alcoy), March 8, 15, and 22, 1923.

Novoa Santos, Roberto. *Physis y psyquis*. Santiago de Compostela, 1922.

Ochoa y Benjumea, José. *El espacio y el tiempo desde Newton a Einstein*. Barcelona: Bazar Ritz, 1924.

Ormaza, Fernando de. "Poesía relativista." *Madrid Científico*, 30 (1923), 36.

———. "Carta abierta." *Madrid Científico*, 30 (1923), 69.

Ortega y Gasset, José. "El sentido histórico de la teoría de Einstein." In *El tema del nuestro tiempo*, 18th edition, pp. 149–168. Madrid: Revista de Occidente, 1976.

———. "Mesura a Einstein." *El Sol*, March 10, 1923. Reprinted in *El tema del nuestro tiempo*, 18th edition, pp. 189–193.

———. "Con Einstein en Toledo." *La Nación*, April 15, 1923. Reprinted in *El tema del nuestro tiempo*, 18th edition, pp. 195–202.

Pahissa y Jó, Jaime. *Idea de la teoría de la relatividad de Einstein*. Barcelona: La Publicidad, 1921.

Palacios, Julio. *El principio de indeterminación*. Discurso leído en el acto de su recepción por. . . . Toledo: Academia de Ciencias Exactas, 1932.

———. *Relatividad: Una Nueva Teoría*. Madrid: Espasa-Calpe, 1960.

———. "Terradas, físico." In *Discursos pronunciados en la sesión necrológica en honor de . . . Esteban Terradas*, pp. 13–21. Madrid: Real Academia de Ciencias Exactas, 1951.

Paniagua, Enrique de. "Un comentario sobre relatividad general." *Madrid Científico*, 30 (1923), 83–85.

———. "La relatividad y la realidad." *Madrid Científico*, 30 (1923), 193–195.

Pellicer, Francisco. "Revolución científica y revolución económica." *Redención* (Alcoy), March 22, 1923.

Pemartín, José. "La física y el espíritu." *Acción Española*, 3 (1932), 595–604; 4 (1933), 27–37, 131–146, 248–256, 347–356, 449–458.

Peña, Fernando. "Bosquejo de la teoría de la Relatividad." In William Watson, *Curso de Física*, pp. 867–886. Barcelona: Labor, 1925.

Pérez de Ayala, Ramón. "El espacio-tiempo." *La Prensa*, November 16, 1930.

Pérez del Pulgar, José Antonio. *De momento philosophico theoriae relativitatis.* Madrid, 1924.

———. "Portée philosophique de la Théorie de la Relativité." *Archives de Philosophie*, 3.1 (1925), 106–142.

———. "El valor filosófico del relativismo: Einstein y Santo Tomás." *Razón y Fe*, 78 (1927), 503–511.

———, and Vicente Burgaleta. "Observaciones sobre la mecánica de Einstein-Minkowski." *Anales ICAI*, 2 (1923), 480–494; 3 (1924), 485–496.

———, and Joaquín Orland. *Introducción a la filosofía de las ciencias físico-químicas.* Liège: Ediciones ICAI, 1934–35.

Pi i Sunyer, August. "Filosofia i ciència experimental." In *Conferències filosofiques*, pp. 209–243. Barcelona: Ateneu Barcelonès, 1930.

Piñerúa Alvarez, Eugenio. "Nociónes acerca de la Teoría de la Relatividad." *La Farmacia Española*, 55 (1923), 241–243ff.

Piracés, Agustin. *Alberto Einstein en Barcelona.* Barcelona, 1955.

Pla Cargol, J[oaquim]. *La teoría de la relatividad.* Gerona: Dalmau Carles, Pla, 1923.

Plans y Freyre, José Maria. "Sobre el movimiento hiperbólico de Born y la cinemática relativista." *Revista de la Academia de Ciencias de Zaragoza*, 3 (1918), 115ff.

———. "Nota sobre la forma de los rayos luminosos en el campo de un centro gravitatorio según la teoría de Einstein." *Anales de la Sociedad Española de Física y Química*, 18, pt. 1 (1920), 367–372.

———. "Algunas ideas sobre la relatividad." *Ibérica*, 13 (1920), 377–380.

———. "Proceso histórico del cálculo diferencial absoluto y su importancia actual." In *Asociación Española para el Progreso de las Ciencias, Congreso de Oporto* [1920], I, 23–43. Madrid, 1921.

———. *Nociones fundamentales de mecánica relativista.* Madrid: Real Academia de Ciencias Exactas, 1921.

———. *Algunas consideraciones sobre los espacios de Weyl y Eddington y los últimos trabajos de Einstein.* Madrid: Real Academia de Ciencias Exactas, 1924.

⸺. *Nociones de cálculo diferencial absoluto.* Madrid: Real Academia de Ciencias Exactas, 1924.

⸺. "Bosquejo histórico y estado actual de la mecánica celeste." *Ibérica,* 23 (1925), 46–48ff.

⸺. "El experimento de Miller y la teoría de la relatividad." *Ibérica,* 27 (1927), 169–171.

⸺. "Nuevas repeticiones del experimento de Michelson." *Ibérica,* 28 (1927), 94–95.

⸺. "Sobre la teoría del campo único de Einstein." *Revista Matemática Hispano-Americano,* 2nd series, 6 (1931), 1–14.

Poto, Mariano. "Einstein y su teoría." *El Liberal,* March 1, 1923.

Prats, Ramon. "Einstein." *Gran Enciclopedia Catalana,* VI, 489–490.

Prieto Delgado, Luis. "Relatividad y racionalidad." *Razón y Fe,* 123 (1941), 279–294ff.

Puig, Ignacio. "Materia y energía." *Madrid Científico,* 34 (1927), 193–196.

Puig Adam, Pedro. "Resolución de algunos problemas elementales en mecánica relativista restringida." *Revista de la Real Academia de Ciencias Exactas,* 20 (1922), 161–216.

Pulido, Pablo. "Einstein en España: Algunos apuntes relacionados con la teoría de la relatividad." *El Auxiliar de la Ingeniería y Arquitectura* (Madrid), 3 (1923), 84–86.

Rafagas [pseudonym]. "Lecciones de humildad." *Heraldo de Aragón,* March 14, 1923.

Rafael, Enrique de. Review of Louis Rougier, *La matérialisation de l'énergie* (Paris, 1919), *Ibérica,* 13 (1920), 192.

⸺. Review of E. Freundlich, *Los fundamentos de la teoría de la gravitación de Einstein* (Madrid-Barcelona, 1920), *Ibérica,* 15 (1921), 63.

⸺. Review of Einstein, *La théorie de la relativité restreinte et généralisée* (Paris, 1921), *Ibérica,* 15 (1921), 288.

⸺. Review of S. Corbella Alvarez, *La teoría de la relatividad de Einstein al alcance de todos* (Barcelona, 1921), *Ibérica,* 16 (1921), 96.

⸺. Review of M. Schlick, *Teoría de la relatividad* (Madrid: Calpe, 1921), *Ibérica,* 16 (1921), 96.

⸺. Review of J. M. Plans, *Nociones fundamentales de mecánica relativista* (Madrid, 1921), *Ibérica,* 16 (1921), 351.

⸺. "De relatividad (Apuntes con ocasión de las conferencias de

E. Terradas en el 'Institut')." *Ibérica*, 15 (1921), 89–91, 218–221, 376–379.

———. "Nociones de mecánica clásica y relativista (Conferencias semanales en el ICAI en el curso 1921–1922)." *Anales de la Asociación de Ingenieros del Instituto Católico de Artes e Industrias*, 1 (1922), 20–26ff.

———. "La teoría de la relatividad." *Razón y Fe*, 64 (1922), 344–359.

———. "El profesor Alberto Einstein en Madrid." *Anales de la Asociación de Ingenieros del Instituto Católico de Artes e Industrias*, 2 (1923), 160–164.

———. "Juventud y formación científica de Terradas." *Discursos pronunciados en la sesión necrológica en honor de . . . Esteban Terradas*, 3–11.

Real Academia de Ciencias Exactas, Físicas y Naturales. *Discursos pronunciados en la sesión solemne que se dignó presidir su majestad el Rey el día 4 de marzo de 1923, celebrada para hacer entrega del diploma de académico corresponsal al profesor Alberto Einstein.* Madrid, 1923.

———. *Discursos pronunciados en la sesión necrológica en honor de . . . Esteban Terradas e Illa.* Madrid, 1951.

[Révésc, Andrés]. "El profesor Einstein en Madrid." *ABC*, March 2, 1923.

Rey Pastor, Julio. "La teoría de la relatividad." *Estudios* (Buenos Aires), 23 (1922), 219–224. Summarized in *Madrid Científico*, 30 (1923), 49–51.

———. "Acerca de la relatividad del espacio." *La Nación*, January 27, February 24, 1924.

———. "Sobre enseñanza técnica y espíritu de cuerpo." *Madrid Científico*, 32 (1925), 337–340.

———. "Esteban Terradas, su vida y su obra." In *Discursos pronunciados en la sesión necrológica en honor de . . . Esteban Terradas*, pp. 35–63. Madrid: Real Academia de Ciencias Exactas, 1901.

Reyes, Alfonso. "Einstein en Madrid." In *Simpatías y diferencias*, 2 vols., II, 92–95. Mexico: Porrúa, 1945.

Riera i Tuebols, Santiago. "Einstein, un home d'excepció." *Avui*, March 8, 1979.

———. "Rafael Campalans, enginyer i polític." *L'Avenç*, no. 16 (May 1979), 6–11.

———. "Albert Einstein: La relativitat del món físic." *L'Avenç*, no. 22 (December 1979), 6–13.

Rigel (pseudonym). "Einstein y la relatividad." *El Heraldo de Madrid,* March 1, 1923.

———. "Más sobre la teoría de la relatividad." *El Heraldo de Madrid,* March 2, 1923.

———. "Algunas consecuencias de la teoría de la relatividad." *El Heraldo de Madrid,* March 3, 1923.

———. "Impugnación a la relatividad." *El Heraldo de Madrid,* March 14, 1923.

Ríos, Sixto, Luis A. Santaló, and Manuel Balanzat. *Julio Rey Pastor, matemático.* Madrid: Instituto de España, 1979.

Rius, Antoni. "Albert Einstein." *Revista del Centre de Lectura* (Reus), 5 (1923), 83–87.

Robles Degano, F. "La relatividad." *El Siglo Futuro,* March 21, 1923.

Roca i Rosell, Antoni. "Alguns aspectes de la història de la física a Catalunya (1900–1930)." *Butlletí de la Societat Catalana de Ciències Físiques, Químiques i Matemàtiques,* 2nd epoch, 1 (1977), 37–45.

———. "Einstein." In *Diccionari de les ciències de la societat als Paisos Catalans,* pp. 168–169. Barcelona, Edicions 62, 1979.

———. "La incidència del pensament d'Einstein a Catalunya (1908–1923)." In *Cententari de la naixença d'Albert Einstein,* pp. 165–184. Barcelona: Institut d'Estudis Catalans, 1981.

———. "Esteve Terradas (1883–1950) en la renovació de la comunitat científica catalana." *Ciència,* 3 (1983), 316–320.

———. "El debat sobre la relativitat a Catalunya (1908–1923)." *Actas II Congreso de la Sociedad Española de Historia de las Ciencias,* 3 vols., II, 325–339. Zaragoza, 1984.

———. "La llegada de la 'Gran Ciencia' a España. Las aportaciones de Esteban Terradas Illa (1883–1950)." *Mundo Científico,* 4 (1984), 908–915.

———. "J. Comas Solá, astrónomo de posición." *Mundo Científico,* 6 (1986), 290–303.

———, and Thomas F. Glick. "Esteve Terradas (1883–1950) i Tullio Levi-Civita (1873–1941): Una correspondència." *Dynamis,* 2 (1982), 387–402.

Rodés, Luis. "De los cuerpos reales al éter hipotético." *Razón y Fe,* 30 (1911), 73–86ff.

———. "El principio de Doppler-Fizeau en su relación con la ley de Kirchoff." *Revista de la Real Academia de Ciencias Exactas,* 17 (1918–19), 471–487.

Rodés, Luis. "Las estrellas a favor de Einstein." *Madrid Científico*, 30 (1923), 199–200.

Rodríguez, Angel. *Sobre la teoría de la relatividad propuesta por el Dr. Einstein*. Madrid, 1924.

Rodríguez, Teodoro. *Relatividad, modernismo, matematicismo*. Barcelona: U. G. de E., 1924. Originally published in *La Ciudad de Dios*, 135 (1923), 42–67ff.

Rosich, Juan. *De las hipótesis y teorías en las ciencias físicas. Algunos antecedentes a la teoría de la relatividad*. Tarrasa: Escuela Industrial, n.d.

Rossi, Paolo. Review of Luis Urbano, *Einstein y Santo Tomás*, (Madrid-Valencia, n.d.), *Rivista di Filosofia Neo-Scolastica*, 20 (1928), 128–132.

Royo-Villanova y Morales, Ricardo. "La teoría de Einstein." *El Norte de Castilla*, June 12, 1935.

———. "La crisis de la ciencia." *El Siglo Médico*, 98 (1936), 58–70.

Sagarra, Josep Maria de. "Einstein." *La Publicitat*, March 4, 1923.

Salaverría, José María. "Las originalidades einsteinianas." *ABC*, March 10, 1923.

Sales, Joan. *Cartes a Marius Torres*. Barcelona: Club Editor, 1976.

Sánchez Peguero, C. "Un aspecto minísculo de la relatividad." *El Noticiero*, March 13, 1923.

Sánchez Pérez, José Augusto. "Necrología del R. P. Enrique de Rafael Verhulst." *Revista de la Real Academia de Ciencias Exactas*, 49 (1955), 213–222.

Sánchez Ron, José Manuel. *El origen y desarrollo de la relatividad*. Madrid: Alianza, 1983.

———. "The Introduction of Relativity and Quantum Mechanics in Spain: Some Facts and Open Problems" (in press).

Sánchez Ron, José Manuel, and Thomas F. Glick. "La oferta de una catedra extraordinaria a Albert Einstein por la Universidad Central: Madrid 1933." *Actas II Congreso de la Sociedad Española de Historia de las Ciencias*, 3 vols., II, 427–436. Zaragoza, 1984.

———. *La España posible de la Segunda República: La oferta de una Catedra Extraordinaria a Albert Einstein (Madrid, 1933)*. Madrid: Editorial de la Universidad Complutense, 1983.

Schlick, M. *Teoría de la relatividad*. Manuel García Morente, trans. Madrid: Calpe, 1921.

Selfa Mora, Rafael. "La sed intelectual." *El Luchador*, March 14, 1923.

Sellés, Manuel A., "La teoría de la relatividad de Julio Palacios." *Actas*

II Congreso de la Sociedad Española de Historia de las Ciencias, 3 vols., II, 437–452. Zaragoza, 1984.

Silván, Graciano. "La actualidad científica." *El Noticiero*, March 14, 1923.

Sittert, Julius de. "Teoría general de la relatividad y el espectro solar." José Sagristá, trans. *Revista de la Sociedad Astronómica de España y América*, 12 (1922), 8–12.

Soldevila, Carles. "La popularitat d'Einstein." *La Publicitat*, February 25, 1923.

Tallada y Ormella, Fernando. "Fundamentos del principio de relatividad." *Técnica. Organo Oficial de la Asociación de Ingenieros Industriales de Barcelona*, 45 (1922), 237–244.

———. "Einstein en Barcelona." *La Vanguardia*, March 4, 13, and 24, 1923.

———. "El método axiomático en las ciencias físicas." *Memorias de la Real Academia de Ciencias y Artes de Barcelona*, 3rd epoch, 22 (1931–32), 303–318.

Tato Puigcerver, J. J. "Una nota sobre la *Revista General de Marina* y la recepción de la relatividad en España." *Llull: Boletín de la Sociedad Española de Historia de las Ciencias*, vol. 3, no. 1 (October 1980), 137–138.

Terradas, Esteve. "Sobre el principi de relativitat." *Arxius de l'Institut de Ciències*, vol. 1, no. 2 (1912), 84–94. Reprinted in *Ciència*, 3 (1983), 510–513.

———. "Relatividad." *Enciclopedia Universal Ilustrada*, 50 (1923), 455–512.

———, and Ramón Ortiz Formiguera. *Relatividad*. Buenos Aires: Espasa-Calpe, 1952.

Tous y Biaggi, José. "El principio de contradicción en la geometría no euclídea y en el principio de relatividad." *Memorias de la Real Academia de Ciencias y Artes de Barcelona*, 3rd epoch, 20 (1926), 17–42.

Ubach, José. *La teoría de la relatividad en la física moderna—Lorentz, Minkowski, Einstein*. Buenos Aires, 1920.

Ugarte, Nicolás de. "Informe de la Real Academia de Ciencias Exactas, Físicas y Naturales sobre la Memoria presentada al concurso de premios del año 1919, y cuyo lema es 'La Science est essentiellement mobile, et n'est formée que d'approximations succesives.'" *Revista de la Real Academia de Ciencias Exactas*, 19 (1920–21), 234–243.

Ugarte, Nicolás de. "Las teorías relativistas." *Madrid Científico*, 31 (1924), 178–179.

Ugarte de Ercilla, E[ustaquio]. "Exposición y refutación de la relatividad." *Razón y Fe*, 73 (1925), 426–428.

Urbano, Luis. *Einstein y Santo Tomás: Estudio crítico de las teorías relativistas*. Madrid-Valencia: Ciencia Tomista, n. d. [1926].

———. "Einstein y Santo Tomás. Las teorías relativistas acerca del movimiento y la doctrina del Angélico Doctor." *Ciencia Tomista*, 32 (1928), 321–344ff.

Vecino, Jerónimo. "La teoría de la relatividad de Einstein." *El Heraldo de Aragón*, March 14, 1923.

Velasco de Pando, Manuel. *Relatividad general y restringida*. Bilbao: Asociación de Ingenieros Industriales, 1924 [2nd edition, 1926].

Vera, Francisco. "La primera conferencia de Einstein: Relatividad restringida." *El Liberal*, March 4, 1923.

———. "La segunda conferencia de Einstein: Relatividad general." *El Liberal*, March 6, 1923.

———. "La tercera conferencia del profesor Einstein: Consecuencias relativistas." *El Liberal*, March 8, 1923.

———. "Sarampion relativista." *El Liberal*, March 14, 1923.

———. "El doctorado 'honoris causa' y otras grandes menudencias." *El Liberal*, March 16, 1923.

———. *El hombre bicuadrado*. Madrid: Páez, 1926.

———. *Espacio, hiperespacio y tiempo*. Madrid: Páez, 1928.

Vilamitjana y Masdevall, Ramón. "¿Teoría de la relatividad?" *Técnica*, 45 (1922), 92–94.

———. "La cinemática relativista." *Técnica*, 45 (1922), 124–126.

Vizuete, Pelayo. *Einstein y el misterio de los mundos*. 2 vols. Madrid: Arte y Ciencia, 1923–24.

Wahr [pseudonym]. "La teoría de la relatividad." *ABC*, January 19, 1922ff.

Wulf, Theodor. *La teoría de la relatividad de Einstein*. Joaquín María de Barnola, trans. Barcelona: Viñals de Sarriá, 1925.

———. *Física*. Eduardo Alcobé, trans. Barcelona: Casals, [1929?].

X[irau] P[alau], J[oaquim]. "Les conferències del profesor Einstein." *La Publicitat*, March 4, 1923.

Zubiri, Xavier. *Naturaleza, historia, Dios*. 3rd edition. Madrid: Editora Nacional, 1955.

BIBLIOGRAPHY

3. Einstein and The Comparative Reception of Relativity

Bernstein, Jeremy. *Einstein*. New York, Viking, 1973.

Biezunski, Michel. "La diffusion de la théorie de la relativité en France." Doctoral dissertation, University of Paris, 1981.

———. "Einstein à Paris." *La Recherche*, 13 (1982), 502–510. Spanish translation: "Einstein en Paris." *Mundo Científico*, 2 (1982), 592–600.

Carter, Paul A. *Another Part of the Twenties*. New York: Columbia University Press, 1977.

Clark, Ronald. *Einstein: The Life and Times*. New York: World, 1971.

Crelinsten, Jeffrey. "Einstein, Relativity and the Press: The Myth of Incomprehensibility." *The Physics Teacher*, 18 (1980), 115–122.

———. "Physicists Receive Relativity: Revolution and Reaction." *The Physics Teacher*, 18 (1980), 187–193.

———. "William Wallace Campbell and the 'Einstein Problem': An Observational Astronomer Confronts the Theory of Relativity." *Historical Studies in the Physical Sciences*, 14 (1983), 1–91.

Dirac, P.A.M. "The Early Years of Relativity." In Holton and Elkana, eds. *Albert Einstein: Historical and Cultural Perspectives*, pp. 79–90.

Dixon, Peter Morris. *Popular Criticisms of Relativity Around 1920*. Honors thesis. Harvard College, 1982.

Earman, John, and Clark Glymour, "Relativity and Eclipses: The British Eclipse Expeditions of 1919 and Their Predecessors." *Historical Studies in the Physical Sciences*, 11 (1980), 49–85.

Eddington, A. S. *Space, Time and Gravitation: An Outline of the General Relativity Theory*. Cambridge, Eng.: Cambridge University Press, 1920.

Einstein, Albert. *Relativity: The Special and General Theory*. Robert W. Lawson, trans. 9th edition. London: Methuen, 1929 [1st edition, 1920].

Einstein, 1879–1979. Exhibition, Jerusalem, March 1979. Jerusalem: Jewish National and University Library, 1979.

Eisenstaedt, Jean. La relativité générale à l'étiage, 1925–1955." *Archive for History of Exact Sciences*, 35 (1986), 115–185.

Feuer, Lewis S. *Einstein and the Generations of Science*. New York: Basic Books, 1974.

375

Glick, Thomas F., ed. *The Comparative Reception of Relativity*. Dordrecht: D. Reidel, 1987.

Goldberg, Stanley. "In Defense of Ether: The British Response to Einstein's Special Theory of Relativity, 1905–1911." *Historical Studies in the Physical Sciences*, 2 (1970), 89–125.

Graham, Loren R. "The Reception of Einstein's Ideas: Two Examples from Contrasting Political Cultures." In Holton and Elkana, eds., *Albert Einstein, Historical and Cultural Perspectives*, pp. 107–136.

Henderson, Linda Dalrymple. *The Fourth Dimension and Non-Euclidean Geometry in Modern Art*. Princeton, N.J.: Princeton University Press, 1983.

Holton, Gerald, and Yehuda Elkana, eds. *Albert Einstein: Historical and Cultural Perspectives*. Princeton, N.J.: Princeton University Press, 1982.

Kaneko, Tsutomu. "Einstein's View of Japanese Culture." *Historia Scientiarum*, no. 27 (1984), 51–76.

Kirsten, Christa, and Hans-Jürgen Treder. *Albert Einstein in Berlin*. 2 vols. Berlin: Akademie Verlag, 1979.

Langevin, Jean, and Michel Paty, eds. *Le séjour d'Einstein en France en 1922*. Strasbourg: Université Louis Pasteur, 1979.

Moyer, Donald Franklin. "Revolution in Science: The 1919 Eclipse Test of General Relativity." In Arnold Perlmutter and Linda F. Scott, eds., *On the Path of Albert Einstein*, pp. 55–101. New York: Plenum, 1979.

Nisio, Sigeko. "The Transmission of Einstein's Work to Japan." *Japanese Studies in the History of Science*, 18 (1979), 1–8.

Okamoto, Ippei. "Albert Einstein in Japan, 1922." Kenkichiro Koizumi, trans. *American Journal of Physics*, 49 (1981), 930–940.

Pyenson, Lewis. *The Young Einstein: The Advent of Relativity*. Bristol-Boston: Adam Hilger, 1985.

Sayen, Jamie. *Einstein in America: The Scientist's Conscience in the Age of Hitler and Hiroshima*. New York: Crown, 1985.

Schwarz, Boris. "Musical and Personal Reminiscences of Albert Einstein." In Holton and Elkana, eds., *Albert Einstein: Historical and Cultural Perspectives*, pp. 409–416.

Swenson, Loyd S., Jr. *The Ethereal Aether: A History of the Michelson-Morley-Miller Aether-Drift Experiments, 1880–1930*. Austin: University of Texas Press, 1972.

Index

ABC, 130, 242, 248, 329
"Abecé" (columnist), 273
Abel, Niels Henrik, 18
Absolute differential calculus, 64–70. *See also* Relativity: General relativity
Academic freedom, 6, 7
Academy of Exact Sciences (Madrid), 38, 73, 186, 264, 316, 329, 332; prizes, 44, 45, 68, 223; session honoring Einstein, 126–129, 325, 330
Academy of Military Engineers (Guadalajara), 13
Academy of Sciences (Barcelona), 27, 103, 105, 121–122, 152, 219, 327
Academy of Sciences (Zaragoza), 146, 149, 331
Adellac, Miguel (journalist), 266–267
Aether, 35, 38, 55, 87, 91, 158, 172, 174, 185n, 198, 217, 218, 220, 281; aether drag, 318; aether-drift experiments, 35, 36, 154–157; aether wind, 93; Alcobé on, 308; and English physics, 75; Comas Solà on, 150–152; *Enciclopedia Universal Ilustrada* article on, 248–249; mechanical model of, 222, 235
Aguilar, Florestán (odontologist), 126, 130, 139
Aguilera, Juan, 320
Alcobé, Eduard (physicist), 308–310
Alfonso XIII, 326, 331; at Academy of Sciences, 126, 129, 138, 325, 330

Alvarez Ude, José G. (mathematician), 62, 63, 66
American Committee for Spanish Freedom, 317
Annalen der Physik, 33
Anti-Mathematicism, 176, 177–178, 179, 220–221, 253, 287
Anti-Semitism, 92, 95, 137, 264, 305, 313
Antón de Olmet, Luis (writer), 291
Aparici, Rafael (engineer), 223
Appell, Paul, 45
Apráiz, Félix (engineer), 315
Araquistáin, Luis (journalist), 243–244
Araujo, Roberto (mathematician), 20
Arcimis, Augusto T. (meteorologist), 50
Assía, Augusto (journalist), 317–318
Association of Engineers (Madrid), 138; Einstein's lecture, 141–142, 329
Association of Industrial Engineers (Barcelona), 202
Association of Industrial Engineers (Bilbao), 201
Astronomers: and relativity, 51–55
Astronomy: in Spain, 48–55
Athenaeum (Madrid), 40, 51, 142, 229, 234, 282n, 329
Ayala, Francisco (novelist), 282
Azcárate, Gumersindo de (educator), 10
Aznar, Manuel (journalist), 262

Baeza, Ricardo (journalist): interviews Einstein, 102, 158

377

Bagaria, Lluís (cartoonist), 293, 294, 297

Balmes, Jaime (philosopher), 196, 235

Banús, Carlos (engineer), 222–223

Barga, Corpus (journalist), 94, 264, 289

Baroja, Pío (novelist), 137, 300; *Las veleidades de la fortuna*, 258–260

Bartrina i Capella, Josep M. (mathematician), 18

Basalla, George, 292

Bataillon, Marcel, 166

Bauer, Ignacio (physician), 126, 330

Bauzá, Felipe (navigator), 3

Bechert, Karl, 59

Benito, Enrique de (journalist), 179, 265

Bentaból, Horacio (engineer), anti-relativity campaign, 225–230, 291, 309

Bergson, Henri, 94, 175, 211, 260; as popularizer, 245; on simultaneity, 168–170

Berkeley, George, 183

Berson, Arthur, 50

Betancourt, Agustín de (engineer), 3

Bethe, Hans, 321

"Betibat" (columnist), 179, 267–268

Biezunski, Michel, 181, 299; and Einstein myth, 76, 77; on *gens du monde*, 258; on "semantic slides," 81, 276

Blanche, Jacques Emile, 277

Bohr, Niels, 42, 71, 73

Bolívar, Ignacio de (entomologist), 9, 10, 126

Bolyai, János, 67

Born, Max, 303

Bouasse, Henri, 226

Brentano, Franz, 138

Broglie, Louis de, 153

Brooklyn Tablet, 317

Brownian movement, 127

Buchanan, John, 50

Buen, Odón de (marine biologist), 142–143, 144

Burgada i Julià, Joan (editor), 122, 241–242

Burgaleta, Vicente (engineer), 131; on "relativity paradox," 208–209; on special relativity, 132, 221–222

Cabrera, Blas (physicist), 21, 28, 29, 30, 31, 32, 33, 57, 63, 65, 66, 67, 123, 126, 130, 131, 139, 173, 174, 192, 202, 247, 249, 309, 320, 325, 333; on anti-relativists, 154; on Einstein's popularity, 262; Einstein visits laboratory of, 124, 125, 264; on general relativity, 41–42; and Lorentz, 72; in Munich, 72; and Ortega y Gasset, 160; and Pierre Weiss, 58–59; on Plans, 44; and relativity, 38–44; on Spanish science, 128, 144, 145, 268–269; on special relativity, 38, 41, 127

Calder, Nigel, 88

Calderón, Laureano (chemist), 5, 6

Calderón, Salvador (geologist), 6

Calleja, Julián (anatomist), 123, 126, 139

Callendar, Hugh, 50

Cámara, Sixto (mathematician), 66

Camba, Julio (journalist), 253, 260, 290–291, 292

Cambó, Francesc (politician), 111, 114

Campalans, Rafael (engineer), 57, 58, 108n, 117, 325, 328; and Einstein, 115–116, 118, 119, 302–303; supper for Einsteins, 120–121, 328

Campana de Gracia, La, 106, 117

Campo, Angel del (chemist), 141

Cánovas de Castillo, Antonio (politician), 5

Cantero Villamil, Federico (engineer), 223–224

Cantor, W., 319
Caro Baroja, Julio (anthropologist), 5
Carracido, José R. (chemist), 10, 26, 29, 126, 130, 135, 138, 139, 141, 230, 330; Academy of Science address, 127
Carrasco, Pedro (astronomer), 51–55, 59, 66, 123, 124, 125, 333; and general relativity, 52–54
Cartan, Elie, 94, 250
Casares Gil, José (chemist), 10, 128
Casares Roldán, José, 318
Castillejo, José de (educator), 4, 9
Castro, Fernando de (philosopher), 4
Castro, Miguel de (journalist), 254
Catalán, Miguel (physical chemist), 32; and Sommerfeld, 59, 60
Cauchy, Augustin-Louis, 18, 19
Cavendish, Henry, 204
Central University. See University of Madrid
Chasles, Michel, 18
Chavaneau, François, 3
Chemists: and relativity, 232–235
Ciudad de Dios, 174
Civil discourse, 4, 16, 17, 131, 179–180, 272, 303, 322; breakdown of, 180, 308–314; defined, xi; emergence of, 8–11, 15
Clarín (Leopoldo Alas, novelist), 13–14
"Clarover" (columnist), 179
Cleveland Plain Dealer, 90, 91
Collège de France, 93, 94, 201, 289
College of Physicians (Madrid), 123, 126, 330
Colominas Maseras, Joan (journalist), 252
Comas Solà, Josep (José, astronomer), 21, 48, 49, 55, 104, 122, 247, 271, 311, 312; on aether-drift experiments, 155–156; and Benta-

ból, 230; emissive-undulatory theory, 150–153; on Michelson-Morley experiment, 153, 155; and popular disillusion with relativity, 255–256; on special relativity, 151
Committee of Intellectual Cooperation, 116
Consejo Superior de Investigaciones Científicas, 314–315
Corbella Alvarez, Salvador (engineer), 201–202, 250
Correa, M., 205
Coso, Marcial del (journalist), 148, 291
Cossío, Manuel B. (art historian), 130, 136, 137
Crelinsten, Jeffrey, 70–71

Dalí, Salvador (artist), 277–281
Darwin, Charles, 125, 175, 176; proscribed, 315, 322; Spanish translations of, 84
Darwinism, 5, 10–11, 16, 171, 181, 278, 292; and Catalan bourgeoisie, 111–112; compared with relativity, 272; as model for revolution in science, 175; natural selection, 185; popularization of, 245; and "Scientific Middle Class," 191–192; Spanish reception of, 83–85
Daudet, Léon, 116
Debate, El, 11n, 179, 265, 332; on chair offered to Einstein, 304–305; coverage of Einstein's trip, 243
Descartes, René, 47, 183, 211, 212, 280
Diario de Barcelona, 122, 241
Diario de Lérida, 267
Dick, Marcel, 79
Diluvio, El, 113
Dingle, Herbert, 319
Dirac, P.A.M., 199–200
Dorronsoro, Bernabé (chemist), 27
Dreisch, Hans, 165

Drude, Paul, 33
Duhem, Pierre, 196
Durán, Miguel-Emilio (journalist),
    106–107
Dyson, Frank, 53, 59

Eberty, Felix, 159
Echegaray, José de (mathematician),
    10, 17, 18, 192; as popularizer,
    245, 258
Eclipse: of 1905, 49–50, 228; of
    1919, 43, 45, 52–53, 74, 75–76,
    204, 213, 218, 243, 249, 347; of
    1922, 155, 250; of 1923, 142–43
Eddington, Arthur, 47, 53, 68, 69,
    258, 354; impact on engineers,
    200; invoked by Einstein, 128; and
    relativistic paradoxes, 70, 75–76,
    206–209, 215n; and religion, 180;
    in Spain (1932), 72; and Spanish
    polemics, 204–208, 299
Edison, Thomas A., 273; compared
    with Einstein, 299–300
Education, in nineteenth-century
    Spain, 5–14
Einstein, Albert: ability to "count"
    music, 79–80; on absolute rota-
    tions, 133; addresses anarcho-
    syndicalists, 107–111, 328; appeal
    to conservatives, 308; attends
    theater, 124, 148; and Bergson,
    94, 168–169; at Cabrera's labora-
    tory, 124, 125, 264; and Campa-
    lans, 115–116; and Catalan nation-
    alism, 111–116; in Cleveland, 89–
    91; compared with Cambó, 114;
    compared with Edison, 299–300;
    consequences of general relativity,
    347–348; conversation with Zu-
    biri, 307–308; "Curiosity Files,"
    91–92; on deformation of solids in
    general relativity, 134; denies rela-
    tivity is revolutionary, 108–110,
    328, 346; on dualism in physics,

139; and eclipse of 1923, 142–143;
    elected to Barcelona Academy,
    120–121; elected to Madrid Acad-
    emy, 129; on El Greco, 137; on
    fourth dimension, 216, 261; as
    hope of Spanish science, 239; in-
    terviews with, 90–91, 100, 102,
    110, 118, 126, 158; invited to
    Spain, 85, 100–102; on Japan, 98;
    on Kant, 336; lectures in French,
    93, 101, 135, 243, 244, 265, 333;
    and Levi-Civita, 64; on light
    quanta, 128; and "Manifesto of
    the 93," 108, 109, 110, 266, 268,
    282, 328; at Marqueses de Villa-
    vieja's tea, 130, 330; at Mathemat-
    ical Society, 132–134; on Max-
    well, 144; moral stature of, 265–
    266; at Museo del Prado, 326;
    myth of, 74–81, 299–301; offered
    chair at Madrid, 303–304; on
    overspecialization, 137–138; and
    perception of science in Spain,
    268–275; persona of, 77, 78; phi-
    losophy of, 158; physical charac-
    teristics of, 263–264; on physics,
    307–308; plays violin, 103, 107,
    131, 148; as political symbol, 266–
    268; popularity of, 261–262, 299;
    on popularization of relativity,
    251; on principle of equivalence,
    340; quality of voice, 265; receives
    honorary degree, 139–141, 326; at
    Refectorium, 117; on relativity,
    158; resigns from Committee of
    Intellectual Cooperation, 116; and
    Rodríguez Bachiller, 134; on sar-
    danas, 119–120; on simultaneity,
    336–337; on space-time, 338; and
    Spanish Civil War, 316–317; and
    Spanish physicians, 126, 139; and
    Terradas, 118, 281; in Toledo,
    136–137, 325–326; travel diary,
    325–326; visit to Poblet, 117; vis-

its Cajal, 134–135, 325; on Zaragoza, 146
—Madrid Lectures: First (special relativity), 125–126, 333–339 (text); Second (general relativity), 135–136, 339–348 (text); Third (latest results), 138–139, 348–355 (text); Fourth (philosophical consequences), 143–144
—Travels: Argentina (1925), 299; Barcelona (1923), 100–122, 325, 327–329, 331; Japan (1922), 95–98; Madrid (1923), 123–145, 325–326, 329–331; Palestine (1923), 98–99; Paris (1922), 92–95; United States (1921), 76, 88–92; Zaragoza (1923), 145–149, 331
—Works: Relativity: The Special and General Theory, 168; La teoría de la relatividad al alcance de todos, 123–124, 172, 196n, 231, 250, 289, 333
Einstein, Elsa, 78, 274–275
Elhuyar, Juan José (mineralogist), 3
Enciclopedia Universal Ilustrada, 36; relativity in, 248–249, 309–310
Engineering: and ideology, 14–15
Engineering schools, 12; courses on relativity, 195–200; and mathematics, 192–194; relativity in textbooks, 197–198
Engineers: military, 139, 222; and relativity, 85, 86, 188, 200–209, 219–225
En Patufet, 106, 273, 289–290
Enriques, Federigo, 22
Epoca, La, 243
Escalante, Gil de (journalist), 130
Escofet, Josep (editor), 226n, 251; on Spanish science, 270–271
Escola d'Enginyers Industrials (Barcelona), 13, 197, 219
Escriche i Mieg, Tomàs (inventor), 122

Escuela de Caminos (Madrid), 13, 17, 18, 192, 199; relativity course, 196–197
Escuela de Ingenieros de Montes, 13, 198
Escuela de Minas (Madrid), 13
Escuela Naval Militar: relativity at, 198
Esquella de la Torratxa, L', 106, 114, 287–289
Evershed, John, 50, 53
Exposition of Lunar Studies (Barcelona), 50–51
Eza, Vizconde de, 130

Fabra Rivas, Antonio (journalist), 118
Faculties of Science, 11–12
Fajans, Kasimir, 71, 72
Falconer, Bruce, 92
Faraday, Michael, 280
Farmacia Española, La, 232
Fernández, Obdulio (chemist), 63
Fernández Ascarza, Victoriano (astronomer), 10
Fernández Baños, Olegario (mathematician), 20, 59, 73
Fernández Barreto, Antonio (soldier), 200–201
Fernández Bordas, Antonio (violinist), 131, 326
Fernández Flores, Wenceslao (humorist), 110, 260, 266, 272, 291
Ferrer, Francesc (educator), 112
Flammarion, Camille, 48, 247
Fock, V. A., 180
Fontanilla, Ruperto (mathematician), 66
Fontseré, Eduard (meteorologist), 27, 62
Forestry School. See Escuela de Ingenieros de Montes
Foucault, Jean, 151
Fourth dimension, 216–217, 261

Fowler, Alfred, 50, 59
Franco regime: and science, 314–315
Freud, Sigmund, 176; proscribed, 315, 322; Spanish translations of, 84
Freudian psychology, 16, 262n, 278, 279, 281n, 282. *See also* Psychoanalysis
Freundlich, Erwin, 158, 250
Fuchs, Klaus, 321

Gale, H. H., 155
Galilei, Galileo, 143, 254, 262, 346; principle of inertia, 334, 335
García, Juan, 316
García Alix, Antonio (politician), 8
García de Galdeano, Zoel (mathematician), 18, 19, 23
García Morente, Manuel (philosopher), 130, 139, 157, 250, 282; on relativity, 158–159
García Rodejo, Vicente (physicist), 31
Garrabou, Ramon (historian), 14
Gascón y Marín, Conde de Grove, 333
Gauss, Carl Friedrich, 18, 343, 344, 345
Gaziel [Agustí Calvet i Pascual] (editor), 281
Geometrical practitioners: and relativity, 236–237, 238
Gianfranceschi, Giuseppe, 310
Gil, Rodrigo (astronomer), 143n
Gimeno, Amalio (physician, politician), 9, 10, 125, 272, 333
Giner de los Ríos, Francisco (educator), 5
Goicoechea, José María (chemist), 174, 233–235
Gómez de Nicolás, Tomás (journalist), 253
Gómez de la Serna, Ramón (writer): on Einstein, 130, 265; *El dueño del*

átomo, 30; *tertulia* of, 282–283
Gómez Ocaña, José (physiologist), 28, 56
González de Linares, Augusto (marine biologist), 5, 6
González Martín, Ignacio (physicist), 141
González Quijano, Pedro M. (mathematician), 63, 131, 194, 208
Graham, Loren, 180
Graña, Manuel (journalist), 243
Grossmann, Marcel, 64, 68
Güell i López, Eusebi (architect), 230–232

Hadamard, Jacques, 25, 94, 116, 250
Haeckel, Ernst, 175
Hebrew University, 98, 303
Heisenberg, Werner, 306; uncertainty principle, 306, 310. *See also* Indeterminacy
Heraldo de Aragón, 147, 245
Heraldo de Madrid, 248, 269
Hernández Pacheco, Eduardo (geologist), 126
Hernando, Teófilo (internist), 130, 139
Herrera, Angel (editor), 332
Herrera, Emilio (engineer), 50, 66, 125, 131, 139, 155, 198, 214, 215, 238, 280, 300; on "fourth dimension" (hyperspace), 216–217, 225, 237; on general relativity, 216–219; on relativity and intuition, 209–211, 321; and "relativity paradox," 206–209; special relativity, 210, 218–219, 320–321; and weight of light polemic, 203–206, 299
Herzfeld, Karl, 72
Hilbert, David, 211, 212, 221
Hinton, Charles Howard, 216
Honigschmid, Otto, 71, 72
Huertas, Ataulfo (priest), 170–171,

174, 179, 322; on Bergson, 169–170; critique of Teodoro Rodríguez, 177–178
Hume, David, 158, 183
Humorists: and relativity, 106, 286–299
Hunolt, Emilio (chemist), 248
Huxley, T. H., 175

Ibeas, Bruno (priest), 171, 179, 232
Ibérica, 158, 171; relativity in, 249–250
Illustrated London News, 204
Imparcial, El, 126, 139
Indeterminacy, 305, 312, 318
Industrial Engineering School (Barcelona). See Escola d'Enginyers Industrials
Industrial School (Escola Industrial, Barcelona), 61, 104, 115, 117, 119
Infeld, Leopold, 261, 303
Iñiguez Almech, José María (lawyer), 147n
Institución Libre de Enseñanza, 5–6, 10, 50
Institut d'Estudis Catalans, 34, 37, 327, 328
Institute of Physical Research (Madrid), 9, 29, 30–32, 43, 56
Institute of Radioactivity (Madrid), 63
Institute-School (Instituto-Escuela, Madrid), 9, 11
Instituto Católico de Artas e Industrias [ICAI] (Madrid), 14, 141, 193, 198, 221; calculus instruction, 194; relativity course, 195–196
Intelligentsia: and relativity, 80–81, 95, 258
Interferometer, 28, 36, 155, 156
International Education Board, 32, 59
International Physical Congress

(1927), 309–310
International Writers' Congress (1937), 316
Intuition: and science, 155, 210–211, 305
Ishiwara, Jun, 95, 96
Izaguirre, R. (physical chemist), 204–205

James, William, 158
Jardí, Ramon (physicist), 37n, 121–122, 152
Jesuits, 14, 46, 49, 62, 170, 171, 176, 182, 193, 249, 305. See also Ibérica; Razón y Fe
Jiménez, Enrique (mathematician), 193–194
Jiménez Fraud, Alberto (educator), 130
Jiménez Rueda, Cecilio (mathematician), 23, 66, 126
Jiménez Vicente, Inocencio, 11n
Jouffret, Esprit Pascal, 216
Junta para Ampliación de Estudios, 9–11, 19, 20, 30, 31, 56, 58–59, 61 130, 269, 273n; dissolved by Franco, 314; and Einstein's trip, 101. See also Study abroad
Jurdant, Beaudouin, 82, 86

Kamerlingh Onnes, Heike, 60
Kant, Immanuel, 162, 182, 183, 211, 213, 259
Klein, Felix, 221
Kocherthaler, Julio, 124, 126, 130, 136, 325
Kocherthaler, Kuno, 131, 136, 325
Kocherthaler, Lina, 124, 126, 130, 136, 326
Krause, Karl, 4; Krausism, 5
Kuwaki, Ayao, 95

Laboratories, 13, 26–30; crisis of, 29; equipment, 28

Laboratory of Automation (Madrid), 28
Lacan, Jacques, 279
La Cierva, Juan [father] (politician), 48, 290
La Cierva, Juan [son] (inventor), 11n
Ladd-Franklin, Christine, 18
Lafora, Gonzalo R. (neurologist), 130, 282
Lafuente, Antonio (historian), 185
La Llave, Joaquín de (engineer), 66, 139
Lallemand, Charles, 248
Lamo, Regina (journalist), 113
Lana Serrate, Casimir (engineer), 61, 102, 117, 118, 325
Landerer, José Joaquín (astronomer), 48, 49
Langevin, Paul, 93, 94, 95, 155; "Langevin's paradox," 93, 319
Larmor, Joseph, 74
Lassaleta, Bernat (chemist), 117, 121
Laub, Jakob, 33, 63
Laue, Max von, 34, 69
Lavoisier, Antoine-Laurent, 215
Ledesma Ramos, Ramiro (philosopher), 70, 313; on relativity, 164–166
Lémeray, E. M., 174
Le Normand, Henri, 201, 232
LeRoy, Edouard, 211
Leverrier, Urbain Jean Joseph, 164
Levi-Civita, Tullio, 20, 21, 22, 25, 37, 67, 187, 199; and Einstein, 64; invoked by Einstein, 128; and second-order tensor, 352–353; visit to Spain (1921), 64–66
Levy, Hyman, 224
Lewis, G. N., 181
Liberal, El, 125, 243, 268
Lindemann, Carl, 18
Lliga Regionalista, 111, 113, 114
Lobatchevsky, N. I., 67
Lodge, Oliver, 75, 181, 247

Loeb, Jacques, 160
London Times, 70, 109
López Soler, Juan (mathematician), 131
Lorente de Nó, Fernando (mathematician), 20, 21, 65, 123, 172, 196n, 231, 250, 333; on book by Plans, 68
Lorente Pérez, José M. (mathematician), 20
Lorentz, Hendrik Antoon, 160, 180, 248; contraction, 35, 41, 88, 168, 172, 204, 227, 231, 233, 243, 308, 342; electron theory, 134; in Spain (1925), 72, 73; transformations, 41, 150, 203, 219, 249, 337, 345
Lorenzo Pardo, Manuel (engineer), 63n, 149
"Lotario" (columnist), 229, 291
Lozano, Eduardo (physicist), 27, 141
Lozano Rey, Edmundo (zoologist), 141
"Lucanor" (columnist), 244–245, 258
Lucía Ordóñez, Pedro (engineer): course on relativity, 196–197; on science and intuition, 212–214
Lucini, Manuel (engineer), 131, 132, 315; on "relativity paradox," 208

Mach, Ernst, 158, 202; on classical mechanics, 348–350; "Mach's principle," 133; physiology, 235, 236
Machado, Antonio (poet), 281
Macià, Francesc (politician), 111, 114
Madariaga, José M., 26
Madrid Científico, 206–208, 209, 214, 238
Madrid Moreno, José, 141
Maeztú, María de, 139

Maeztu, Ramiro de (writer), 111, 130, 266

Magie, W. F., 181

Mañas y Bonví, J. (physicist), 197–198

Mancomunitat of Catalonia, 35, 103, 111, 327; laboratories, 27

Marañón, Gregorio (endocrinologist), 130, 215, 281n; on reception of scientific ideas, 284–285; on structure of disciplines, 185–186

March, Juan (financier), 313–314

Mariscal de Gante, Jaime (journalist), 254, 274

Maritain, Jacques, 168

Martín Escobar, Teófilo (mathematician), 20, 21, 63, 332

Martínez Risco, Manuel (physicist), 28, 59, 60, 282

Masriera, Miguel (physical chemist), 287; on Bergson-Einstein exchange, 168–169; critique of Ortega, 163

Mataix Aracil, Carlos (mathematician), 63, 197

Mathematical Laboratory. See Mathematical Seminar

Mathematical Seminar (Madrid), 19, 20, 23, 24, 26, 46, 56, 131, 198, 306

Mathematical Society (Madrid), 21, 23, 24; meeting with Einstein, 131, 132–134, 325; relativity sessions, 131

Mathematics: and engineering education, 192–194; relations with Italian mathematics, 20–23, 65, 66, 186; in Spain, 17–26, 65–71

Maura, Antonio (politician), 125, 332

Maurín, Joaquín (politician), 108

Maurras, Charles, 318

Maxwell, James Clerk, 280, 310, 354

Maynés, Enric (politician), 113

Medio, Pedro Nolasco de (priest), 171, 179

Mendoza y Ríos, José (navigator), 3

Menéndez Ormaza, J. (journalist), 274

Menéndez y Pelayo, Marcelino (literary historian), 10

Meyer, Walter, 303

Meyerson, Emile, 94

Michelson, Albert, 34n, 36, 155

Michelson-Morley Experiment, 35, 36, 53, 96, 151, 153, 155, 156, 160, 172, 220, 231, 335; misinterpretations of, 204, 233n

Mier y Mirva, Eduardo (seismologist), 28

Miller, Dayton, 36, 91, 156–157, 311

Millikan, Robert, 89

Minkowski, Hermann, 249, 313, 338

Mira y López, Emilio (psychiatrist), 262n

Miral, Domingo (philosopher), 146

Modernization: and science, xi, 15, 17

Mola, Emilio (general), 313

Moles, Enric [Enrique] (chemist), 32, 40, 63; and laboratory crisis, 29; in Germany, 72; on language, 57

Monsieur Homais, 245, 259, 260, 286

Moreno-Carraciolo, Manuel (engineer), 27, 66; on Institute of Physical Research, 30–32; as popularizer of relativity, 248, 285

Moreux, Abbé Th., 201

Mori, Arturo (journalist), 260–261, 291

Morley, Edward W., 156

Moyano Plan (1857), 4, 8, 12

Museum of Natural Sciences, 3, 11, 130

Narváez, Ramón María (general), 4
Nature, 206, 207, 262
Naval School. *See* Escuela Naval
　Militar
Navarro, Benjamín (priest), 170; on
　relativity, 173–174
Negrín, Juan (physiologist), 28, 283
Netto, Eugen, 33
Newall, Hugh, 59
Newton, Isaac, 39, 70, 143, 204,
　262, 280, 310; on gravitation, 342;
　theory of emission, 151
New York Times, 89
Nicolai, Georg, 282–283
Nietzsche, Friedrich Wilhelm, 312
Noble, H. R., 157
Non-Euclidean geometry, 220, 221,
　229, 231, 237, 238, 259, 280n
Nordmann, Charles, 95, 109, 232,
　247, 250
Noticiero Universal, 242
Novoa Santos, Roberto (patholo-
　gist), 235–236

Obermaier, Hugo, 130, 300
Observatories: Ebro (Tortosa), 14,
　49, 62, 171, 249; Fabra (Barce-
　lona), 49; Royal (Madrid), 12, 48–
　49, 51, 62; San Fernando (Cádiz),
　49
Ochoa y Benjumea, José (engineer),
　202
Octavio de Toledo, Luis (mathema-
　tician), 66, 125, 141, 333
Orovio, Manuel de (politician), 4, 5
Ors, Eugeni d' (philosopher), 281,
　285; on German science, 274; on
　relativity, 166
Ortega y Gasset, José (philosopher),
　11n, 17, 130, 145, 157, 214, 277,
　326; criticizes Einstein's politics,
　316; on Einstein's popularity, 301;
　on empiricism in physics, 162; on
　Newton and Galileo, 161; per-

spectivism, 163; on relativity, 138,
　144, 159–164; *tertulia* of, 214, 282;
　in Toledo with Einstein, 136–138
Orts, José María (mathematician),
　20; on book by Plans, 69
Orueta, Domingo (geologist), 11n
Ostwald, Wilhelm, 233
Ostwald, Wolfgang, 72

Pahissa i Jo, Jaume (composer), 107,
　230; on role of mathematics in sci-
　ence, 246
Painlevé, Paul, 93, 94, 95, 154, 201
Palacios, Julio (physicist), 59, 60, 66,
　131, 141, 333; countertheory of
　relativity, 318–321; on indetermi-
　nacy, 318; on Terradas, 34
Paniagua, Enrique de (engineer),
　223
Pellicer, Francisco (journalist), 109
Pemartín, José (writer), 184; critique
　of Ortega, 164
Peña, Fernando (mathematician),
　21, 131, 198
Pérez de Ayala, Ramón (writer),
　303, 317
Pérez del Pulgar, José Antonio (en-
　gineer), 63, 71, 154–155, 193, 305;
　on special relativity, 221–222, 316
Pestaña, Angel (syndicalist), 108,
　328
Petit Parisien, Le, 268
Philosophy of science, 209–216, 236
Physicians: and relativity, 235–236,
　260
Physicists: and relativity, 32–48
Physics: in Spain, 26–32
Pi Sunyer, August (physiologist),
　236
Pi Sunyer, Carles (politician), 104
Pidal, Marqués de (politician), 7
Pierce, Charles Sanders, 19
Piña de Rubíes, Santiago (physicist),
　31, 32

Pineda, Pedro (mathematician), 21, 58

Piñerúa Alvarez, Eugenio (chemist), 232–233

Pittaluga, Gustavo (parasitologist), 126, 130, 139

Planck, Max, 34, 42

Plans, Josep María (mathematician), 21, 42, 63, 66, 124, 125, 131, 140, 141, 170, 174, 181, 192, 202, 208, 249, 332, 333; on absolute differential calculus, 67; on aether-drift experiments, 156–157; on general relativity, 132–133; on Lorentz, 73; and relativity, 44–48; religious views, 46–47; and scientific communication, 186–187; on Spanish mathematics, 25; as teacher, 45; translator of Freundlich, 158, 250 —Works: *Nociones de cálculo diferencial absoluto*, 68–70, 200; *Nociones fundamentales de Mecánica relativista*, 44, 223, 250

Plaskett, John S., 311

Poincaré, Henri, 181, 196; and intuition, 211

Poto, Mariano (journalist), 268, 272

Poynting, John Henry, 132

Prat de la Riba, Enric (politician), 111

Prestige: and relativity, 86, 199, 225

Prieto, Luis, 316

Primo de Rivera, Miguel (general), 184

Progreso Matemático, El, 23

Proust, Louis Joseph, 3

Proust, Marcel, 277

Psychoanalysis, 292; Spanish reception, 83–85, 259, 260

Publicitat, La, 105, 106, 112

Puente, José de la (physicist), 31

Puente, Santos María de la (engineer), 199

Puig Adam, Pere [Pedro] (mathematician), 20, 21, 24, 46, 198

Pulido, Angel (physician), 126

Pulido, Pablo (practical geometer), 237

Quantum theory, 34, 39, 40, 278, 280, 305, 312

Rafael, Enric [Enrique] (mathematician), 35, 36, 43, 44, 47, 62, 131, 132, 170, 181, 184, 307, 333; as book reviewer, 249–250; course on relativity, 195–196, 221; on Kant and Einstein, 182; on philosophy of science, 196; on Poinsot's motion, 133–134; on relativity, 171–173; on Righi, 156; on Schlick, 158; on Terradas, 32

"Rafagas" (columnist), 251

Ramón y Cajal, Santiago (neurologist), 9, 10, 28, 254; meets Einstein, 134–135, 325

Rankin, John, 317

Rathenau, Walter, 95

Razón y Fe, 171, 176, 316

Recasens, Sebastià (obstetrician), 139

Redención, 109

Refectorium (restaurant), 117, 325

Reichenbach, Hans, 298n

Relativism, 184, 294, 334; confused with relativity, 81, 140

Relativity: in Argentina, 283; comparative reception of, 180–186; confusion between special and general theories, 76, 87; diffusion of, 82–86; in engineering schools, 195–200; incomprehensibility of, 76, 79, 80, 86, 89, 94–95, 97, 110, 145, 225, 239, 240, 250–262, 287, 292, 293–294; moral consequences of, 311–312; philosophical consequences of, 105, 143–144, 209–216; and politics, 184; popular ac-

Relativity (*cont.*)
cess to, 283–287; popularization
of, 85–88, 148, 200–209, 238–301;
press coverage of, 105–106, 126,
240–245, 253, 260, 330; principle
of, 334–335; and relativism, 81;
and religion, 170–180, 181, 184;
and Spanish philosophers, 157–
170; of time, 289–290, 319, 320;
utility of, 273
—Special relativity, 33, 38, 41, 44,
46, 75, 95, 180–181, 220, 221–222,
231, 279, 309; in England, 181;
experimental confirmation of,
127; in France, 180; in Germany,
180; Herrera on, 320–321; and intuition, 210, 280; invariance of
speed of light, 87, 127, 132, 153,
206, 210, 219, 220, 222, 228, 316,
335; kinematics, 55, 87; misconceptions of, 280; popularization
of, 203; simultaneity, 88, 94, 105,
168–169, 219, 318–319, 320, 336–
337; in United States, 181. *See also*
Lorentz, Hendrik Antoon
—General relativity, 41–42, 45, 52–
54, 74, 132–133, 135, 220–221;
and absolute differential calculus,
64–70, 183, 225, 253, 254, 333;
absolute rotation, 133; cosmology, 178; deflection of light rays,
75, 154, 347; observational confirmations of, 75, 107, 174, 228, 248;
perihelion of Mercury, 43, 45, 75,
107, 154, 222, 228, 246, 298n,
347; principle of equivalence, 156;
rotating disk, 342, 343; and scholastic cosmology, 182–183; theoretical imperfections, 348–355.
See also Eclipse of *1919*
Residencia de Estudiantes (Madrid),
9, 38, 130, 144, 158, 159, 166,
326, 329; laboratories, 9n, 27, 29
Révéscz, Andrés (journalist), 110,

130, 131, 134, 251
Revista Calasancia, 170, 173, 177,
234
*Revista Matemática Hispano-Ameri-
cana*, 23, 332
Rey Pastor, Julio (mathematician),
17, 21, 25, 58, 63, 65, 66, 71, 118,
198, 199, 203, 215; and engineering curriculum, 194; *Introducción a
la matemática superior*, 24; and invi-
tation to Einstein, 100, 101; on
Italian mathematics, 22–23; on
mathematics and relativity, 70; on
Spanish mathematics, 19; on
study abroad, 60–61; on Zubiri,
306
Reyes, Alfonso (writer), 166
Reyes Prosper, Ventura (mathemati-
cian), 18
Ricci-Curbastro, Gregorio, 64, 67,
69
Riemann, Bernhard, 18, 67, 193,
352, 353; tensor, 354
"Rigel" (columnist), 248
Righi, Augusto, 53, 156
Río, Andrés Manuel de (chemist), 3
Ríos, Fernando de los (politician),
303
Ritz, Walter, 150, 153
Rius, Antoni (chemist), 256–257,
258
Robles Degano, Felipe (priest), 171,
179
Roca, Antoni (historian), 35
Rocasolano, Antonio de Gregorio
(biochemist), 72, 145, 281n; and
Brownian motion, 146
Rodés Campdera, Lluis (astrono-
mer), 62
Rodríguez, Angel (astronomer), 170
Rodríguez, Teodoro (priest), 170;
on relativity, 174–178
Rodríguez Bachiller, Tomás (mathe-
matician), 20, 132, 193, 198, 199;

abstracts of Einstein's lectures, 243, 332–355; on Bentaból, 229; instructed by Einstein, 134; on Palacios, 319; on Plans, 45
Rodríguez Carracido, José. *See* Carracido, José R.
Rolland, Romain, 265
Romanones, Conde de (politician), 8
Romestead, Christian, 277
Rosich, Joan (engineer), 214–215
Roso de Luna, Mario (theosophist), 237
Rossi, Paolo, 172n
Rotch, A. Laurence, 50
Royal Botanical Garden (Madrid), 3, 11
Royal Society (London), 3, 74
Royo-Villanova, Antonio (politician), 116
Royo-Villanova, Ricardo (physician), 145, 147, 308; as anti-relativist, 310–313
Rutherford, Ernest, 261

Sacristán, José M. (neurologist), 130, 282
Sagarra, Josep Maria (writer), 105, 106, 147, 251; on Einstein, 112–113
Sagnac, Georges, 93, 155
Salaverría, José María (writer), 130, 272, 273
Saldaña, Angel (mathematician), 21
Salmerón, Nicolás (philosopher), 4
Salvatella, Joaquim (politician), 125, 130, 144–145, 333
Sánchez Peguero, C. (journalist), 286
Sànchez Pérez, José (mathematician), 66
Sànchez de Toca, Joaquín (politician), 7, 10, 11
Sans i Guitart, Pau (engineer), 14

Sanz del Río, Julián (philosopher), 4
Sauer, Emil von, 148
Schlick, Moritz, 158, 212, 250
Schnabel, Artur, 79
Schrödinger, Erwin, 261
Schwarz, Boris, 79
Scientia, 22, 153
Scientific communication, 32, 37, 55–65, 71–73, 186–187; circles of affinity, 282–283; language and, 56–58, 135–136
Scientific community, structure of, 39
Scientific disciplines, structure of, 71, 185–186
Scientific meetings, 62–63
"Scientific Middle Class," 66, 107, 188–192, 230, 238, 260, 283; and Darwinism, 191–192, 245; defined, 188; and popularization of relativity, 246, 248, 250. *See also* Engineering; Engineers; Monsieur Homais
Scientific popularization, 81–88; demand for, 82, 86
Seelig, Carl, 116
Selfa Mora, Rafael (philosopher), 257, 258, 260
Severi, Francesco, 22; trips to Spain, 72
Shankland, Robert, 157
Siglo Futuro, El, 179, 267
Simarro, Luis (neurologist), 10
Sitter, Willem de, 133, 153, 183, 311
Sittert, Julius van, 248
Sociedad Española de Física y Química: *Anales*, 32, 57; membership of, 189–191
Société Française de Philosophie, 93, 94, 168, 264
Sol, El, 47, 94, 118, 145, 158, 204–205, 238, 268, 272, 290, 293, 299
Soldevila, Carles (writer), 106, 252–253, 258

Sommerfeld, Arnold, 40, 42, 43, 59, 60, 71, 72

Spanish Association for the Progress of Science: membership of, 189–191; Oporto meeting (1921), 67, 156, 209, 218; Zaragoza meeting (1908), 33, 38

Staudt, Christian von, 18

Steinmetz, Charles P., 193

Study abroad, 58–62; and scholarly productivity, 59–60

Suárez, Francisco (philosopher), 183

Suárez Somonte, Ignacio (mathematician), 66

Subirá, José, 11

Subjectivism, 175, 177, 179, 221

Surrealism, 280

Szilard, B., 63

Tallada, Ferran [Fernando] (mathematician), 105, 106, 121, 197, 219; as popularizer of relativity, 202–203, 247

Terradas, Esteve [Esteban] (mathematician, physicist), 22, 42, 43, 47, 57, 63, 65, 67, 103, 117, 118, 152, 170, 181, 184, 192, 202, 208, 219, 305, 307, 325; on aether, 248–249; on aether-drift experiments, 36; articles in Enciclopedia Universal Ilustrada, 248–249; conversations with Einstein, 118, 281; Einstein's regard for, 118; and invitation to Einstein, 100, 102; on Plans, 46–47; and relativity, 32–38, 182–183, 197–198, 315; and scholastic philosophy, 182–183; and scientific communication, 187; on Spanish mathematics, 25, 71; as teacher, 34; on Walter Ritz, 153

Tertulias, 82, 282–283, 285, 286

Tetrode, Hugo, 128

Textbooks: mathematics, 19; relativity in, 197–198

Thirring, H., 169

Thompson, S. P., 205

Thomson, William [Lord Kelvin], 181

Tinoco, José (astronomer), 62

Tirpitz, Admiral von, 117

Tofiño, Vicente (navigator), 3

Tolman, R. C., 181

Tono [Antonio de Lara Gavilán] (cinemast), 78

Toreno, Conde de (politician), 6

Torres Quevedo, Leonardo (inventor), 10, 28, 29, 64, 126

Torroja i Caballé, Eduard (mathematician), 18

Torroja i Miret, Antoni (mathematician), 62, 231

Torroja i Miret, Eduard (engineer), 126

Torroja i Miret, Josep Maria (engineer), 11n

Tous i Biaggi, Josep (engineer), 219–221

Trouton, F. T., 157

"Two cultures" journalism, 238, 240

Tyndall, John, 6

Ubach, Josep (astronomer), 171

Ugarte, Nicolás de (engineer), 44, 126, on relativity, 44n, 223

Ugarte de Ercilla, Eustaquio (priest), 176–177, 178

Unamuno, Miguel de (philosopher), 157, 281; on relativity, 166–168

University of Barcelona, 34, 308, 318

University of Madrid, 18, 27, 29, 42, 329; awards Einstein degree, 139–141; offers chair to Einstein, 303–304

University of Zaragoza: invitation to Einstein, 102

Universo, El, 174

Urbano, Luis (priest), 30, 170, 179, 181, 307; on Einstein-Bergson exchange, 169; and privileged frames of reference, 172n; on Teodoro Rodríguez, 176
Uriarte, Gregorio (engineer), 66

Valera Candel, Manuel (historian), 57, 59, 60
Vallhonesta Vendrell, Josep (engineer), 15
Vanguardia, La, 105, 106, 169, 271, 317
Vecino, Jerónimo (physicist), 31, 145, 148; on popularizing relativity, 245–246, 247
Vegas, Miguel (mathematician), 141
Vela, Antonio (astronomer), 51
Velasco de Pando, Manuel (engineer), 200–201
Vera, Francisco (mathematician), 123, 125, 135, 138, 139, 187, 243, 265, 285; on Bentaból, 229; El hombre bicuadrado, 24; on utility of relativity, 273
Veu de Catalunya, La, 103, 113, 120
Vicenti, Eduardo (politician), 10
Vidal i Guardiola, Miguel (politician), 120
Vilamitjana i Masdevall, Ramon (engineer), 219
Villavieja, Marqueses de, 130, 272
Vives i Vich, Pere [Pedro] (engineer), 50
Vizuete, Pelayo (science writer), 250
Volterra, Vito, 19, 21, 22; in Spain (1932), 72

Wachhorst, Wynn, 300
"Wahr" (columnist), 242, 248
War Resisters League, 303
Watson, William, 198
Weight of Light, 203–206, 240, 246, 293, 298
Weiss, Pierre, 58–59
Weizmann, Chaim, 76, 89
Weyl, Hermann, 20, 58, 68, 69, 71, 163, 169, 187; on dualism in physics, 139; and geometrization of physics, 307; invoked by Einstein, 128; Raum, Zeit, Materie, 133; and second-order tensor, 352–353; trip to Spain, 71; and unified field theory, 353–354, 355
Whitehead, Alfred North, 158
Wien, Max, 72
Wulf, Theodor, 176–177, 179; Tratado de Física, 310

X-rays, 27
Xirau, Joaquim (philosopher), 105, 106

Yahuda, A. S., 303

Zeeman, Pieter, 59, 60; Zeeman effect, 26
Zozaya, Antonio (journalist), 260
Zubiri, Xavier (philosopher), 282, 312, 318; on relativity, 307; on role of mathematics in physics, 306
Zúñiga, Toribio (physician), 126

LIBRARY OF CONGRESS CATALOGING-IN-PUBLICATION DATA

Glick, Thomas F.
Einstein in Spain: relativity and the recovery of science
Thomas F. Glick.
p. cm.
Bibliography: p.
Includes index.
ISBN 0-691-05507-6 (alk. paper): $42.00
1. Science—Social aspects—Spain. 2. Relativity (Physics)
3. Einstein, Albert, 1879-1955—Travel—Spain. I. Title.
Q175.52.S7G48 1988
530'.0946—dc19

87-18946
CIP